MW00845272

*Dean Rood*

**The Troubleshooting and Maintenance Guide for Gas Chromatographers**

## 1807–2007 Knowledge for Generations

Each generation has its unique needs and aspirations. When Charles Wiley first opened his small printing shop in lower Manhattan in 1807, it was a generation of boundless potential searching for an identity. And we were there, helping to define a new American literary tradition. Over half a century later, in the midst of the Second Industrial Revolution, it was a generation focused on building the future. Once again, we were there, supplying the critical scientific, technical, and engineering knowledge that helped frame the world. Throughout the 20th Century, and into the new millennium, nations began to reach out beyond their own borders and a new international community was born. Wiley was there, expanding its operations around the world to enable a global exchange of ideas, opinions, and know-how.

For 200 years, Wiley has been an integral part of each generation's journey, enabling the flow of information and understanding necessary to meet their needs and fulfill their aspirations. Today, bold new technologies are changing the way we live and learn. Wiley will be there, providing you the must-have knowledge you need to imagine new worlds, new possibilities, and new opportunities.

Generations come and go, but you can always count on Wiley to provide you the knowledge you need, when and where you need it!

*William J. Pesce*
President and Chief Executive Officer

*Peter Booth Wiley*
Chairman of the Board

*Dean Rood*

# The Troubleshooting and Maintenance Guide for Gas Chromatographers

Fourth, Revised and Updated Edition

WILEY-VCH Verlag GmbH & Co. KGaA

**The Author**

***Dean Rood***
968 Glide Ferry Way Drive
Sacramento, CA 95831
USA



**Library of Congress Card No.:** applied for

**British Library Cataloguing-in-Publication Data:**
A catalogue record for this book is available from the British Library.

**Bibliographic information published by the Deutsche Nationalbibliothek**
The Deutsche Nationalbibliothek lists this publication in the Deutsche Nationalbibliografie; detailed bibliographic data are available in the Internet at http://dnb.d-nb.de.



**Typesetting** Manuela Treindl, Laaber
**Printing** Strauss GmbH, Mörlenbach
**Binding** Litges & Dopf GmbH, Heppenheim
**Wiley Bicentennial Logo** Richard J. Pacifico

Printed in the Federal Republic of Germany
Printed on acid-free paper

**ISBN** 978-3-527-31373-0

## Preface

Even though gas chromatography (GC) is considered a very mature and highly developed technology, advances continue to be made in the areas of hardware, electronics, software and columns. In some cases, these advances have reduced the occurrence of problems and made their detection easier and more certain. In other cases, greater complexity has been introduced with its own set of problems and solutions. Regardless of the age or complexity of the GC instrument, many of the same problems occur and the underlying causes are often the same. In addition, the guidelines and techniques used to care and maintain the instruments and columns are the same.

With this thought in mind, much of the core information in this edition does not differ significantly from the previous one; however, there are a number of noteworthy additions and enhancements. The majority of the figures are new and improved especially in the injector and detector chapters. A complete section on pressure and flow programmable injectors has been added. Due to its popularity and specific requirements, an Appendix on high speed GC using small diameter columns is new to this edition. Column, hardware, carrier gas and sample considerations and issues are presented in a concise and direct format to ensure successful high speed GC applications. Finally, an extensive Appendix on the basics of quantitative GC is new and relatively unique. This Appendix covers important quantitation definitions, calibration curves, the selection and use of quantitation techniques such as internal and external standards, and several standard preparation techniques. Numerous examples are provided to aid in understanding.

The information contained in this book encompasses nearly 25 years of in-depth experience in the field of GC along with the wisdom passed along from 1000's of personal interactions with GC practitioners around the world. It is often practical information mixed with a touch of theory such as presented and discussed within these pages that most often proves to be the most useful and helpful.

Sacramento, CA, March 2007 *Dean Rood*

*The Troubleshooting and Maintenance Guide for Gas Chromatographers, Fourth Edition.* Dean Rood

ISBN: 978-3-527-31373-0

# Contents



*The Troubleshooting and Maintenance Guide for Gas Chromatographers, Fourth Edition.* Dean Rood

ISBN: 978-3-527-31373-0

# Intentions and Introduction

There already seems to be a number of excellent references on gas chromatography (GC), so why this book? Well, there are several reasons. There is a large number of gas chromatographs in use. If is often stated that gas chromatography is the most common instrumental analytical technique in routine use. The availability of easy to operate, affordable and feature laden instruments has made GC a powerful analytical technique accessible to nearly every laboratory.

Commercially available capillary columns of high quality have existed for about 25 years. For a number of reasons, many GC users are not extremely experienced in the practice of capillary gas chromatography. Many of these users do not possess a level of comprehension of the technique that allows them to prevent and solve many of the problems that commonly occur. Much of this comprehension comes from years of experience and the problems that accompany that experience. The combination of accessible instruments and capillary columns along with inexperienced users has created the need for practical information on the care, maintenance and troubleshooting of capillary columns and instruments.

One of the goals of this book is to provide practical information that will maximize both capillary column lifetime and the performance of the gas chromatographic system. The other goal is to provide an efficient and logical troubleshooting guide with the real intention to reduce or prevent performance breakdown problems from occurring. An in-depth knowledge of chemistry and chromatography (and other foreign languages) is not required. This book, in no shape or form, attempts to thoroughly explain every detail about capillary gas chromatography; it is intended as a practical guide so that the urge to hit the GC with a hammer as a last resort does not occur. In-depth technical information about GC techniques, instrumentation, specific applications and other gory details can be found in the books listed in the reference section.

*Many generalizations and simplifications have been exercised to keep the information in a basic and widely digestible form.* Again, this book is intended for the average GC user and not those whose entire life revolves around capillary gas chromatography. The topics covered within these pages are based on the most common problems, questions and misconceptions about capillary gas chromatography. These topics have been assembled and presented in an unique, practical and concise format suitable even for the most inexperienced GC user.

*The Troubleshooting and Maintenance Guide for Gas Chromatographers, Fourth Edition.* Dean Rood

ISBN: 978-3-527-31373-0

References to specific models of GCs and columns from specific manufacturers have been avoided where possible. Any differences are usually minor and often inconsequential in nature. The operating principles, proper techniques and practices, and underlying theory are the same regardless of the instrument or column manufacturer.

# 1
# Introduction to Capillary Gas Chromatography

## 1.1
## What Is Gas Chromatography?

In a broad sense, gas chromatography is a very powerful and one of the most common instrumental analysis techniques in use. When properly utilized, it provides both qualitative (i.e., what is it?) and quantitative (i.e., how much?) information about individual components in a sample. Gas chromatography involves separating the different compounds in a sample from each other. This allows the easy identification and measurement of the individual compounds in a sample. The compounds are separated primarily by the differences in their volatilities and structures. Many compounds and samples are not suitable for gas chromatographic analysis due to their physical and chemical properties.

## 1.2
## What Types of Compounds Are Suitable for GC Analysis?

For a compound to be suitable for GC analysis, it must possess appreciable volatility at temperatures below 350–400 °C. In other words, all or a portion of the compound molecules have to be in the gaseous or vapor state below 350–400 °C. Another characteristic is the compound must be able to withstand high temperatures and be rapidly transformed into a vapor without degradation or reacting with other compounds. Unfortunately, this type of information about a compound is not readily available in references or other sources; however, some estimates and generalizations can be made from the structure of the compounds.

Compound structure and molecular weight can be used as indicators of potential GC analysis suitability. Compounds with very low volatilities are not suited for GC analysis since they do not readily vaporize. Compound boiling points are not always good indicators of volatility. There are many high boiling compounds that can be analyzed by GC. As a general rule, the greater the molecular weight or polarity of a compound, the lower its volatility. Both factors have to be considered. For example, a large, non-polar compound may be more volatile than a small, polar compound. Also, one polar group on a large molecule has less of an influence than one polar group on a small molecule.

*The Troubleshooting and Maintenance Guide for Gas Chromatographers, Fourth Edition.* Dean Rood

ISBN: 978-3-527-31373-0

Hydrocarbons with molecular weights over 500 are routinely analyzed using standard GC systems, and hydrocarbons with molecular weights over 1400 have been easily analyzed using the properly equipped GC and type of column. The presence of polar functionalities such as hydroxyl and amine groups severely decrease compound volatility. Some small molecules such as sugars and amino acids can not be easily analyzed by GC due to the large number of polar groups.

As a rule, inorganic compounds are not suitable for GC analysis. Metals and salts do not possess the required volatility. Many organo-metallics have sufficient volatility for analysis due to the high organic content of these molecules. Most organic compounds are suitable for GC analysis; however, there are many exceptions. Many biomolecules and pharmaceuticals are thermally sensitive and degrade at the temperatures used in gas chromatography. Some compounds react with the materials used in gas chromatographs and columns and can not successfully analyzed by GC. There are no realistic, absolute guidelines that can be used to determine whether a compound can be analyzed by GC. Overall, it has been estimated that only about 10% of all compounds can be analyzed by GC.

## 1.3 The Basic Parts of a Gas Chromatograph

A gas chromatographic system is comprised of six major components: gas supply and flow controllers, injector, detector, oven, column, and a data system (Figure 1-1). In most cases, the injector, detector and oven are integral parts of the gas chromatograph; the column, gases and recording device are separate items and are often supplied by a different manufacturer. All of the components are further described in individual sections or chapters with the exception of the oven and recording devices.

### 1.3.1 Gas Supply and Flow Controllers

High purity gases are supplied from a pressurized cylinder or gas generator. Pressure regulators on the cylinders or generators control the amount of gas delivered to the gas chromatograph. Flow controllers or pressure regulators in the gas chromatograph control the flow of the various gases once they enter the instrument.

The column is installed between the injector and detector. Gas at a precisely controlled flow is supplied to the injector; this gas is called the carrier gas. The carrier gas flows through the injector and into the open tubular column. The gas travels the length of the column and exits through a detector. To function as desired, most detectors require specific gases at the proper flow rates.

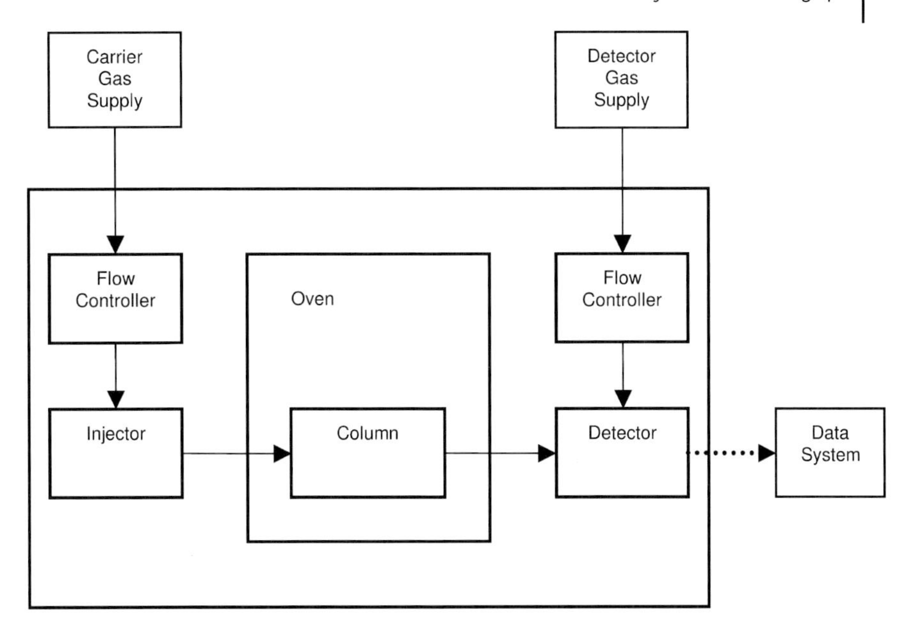


**Figure 1-1** Block diagram of a typical gas chromatograph.
Solid arrows denote gas flow paths and dotted arrows denote electronic signal flow paths.

### 1.3.2
### Injector

The injector introduces the sample into the open tubular column. The injector is a hollow, metal cylinder containing a glass liner or insert. The column is inserted into the bottom of the injector so that the column end resides in the lower region of the glass liner. A liquid, or sometimes a gas, is introduced into the injector through a resealable septum using a small syringe. The injector is heated to 100–300 °C, thus any volatile sample components are rapidly transformed into a vapor. The carrier gas mixes with the vaporized portion of the sample and carries the sample vapors into the column.

An on-column injector deposits the sample directly into the column without a vaporization step and it is used for select types of samples. In some cases, non-syringe techniques utilizing specialized equipment or devices (e.g., purge and trap, headspace, and valves) can be used to introduce a sample into a column.

### 1.3.3
### Capillary Column and Oven

The column resides in an oven whose temperature is accurately controlled. If unimpeded, vaporized compounds move through the column at the same rate as the flowing carrier gas. However, the interior walls of columns are coated with a thin film of polymeric material called the stationary phase. This stationary phase impedes the movement of each compound down the column by a different amount. This behavior is called retention.

The length and diameter of the column, the chemical structure and amount of the stationary phase, and the column temperature all affect compound retention. If all of these factors are properly selected, each compound travels through the column at a different rate. This makes the compounds exit the column at different times. As each compound leaves the column, its presence and amount are measured by the detector.

### 1.3.4 Detector

As each compound exits the column, it enters the detector. The detector interacts with the compounds based on some physical or chemical property. Some detectors respond to every compound while others respond only to a select group of compounds. The interaction generates an electrical signal whose size corresponds to the amount of the compound. The detector signal is then sent to a recording device for plotting.

### 1.3.5 Data System

The recording device plots the size of the detector signal versus the time elapsed since sample introduction into the injector. The plot is called a chromatogram and appears as a series of peaks (Figure 1-2). Except very old recorders, some type of report is provided by the data system.

The most common data recording devices are computer (PC) based. Older GC systems may use an integrator or a strip chart recorder which produce printed versions of the chromatogram and report with little or no data storage and recall capability. PC based data system are extremely powerful and offer numerous data plotting, reporting and storage options, thus their popularity. Most computer data system can also control and automate the operation of the GC.

## 1.4 The Chromatogram

In the ideal situation, each peak in the chromatogram represents a single compound in the sample. It is not unusual for more than one compound in a sample to interact with the column in the same manner, thus each compound has the same retention. This results in a single peak that represents more than one compound (complete co-elution). In some cases, the interactions are very similar, but not identical. This results in two peaks that partially overlap (partial co-elution). Using the proper column and operating conditions minimizes dual

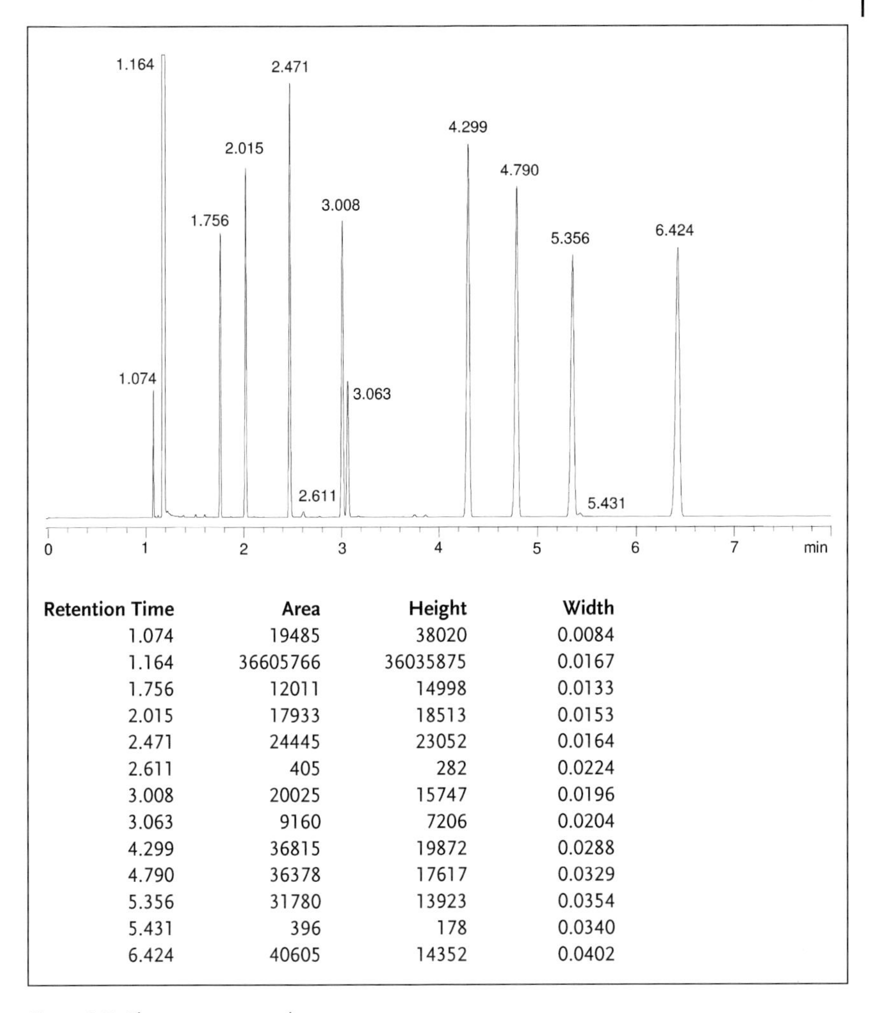


| Retention Time | Area | Height | Width |
|---|---|---|---|
| 1.074 | 19485 | 38020 | 0.0084 |
| 1.164 | 36605766 | 36035875 | 0.0167 |
| 1.756 | 12011 | 14998 | 0.0133 |
| 2.015 | 17933 | 18513 | 0.0153 |
| 2.471 | 24445 | 23052 | 0.0164 |
| 2.611 | 405 | 282 | 0.0224 |
| 3.008 | 20025 | 15747 | 0.0196 |
| 3.063 | 9160 | 7206 | 0.0204 |
| 4.299 | 36815 | 19872 | 0.0288 |
| 4.790 | 36378 | 17617 | 0.0329 |
| 5.356 | 31780 | 13923 | 0.0354 |
| 5.431 | 396 | 178 | 0.0340 |
| 6.424 | 40605 | 14352 | 0.0402 |

**Figure 1-2** Chromatogram and report.

peak identities or overlapping problems, but there are cases where complete separation is not possible.

Each peak in the chromatogram is assigned a retention time. It is the time required for a compound to travel through the column. The data system usually calculates and prints the retention times and size for each peak on the chromatogram or in a table (Figure 1-2); additional information may also be included in the report table. Retention times are usually reported in minutes and the peak size in an unitless area or height value.

Identifying the compounds corresponding to each peak in the chromatogram is accomplished by comparison to a previously generated reference chromatogram. A prepared solution containing known amounts of each compound (commonly called a standard) is analyzed to obtain their respective retention times and peak sizes. Using the same column and GC parameters, the sample is analyzed. If any of the peaks in the sample have the same retention times as those in the standard, there is a good probability that the sample contains one or more of the compounds. If the peaks in the sample do not correspond to those in the standard, the sample does not contain any of the compounds.

To determine the amount of a compound in the sample, the size of its peak is used. The size of a peak is proportional to its amount in the sample or standard. Since the standard contains a known amount of each compound, the peak sizes can be used as a reference. The size of the peak in the sample is compared to the size of the corresponding peak in the standard. A simple ratio is set up for quantitation. For example, if the peak in the sample is two times larger than the peak in the standard, the injected portion of the sample contains two times the amount of the compound than the amount known to be present in the standard.

There are numerous situations where peak misidentification or quantitation errors can occur. Adhering to good GC practices will minimize the occurrence of these types of errors. Additional information on quantitative GC can be found in Appendix F.

## 1.5
## The Mechanism of Compound Separation

How does the column work? What happens inside the column? How do the compounds move through the column? Why do some compounds stay in the column longer than others? How does the sample get into the column? These are some of the most basic questions asked about gas chromatography. Knowing the answers does not automatically make a chromatographer produce better results, but the knowledge is very valuable in solving and preventing problems, selecting columns, and understanding unexpected results. Complicated discussions involving thermodynamics and molecular interactions are necessary to fully answer these questions. Fortunately, comprehension at this level is not necessary to become an excellent chromatographer. A basic understanding of the concepts, and not the intricate details, provides a chromatographer with all of the information necessary to produce the most consistent, trouble free and best results.

### 1.5.1
### A Simple Description of the Chromatographic Process

The separation of a sample into its individual compounds by a capillary GC column can be described by a very simple concept. The sample containing a mixture compounds enters the column and collects in the front of the column (Figure 1-3a). Then the molecules of each compound start to collectively move down the column at a different rate (Figure 1-3b). The fastest moving molecules reach the end of the column first, enters the detector, thus corresponding with the first peak in the chromatogram (Figure 1-3c). The next fastest compound molecules follows,

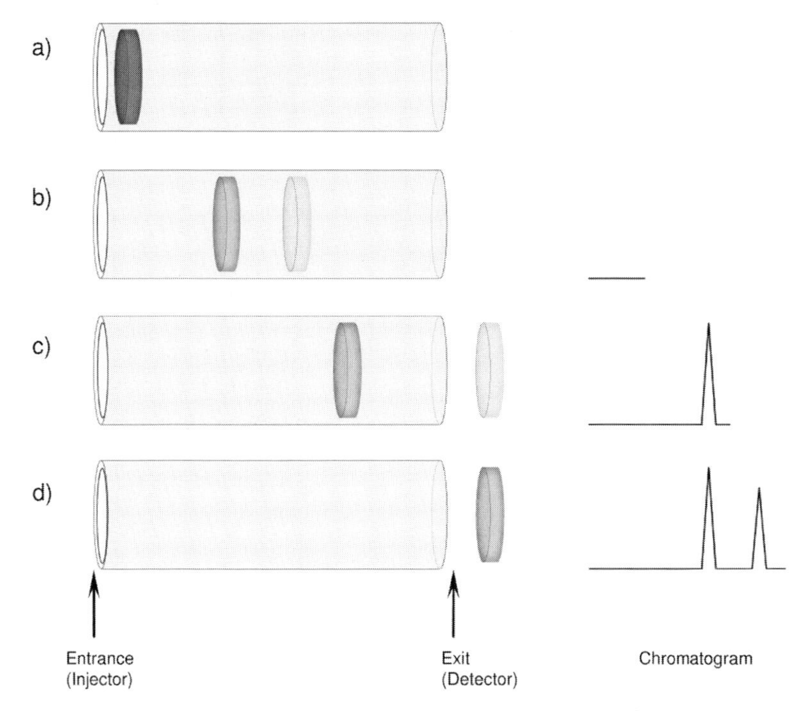


**Figure 1-3** Separation of the sample in the column.

and this process continues until all of the remaining compounds have left the column (Figure 1-3d). Since the compounds each leave the column at different times, they are separated. Any compounds that travel through the column at the same rate are not separated and have the same retention times.

### 1.5.2
### A Detailed Description of the Chromatographic Process

Capillary columns are composed of three distinct parts. The tubing is fused silica (glass) with an outer protective coating. The inner walls are coated with a thin film of polymeric material called the stationary phase. The sample compounds interact with the stationary phase, and this interaction is responsible for the separation properties of the column.

Once in the column, the molecules for each compound distribute between the mobile phase (carrier gas) and the stationary phase (Figure 1-4a). Molecules in the mobile phase move down the column; molecules in the stationary phase do not move down the column (Figure 1-4b). The carrier gas transports the compound molecules down the column. Simultaneously, the molecules are moving in a random motion. Eventually, each molecule comes into contact with the stationary phase. Each one enters the stationary phase when this occurs. For every molecule entering the stationary phase, another one leaves the stationary phase to take

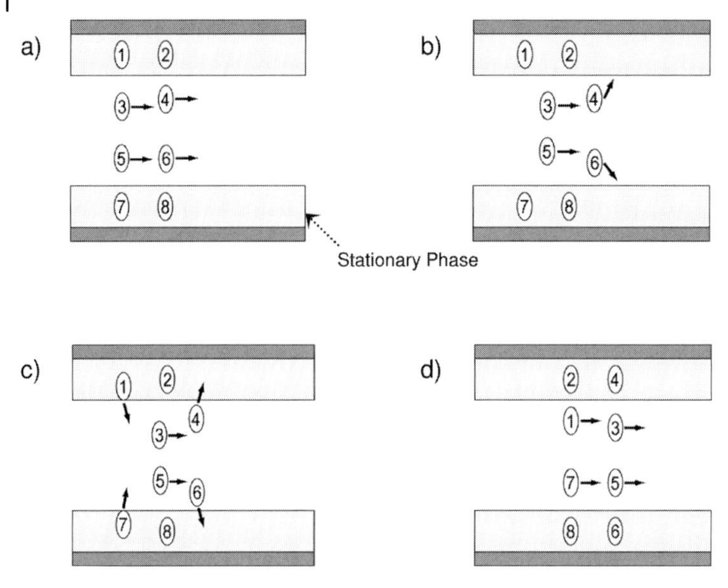


**Figure 1-4** Movement of molecules down the column. Longitudinal cross-section view of a column.

its place in the mobile phase (Figure 1-4c). This maintains the same overall distribution of the molecules between the two phases. The process of exchange between the phases is repeated thousands of times for each molecule. The net effect is the movement of the molecules down the column (Figure 1-4d).

The rate of molecule movement down the column depends on the distribution of the molecules between the stationary and mobile phases. The greater the percentage of molecules in the mobile phase, the faster the molecules travel down the column. This results in a short residence time for the molecules in the column and a short retention time for the corresponding peak. Separation of two compounds occurs when the distribution of their molecules between the stationary and mobile phases are different. If the distributions are the same, co-elution occurs.

The distance or time between the various groups of molecules (with each group representing one compound) as they exit the column determines the amount of separation between the peaks. While this separation distance is important, there is more to chromatography than just separation. The length of column occupied by the molecules for each compound is critical. A narrow band of compound molecules occupying a short length of column is desired. If the width of the molecule bands is narrow, a large separation between the band of molecules is not needed to prevent overlap of the different compound molecules (Figure 1-5a). If the width of the molecule bands is broad, the same amount of separation results in an overlap of the different compound molecules (Figure 1-5b). When the molecule bands are broad, greater separation is need to prevent overlapping of the molecule bands (Figure 1-5c).

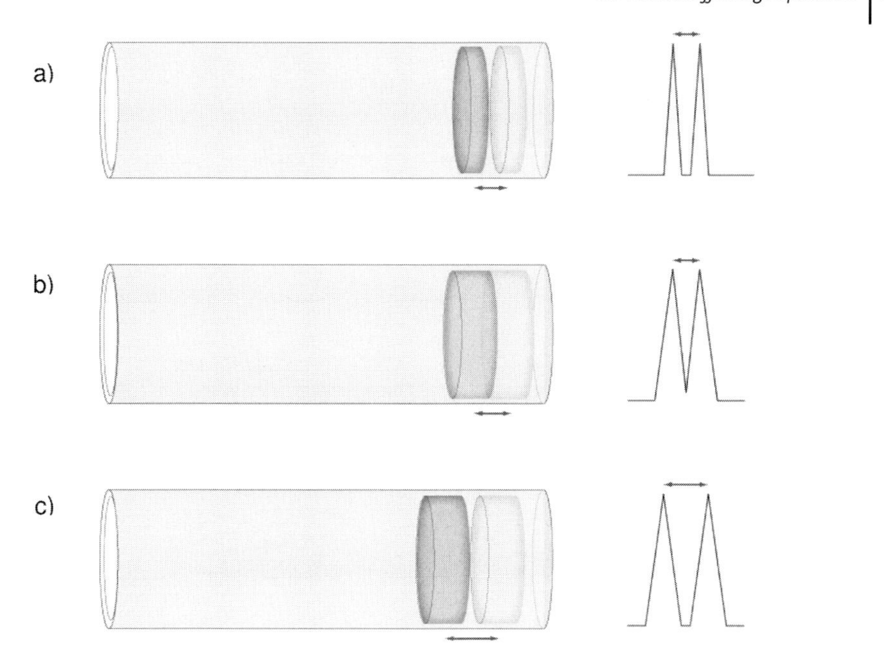


**Figure 1-5** Width of the molecule bands and the effect on the chromatogram.

## 1.6 Factors Affecting Separation

The distribution of compound molecules depends on the stationary phase, the compound and the column temperature. Only these three factors influence the amount of peak separation (i.e., distance or time between the peaks). Other factors such as column dimensions and carrier gas do not have a direct affect on separation.

### 1.6.1 Stationary Phase

Retained compound molecules can be regarded as dissolving in the stationary phase. For a particular compound, its molecules may be more soluble in one stationary phase than in another. In this case, the compound's molecules distribute differently in each stationary phase (Figure 1-6). This means different retention is obtained with each stationary phase. For multiple compounds, the change in retention (i.e., molecule distribution) for each compound is usually different. For this reason, co-eluting compound peaks for one stationary phase often separate with a different stationary phase.

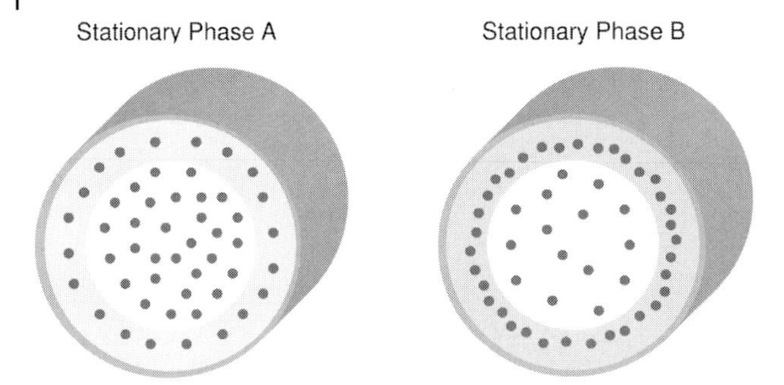


**Figure 1-6** Molecule distribution in the mobile and stationary phases: *change in the stationary phase.* Cross-sectional view of the column.

### 1.6.2 Compound Structure

The molecules for different compounds distribute in a different ratio in the same stationary phase (Figure 1-7). Different compounds have different solubilities in the same stationary phase. If this occurs, peak separation is obtained. If two compounds have the same distribution in the stationary phase, the corresponding peaks do not separate. This is the reason for co-eluting peaks on the same column.

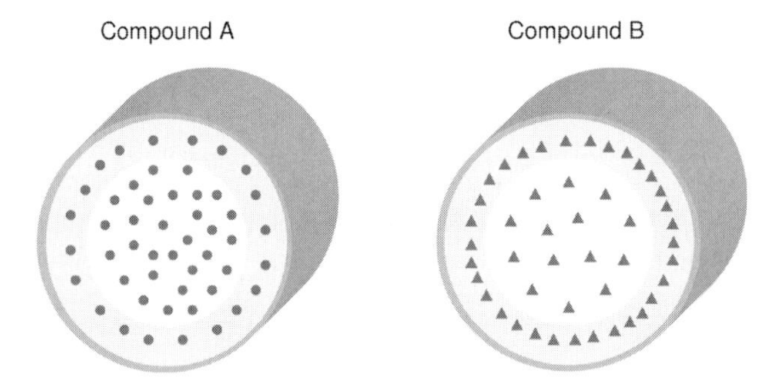


**Figure 1-7** Molecule distribution in the mobile and stationary phases: *change in the compound.* Cross-sectional view of the column.

### 1.6.3 Column Temperature

The distribution of molecules between the stationary and mobile phases depends on column temperature. At higher column temperatures, there are fewer molecules in the stationary phase and more in the mobile phase than at a lower temperature (Figure 1-8). The presence of fewer molecules in the stationary phase results in faster migration down the column and shorter retention times. This accounts for longer retention times at lower column temperatures and shorter retention times at higher column temperatures.

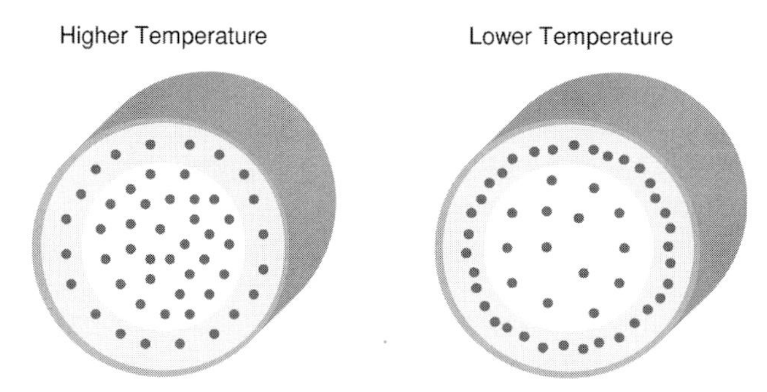


**Figure 1-8** Molecule distribution in the mobile and stationary phases: *change in the column temperature.* Cross-sectional view of the column.

At higher column temperatures, the differences in the distribution of different compound molecules become smaller. Smaller distribution (retention) differences in the various compound molecules result in less separation between the corresponding peaks. In general, this results in a decrease in separation at higher column temperatures (less difference in the molecule distributions) and a separation increase at lower column temperatures (greater differences in the molecule distributions).

# 2
# Basic Definitions and Equations

## 2.1
## Why Bother?

There are a number of terms commonly used to describe various chromatographic and column characteristics, behaviors, and conditions. An understanding of these terms on a basic level is helpful for comparing column performance, determining column quality, troubleshooting, interpreting the test mixture chromatograms, and comprehending the fundamentals of chromatographic theory.

## 2.2
## Peak Shapes

### 2.2.1
### Peak Width (W)

Depending on the compounds, conditions and column, the peaks in a chromatogram have different widths. Peak widths are important indicators and measures of column and system performance. There are two peak width measurements that are commonly used – the width at the base and at half height.

Peak width at the base is measured by extending the peak sides down to the baseline and measuring the width at the baseline (Figure 2-1). This can be uncertain or inaccurate if the baseline is not level, if peaks are partially co-eluting or tailing. Measuring the peak width at half height is usually easier. This involves measuring across the peak half way up its height instead of at the base (Figure 2-1). Peak width at half height is more commonly reported by most computer data systems. Peak widths are reported in time units (usually minutes).

### 2.2.2
### Peak Symmetry

Most calculations assume the peaks are symmetrical (Gaussian). Peak symmetry is determined by bisecting the peak through the apex. The width at 10% height

*The Troubleshooting and Maintenance Guide for Gas Chromatographers, Fourth Edition.* Dean Rood

ISBN: 978-3-527-31373-0

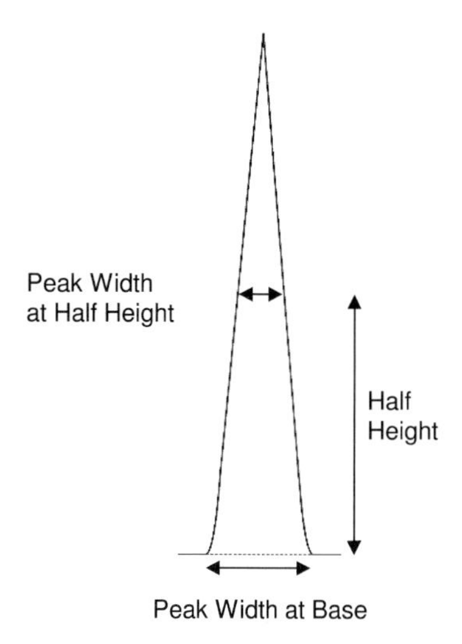


**Figure 2-1** Peak width measurements.

is measured for each half (Figure 2-2). The width of the back half is divided into the width of the front half. A perfectly symmetrical peak has a symmetry factor of 1.00. Values of 0.9–1.1 are common especially with higher compound amounts. Symmetry values outside of this range may indicate peak overloading, a co-eluting peak, activity (tailing) or an injector problem.

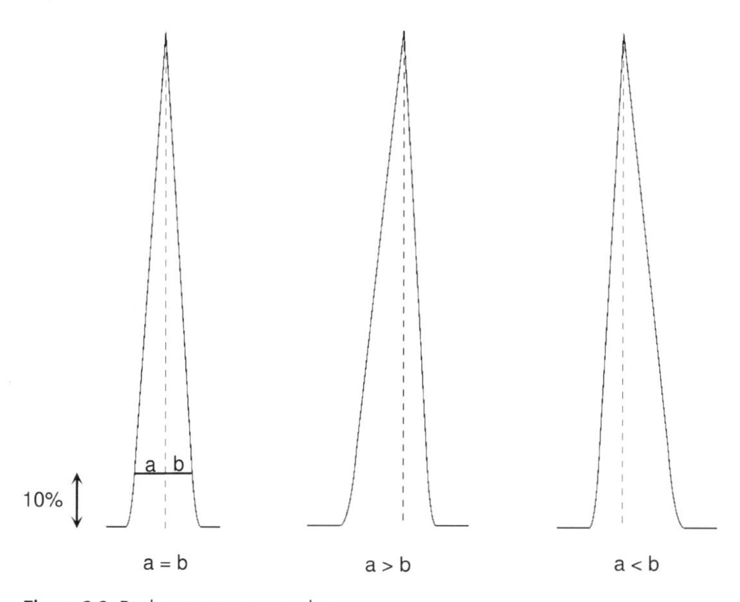


**Figure 2-2** Peak symmetry examples.

## 2.3
## Retention

### 2.3.1
### Retention Time ($t_r$)

One of the most fundamental measurements in GC is peak (compound) retention. Retention time ($t_r$) in minutes is the usual unit of measure, thus it is the default value reported by data systems. Retention time is a measure of how long it takes a compound to travel down the column. It can also be considered the time a compound spends in the column. It is the sum of the time the compound spends in the stationary and mobile phases. A compound's retention time is dependent on the column, column temperature and carrier gas linear velocity. Due to small changes in the GC, retention times may slightly drift over time. Also, run-to-run variations of ±0.05 minutes are not unusual. Larger variations may indicate a potentially serious system problem.

### 2.3.2
### Adjusted Retention Time ($t_r'$)

Compounds move down a column only while they are in the mobile phase (carrier gas). For a given column temperature and carrier gas linear velocity, every compound spends the same amount of time in the mobile phase. The time in the mobile phase is determined by injecting a non-retained compound and measuring its retention time (a list of non-retained compounds can be found in Table 6-1). This retention time is called the column dead time or hold up time and is denoted as $t_m$ or sometimes as $t_o$.

Since every compound spends the same amount of time in the mobile phase, differences in compound retention are due to the differences in the times spent in the stationary phase. If the column dead time ($t_m$) is subtracted from a compound's retention time ($t_r$), a measure of the actual retention of a compound by the stationary phase is obtained. This value is called the adjusted retention time ($t_r'$) and is calculated using Equation 2-1.

$$t_r' = t_r - t_m$$

$t_r$ = retention time
$t_m$ = retention time of a non-retained compound (peak)

**Equation 2-1** Adjusted retention time ($t_r'$).

### 2.3.3
### Retention Factor ($k$)

Another retention measurement is called the retention factor ($k$). It is commonly called the partition ratio or capacity factor, or sometimes written as $k'$. The retention factor is proportional to the time a compound spends in the stationary phase ($t'_r$) relative to the time it spends in the mobile phase ($t_m$). Since all compounds spend the same amount of time in the mobile phase, $k$ is a unitless measure of how long a compound spends in the stationary phase compared to another compound. The retention factor is calculated using Equation 2-2. Compound retention increases as its $k$ value increases. A compound is non-retained if $k = 0$.

$$k = \frac{t_r - t_m}{t_m}$$

$t_r$ = retention time
$t_m$ = retention time of a non-retained compound (peak)

**Equation 2-2** Retention factor ($k$).

The retention factor is a more direct measure of the actual magnitude of compound retention than retention time. A compound with $k = 8$ is twice as retained as a compound with $k = 4$; however, a compound with a retention time of 8 minutes is not twice as retained as a compound with a retention time of 4 minutes. Figure 2-3 shows an example of difference between $k$ and retention time. Retention times can be misleading when measuring column retention. For this reason, retention factors are used in many equations, theoretical discussions and quality control measurements.

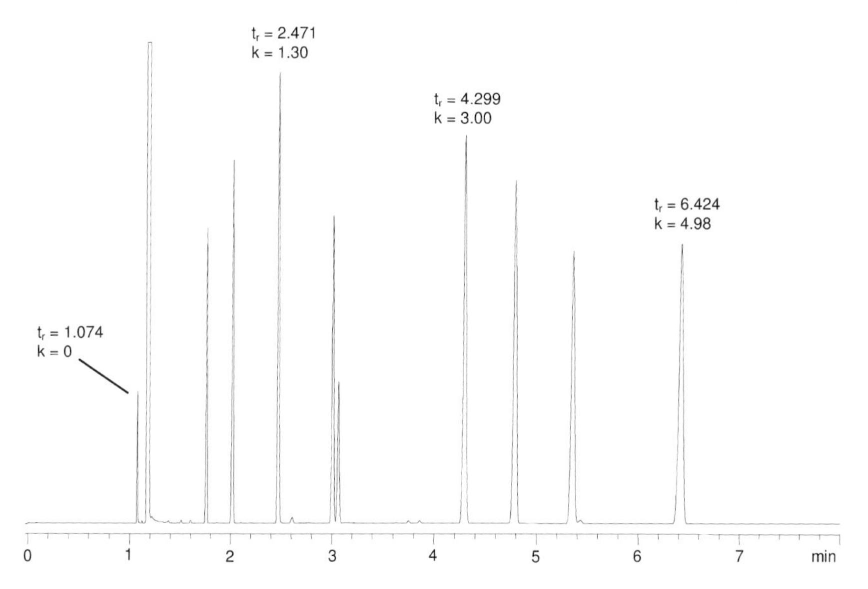


**Figure 2-3** Comparison of $k$ and retention time.

### 2.3.4
### Retention Index (*I*)

Retention index ($I$) is a measure of the relative retention of a compound relative to normal alkanes (straight-chain hydrocarbons) at a given temperature on a particular stationary phase. Retention indices are also known as Kovats indices. Retention indices normalize instrument and column variables so that retention data can be effectively compared for different chromatographic systems. The retention index for a compound is calculated for isothermal temperature conditions using Equation 2-3.

$$I = 100\,y + 100\,(z - y)\left[\frac{\log t'_r(x) - \log t'_r(y)}{\log t'_r(z) - \log t'_r(y)}\right]$$

$t'_r$ = adjusted retention time
$x$ = compound of interest
$y$ = normal alkane with $y$ carbon atoms eluting **before** compound $x$
$z$ = normal alkane with $z$ carbon atoms eluting **after** compound $x$

**Equation 2-3** Retention index ($I$) for isothermal conditions.

The retention index for a normal alkane is equal to its number of carbons multiplied by 100. For example, $n$-dodecane ($n$-$C_{12}H_{26}$) has $I = 1200$ and $n$-tridecane ($n$-$C_{13}H_{28}$) has $I = 1300$. A compound with a retention index of 1225 elutes between $n$-dodecane and $n$-tridecane, but closer to $n$-dodecane under the same test conditions and column.

Retention indices are useful when comparing relative elution orders of various compounds for a given column and conditions. Also, retention indices are good for comparing the retention behavior of two columns with the same description used under the same conditions.

For sample mixtures containing compounds with a wide range of boiling points, using isothermal conditions is not practical. Using Equation 2-3 to calculate retention indices for temperature program conditions results in errors especially if temperature programs starting at high temperatures, covering a wide temperature range or with rapid ramp rates are used. Equation 2-4 is used to calculate retention indices for temperature program conditions. It is important to note that programmed indices require specification of column dimensions, carrier gas type and flow rate and the temperature program.

$$I_T = 100\left[\frac{t_r(x) - t_r(y)}{t_r(z) - t_r(y)}\right] + 100\,y$$

$t_r$ = retention times
$x$ = compound of interest
$y$ = normal alkane with $y$ carbon atoms eluting **before** compound $x$
$z$ = normal alkane with $z$ carbon atoms eluting **after** compound $x$

**Equation 2-4** Retention index ($I_T$) for temperature program conditions.

## 2.4 Phase Ratio (β)

The phase ratio (β) is a unitless value relating the diameter and film thickness of the column (Equation 2-5). Since the phase ratio is unitless, the same units must be used for the column radius and film thickness (usually micrometers). Table 2-1 lists the phase ratios for various sized columns. The phase ratio is used to determine the impact of column diameter and film thickness on retention (Section 2.5).

$$\beta = \frac{r}{2\,d_f} = \frac{d}{4\,d_f}$$

$r$ = column radius (μm)
$d$ = column diameter (μm)
$d_f$ = film thickness (μm)

**Equation 2-5** Phase ratio (β).

**Table 2-1** Phase ratios for common column dimensions.

| Film thickness (μm) | Column diameter (mm) | | | | | |
|---|---|---|---|---|---|---|
| | **0.10** | **0.18** | **0.20** | **0.25** | **0.32** | **0.53** |
| 0.10 | 250 | 450 | 500 | 625 | 800 | 1325 |
| 0.15 | 167 | 300 | 333 | 417 | 533 | 883 |
| 0.18 | 139 | 250 | 278 | 347 | 444 | 736 |
| 0.20 | 125 | 225 | 250 | 313 | 400 | 663 |
| 0.25 | 100 | 180 | 200 | 250 | 320 | 530 |
| 0.33 | 76 | 136 | 152 | 189 | 242 | 402 |
| 0.40 | 63 | 113 | 125 | 156 | 200 | 331 |
| 0.50 | 50 | 90 | 100 | 125 | 160 | 265 |
| 0.83 | | | 60 | 75 | 96 | 160 |
| 1.00 | | | 50 | 63 | 80 | 133 |
| 1.50 | | | | 42 | 53 | 88 |
| 3.00 | | | | 21 | 27 | 44 |
| 5.00 | | | | 13 | 16 | 27 |

## 2.5
## Distribution Constant ($K_C$)

Upon entering the column, compound molecules are distributed between the mobile and stationary phases according to column temperature, the compound and stationary phase structures. The ratio of the concentrations of the compound in the stationary and mobile phases is defined as the distribution constant ($K_C$) (Equation 2-6). Figure 2-4 is a graphical representation of the distribution constant.

$$K_C = \frac{\text{Compound concentration in the stationary phase}}{\text{Compound concentration in the mobile phase}}$$

**Equation 2-6** Distribution constant ($K_C$).

Due to the overwhelming number of possible temperatures, compounds and stationary phases, there are no comprehensive tables or graphs that list $K_C$ values. The distribution constant is most useful in describing and predicting retention changes upon changing column dimensions and temperature.

The distribution constant can be related to retention, column diameter and stationary phase thickness using Equation 2-7. The phase ratio (β) depends only on column diameter and film thickness, and is fixed for a column; $K_C$ is constant for a specific compound, stationary phase and column temperature. This means that $K_C$ and β are independent and are affected by different variables. Also, note that column length, carrier gas type and flow rate are not directly involved in the terms in Equation 2-7.

$$K_C = k\,\beta = k\left(\frac{r}{2\,d_f}\right)$$

$k$ = retention factor
$r$ = column radius (µm)
$\beta$ = phase ratio
$d_f$ = column film thickness

**Equation 2-7** Distribution constant ($K_C$) retention and column relationship.

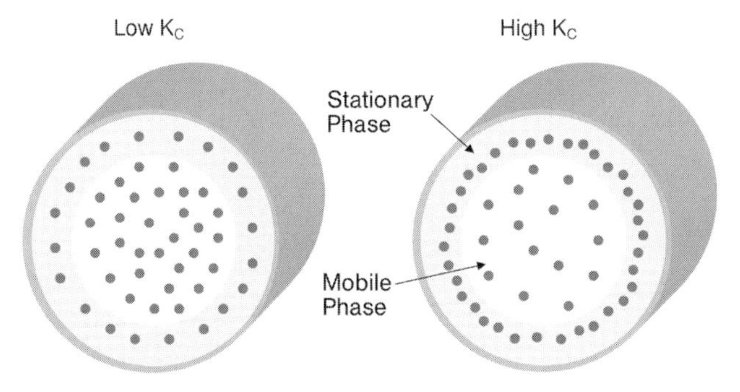


**Figure 2-4** Graphical representation of the distribution constant ($K_C$). Cross-sectional view of a column.

### 2.5.1
### $K_C$ and Column Dimensions

For a given compound, stationary phase and column temperature, a change in column diameter or film thickness causes a change in retention of the compound (see Equation 2-7). This occurs because the distribution constant remains constant for the same column temperature and stationary phase. The retention of the compound shifts depending on whether the phase ratio increases or decreases. Table 2-2 summarizes the effects of changing the film thickness and diameter on retention for the same compound, stationary phase and column temperature (i.e., maintaining a constant $K_C$). If both diameter and film thickness change, the direction of the retention shift depends on the relative change of the two dimensions.

**Table 2-2** Effects of changing column diameter and film thickness on retention.

| Diameter | Film thickness | Phase ratio (β) | Retention ($k$) |
|---|---|---|---|
| Increase | – | Increase | Decrease |
| Decrease | – | Decrease | Increase |
| – | Increase | Decrease | Increase |
| – | Decrease | Increase | Decrease |

These relationship are valid only for constant column temperature, stationary phase and compound.

### 2.5.2
### $K_C$ and Column Temperature

Since the phase ratio (β) is fixed, retention ($k$) is affected by a change in $K_C$. Temperature is related to $K_C$ by an inverse natural log (ln) function. This means that a small change in column temperature correspond to large change in retention. From $K_C = k\,\beta$, retention decreases as column temperature increases (i.e., decrease in $K_C$) and $k$ increases as column temperature decreases (i.e., increase in $K_C$). The distribution constant does not change uniformly with temperature for all compounds. Upon a change in column temperature, the relative change of retention may be different for each peak in the chromatogram.

## 2.6
## Column Efficiency

Symmetrical peaks with the smallest widths are desired in chromatography. Two peaks that are very narrow can be closer together than two broader peaks and still not overlap (see Figure 1-5). All of the peaks in a chromatogram are not the same width. Ignoring injector influences on peak widths, peak widths increase as retention times increases. The effect is more obvious with isothermal temperature conditions. Column efficiency is the relationship between a peak's retention time and its width. This helps to account for the occurrence of broader peaks at longer retention times.

### 2.6.1 Number of Theoretical Plates (*N*)

Column efficiency is expressed as the number of theoretical plates (*N*). A peak in the chromatogram is used to calculate the number of plates using Equation 2-8. The number of plates per meter of column (N/m) is often reported so that columns of different lengths are more easily compared. A higher number of theoretical plates results in thinner peaks at their respective retention times. This is critical when attempting to separate two closely eluting peaks. A column with a low number of plates has more of each peak overlapping with each other. *N* was formerly written as *n*.

$$N = 5.545\left(\frac{t_{\mathrm{r}}}{W_{\mathrm{h}}}\right)^2$$

$$N = 16\left(\frac{t_{\mathrm{r}}}{W_{\mathrm{b}}}\right)^2$$

$t_{\mathrm{r}}$ = retention time (min)
$W_{\mathrm{h}}$ = peak width at half height (min)
$W_{\mathrm{b}}$ = peak width at base (min)

**Equation 2-8** Number of theoretical plates (*N*).

### 2.6.2 Height Equivalent to a Theoretical Plate (*H*)

Another measure of column efficiency is the height equivalent of a theoretical plate (*H*). *H* is usually reported in unit of millimeters (mm). *H* is calculated using Equation 2-9. The shorter each theoretical plate, the greater the number that fits into an unit length of column, thus the greater the number of total theoretical plates per meter. High efficiency columns (i.e., large *N* values) have small values of *H*. *H* was formerly written as *h* or HETP.

$$H = \frac{L}{N}$$

$L$ = column length (mm)
$N$ = total number of theoretical plates

**Equation 2-9** Height equivalent to a theoretical plate (*H*).

### 2.6.3
### Effective Theoretical Plates ($N_{eff}$) and Effective Plate Heights ($H_{eff}$)

A non-retained compound peak has a retention time and peak width; therefore, the theoretical plate number and plate height value are greater than zero. In theory, both values should be equal to zero. This problem is corrected by using the adjusted retention times ($t'_r$) in Equation 2-7 to calculate $N$ which is then used in Equation 2-9 to calculate $H$. The corresponding efficiency values when $t'_r$ is used are called the effective theoretical plate number ($N_{eff}$) and the height equivalent of an effective theoretical plate ($H_{eff}$). If adjusted retention time is used, $N = 0$ and $H = 0$ are obtained. The error introduced by using retention time is small, thus often ignored.

### 2.6.4
### Precautions When Using Theoretical Plates

Care must be exercised when comparing theoretical plate numbers. The number of plates is influenced by the retention of the peak used in the calculation. In theory, the calculated number of theoretical plates should be the same regardless of the retention of the peak used in the calculation. In reality, the number of theoretical plates is dependent on the retention of the peak used to calculate $N$ (Figure 2-5). When calculating $N$ values, a peak with $k > 5$ should be used to ensure valid and consistent numbers. Using a peak with a $k < 5$ may inflate the number of theoretical plates.

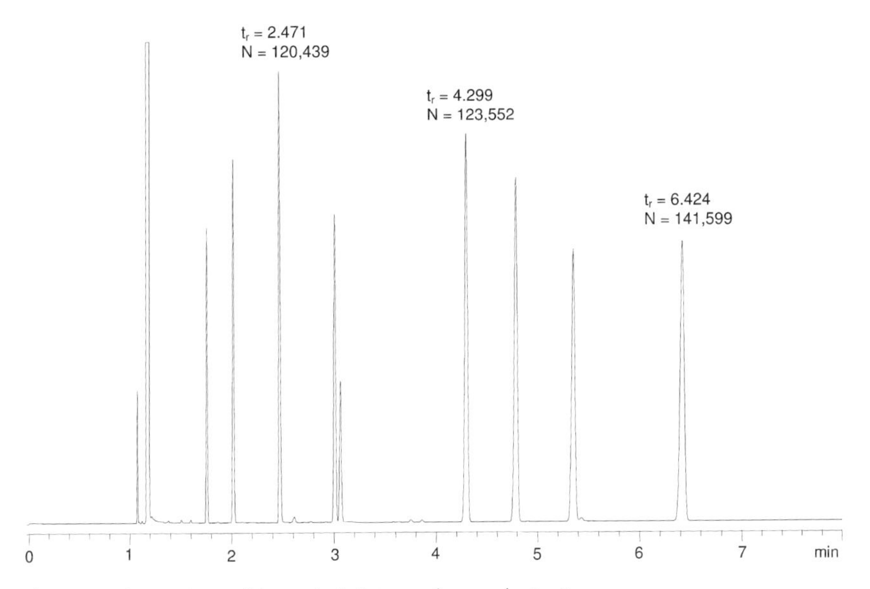


**Figure 2-5** Comparison of theoretical plate numbers and retention.

Another important consideration is that theoretical plate numbers are actually measures of the performance of the GC system and not just the column. Other factors such as injector type and performance, and carrier gas flows can have significant influence on the plate numbers. Instrument settings and conditions need to be identical when comparing column plate numbers. This is especially true when comparing columns from different manufacturers. The $k$ values for the compounds used in the calculation must be the same or the comparison is not valid. In most cases, each column manufacture use their own compounds and test conditions.

The use of temperature programming decreases peak widths. Depending on the temperature program, dramatically different plate numbers are obtained especially compared to isothermal conditions. For the same column and peak retention time, the total number of theoretical plates for temperature program can be 2–5 times higher than for isothermal conditions. Examples can be found by comparing the isothermal and temperature programs chromatograms in Figures 5-1 and 5-2 respectively.

## 2.7 Utilization of Theoretical Efficiency (UTE%)

Another slightly different measure of column efficiency is utilization of theoretical plates (UTE) or coating efficiency (CE). UTE is the comparison between a column's actual efficiency and its theoretical maximum efficiency (Equation 2-10).

$$\text{UTE\%} = \left(\frac{H_{\text{min}}\,(\text{theoretical})}{H\,(\text{actual})}\right) \cdot 100$$

$H$ (actual) = height equivalent of a theoretical plate as calculated

**Equation 2-10** Utilization of theoretical efficiency (UTE%).

Equation 2-11 is often used to calculate the theoretical maximum efficiency ($H_{\text{min}}$).

$$H_{\text{min}}\,(\text{theoretical}) = r\sqrt{\frac{(11\,k^2 + 6\,k + 1)}{3\,(1 + k)}}$$

$r$ = column radius (mm)
$k$ = retention factor

**Equation 2-11** $H_{\text{min}}$ theoretical.

An utilization of theoretical efficiency of 100% means the efficiency is equivalent to that expected for a column with a stationary phase of uniform film thickness and without any physical or chemical flaws. Coating efficiencies less than 100% (which is normal) are departures from this perfect behavior. There is some error (less than ±5%) in the calculation of the theoretical and actual values, thus it is possible to occasionally obtain coating efficiencies slightly greater than 100%. In general, non-polar stationary phases have higher coating efficiencies than polar phases. Typical coating efficiencies are 85–95% for non-polar phases and 60–80% for polar phases. Also, thicker film columns have lower coating efficiencies than the same phase with a thinner film. The less than 100% values are primarily caused by limitations in capillary column manufacturing processes.

Utilization of theoretical efficiency can be thought as a measure of the uniformity of the stationary phase. A more uniform phase results in a column of higher efficiency.

## 2.8 Separation Factor (α)

The separation factor (α) is a measure of the amount of peak separation. It is sometimes called selectivity. The separation factor is easily calculated using Equation 2-12 and is always equal to or greater than 1. If $\alpha = 1$, the peaks have the same retention, thus they completely co-elute. The separation factor does not contain any information about peak widths. It is only a measure of the distance (time) between two peaks.

$$\alpha = \frac{k_2}{k_1}$$

$k_1$ = retention factor of earlier eluting peak
$k_2$ = retention factor of later eluting peak

**Equation 2-12** Separation factor (α).

## 2.9 Resolution (*R*)

Resolution (*R*) is the measure of the amount of separation between two peaks taking the width of the peaks into account. Equation 2-13 is used to calculate the amount of resolution. A resolution of 1.5 represents fully resolved peaks without any baseline or space between the two peaks (Figure 2-6a). Resolution values greater than 1.5 represent peaks with some baseline between them (Figure 2-6b). Resolution values less than 1.5 represent peaks that have some degree of overlap or are partially resolved (Figure 2-6c). If any of the peaks deviate from Gaussian, a slightly inflated resolution number will be obtained and may not completely correspond to the resolution apparent upon visual inspection of the peaks.

Resolution is often given as a percentage value. Resolution expressed as a percentage is useful when peaks are not completely resolved since less than 100% resolution is easy to visualize (Figure 2-6c). Resolution % values are calculated

$$R = 1.18\left(\frac{t_{r2} - t_{r1}}{W_{h1} + W_{h2}}\right)$$

$$R = 2\left(\frac{t_{r2} - t_{r1}}{W_{b1} + W_{b2}}\right)$$

$t_{r1}$ = retention time of the first peak
$t_{r2}$ = retention time of the second peak
$W_{h1}$ = peak width at half height of the first peak
$W_{h2}$ = peak width at half height of the second peak
$W_{b1}$ = peak width at base of the first peak
$W_{b2}$ = peak width at base of the second peak

**Equation 2-13** Resolution ($R$).

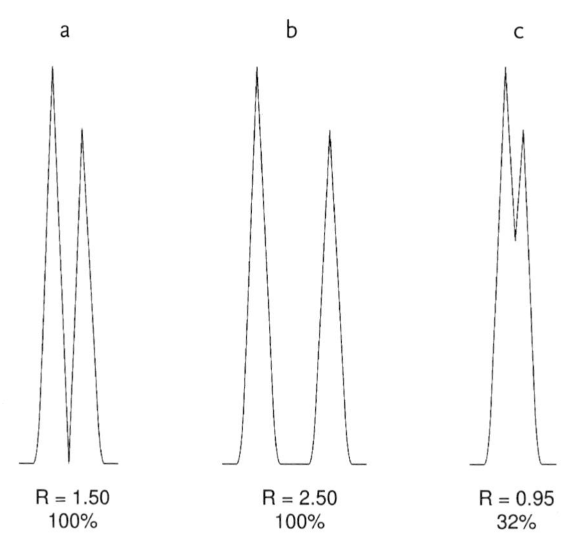


**Figure 2-6** Resolution number examples.

by dividing the depth of the valley between the two peaks (relative to the smaller peak) by the height of the smaller peak. One problem with resolution expressed as a percent is the inability to numerically distinguish between barely baseline and much greater than barely baseline separations. Both situations will be represented as 100% resolution even though the one set of peaks is separated by a substantially greater amount (Figures 2-6a and b).

Peak separation ($\alpha$) means little if peak resolution is not considered. Two broad peaks can have equal or better separation than two narrow peaks, but the narrow peaks are better resolved (Figure 2-7a and b). Less separation can be tolerated if the two peaks are narrow. The two peaks can be closer together and still maintain adequate resolution (Figure 2-7c).

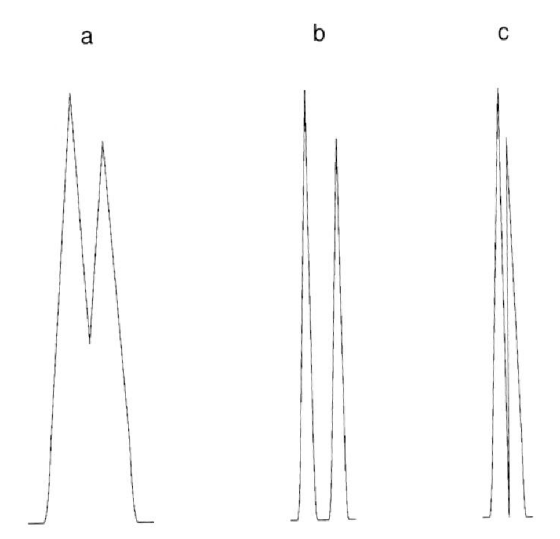


**Figure 2-7** Separation versus resolution examples:
a) separation and incomplete resolution,
b) equal separation and better resolution (compared to a),
c) less separation but complete resolution (compared to a).

In chromatography, the real goal is peak resolution. If resolution is achieved, separation automatically occurs. Instrument setup and operation, and column selection should be primarily focused on obtaining good resolution and less on obtaining separation. For low efficiency situations, more separation is needed to resolve the peaks. For high efficiency situations, less separation is needed to resolve the peaks.

## 2.10 Trennzahl (TZ)

A different measure of resolution is Trennzahl (TZ) or separation number. Like resolution it takes peak widths and the amount of separation into account. Two consecutive *n*-alkanes (hydrocarbons) are used as the peaks. TZ is calculated using Equation 2-14. TZ can be interpreted as the number of peaks (of equal width) than can be placed side by side between the two peaks. The higher the number, the greater the resolution for the two peaks. The TZ number is dependent on the compounds (peaks) used; therefore, the identity of the two hydrocarbons needs to be specified along with the TZ value.

$$TZ = \left( \frac{t_{r2} - t_{r1}}{W_{h1} + W_{h2}} \right) - 1$$

$$TZ = \left( \frac{R}{1.177} \right) - 1$$

$t_{r1}$ = retention time of the first peak
$t_{r2}$ = retention time of the second peak
$W_{h1}$ = peak width at half height of the first peak
$W_{h2}$ = peak width at half height of the second peak
$R$ = resolution

**Equation 2-14** Trennzahl (TZ).

## 2.11
## Column Capacity

Column capacity is the amount of compound that can be introduced into the column without causing excessive peak shape distortion. When too much of a compound is introduced in the column, its peak begins to develop a sloping front and may appear to be asymmetrical. A peak is overloaded when it has a symmetry factor of greater than 1.10 (Figure 2-2). A peak with a symmetry factor slightly above 1.10 often does not appear to be asymmetrical or overloaded. Peak overloading is commonly detected by visual inspection since peak symmetry is rarely measured and monitored in most GC analyses. A shark fin shape is the most common description and appearance of an overloaded peak. The middle peak in Figure 2-2 is an example of an overloaded peak. PLOT columns exhibit peak tailing when overloading occurs.

Column capacities are usually quoted as per peak and not for the entire sample. For example, a column capacity of 150 ng refers to 150 ng per compound or peak and not the mass of the injected sample. Column capacity does not refer to the volume of sample. The maximum sample volume is primarily a function of the injector, column dimensions, sample solvent, initial column temperature and retention factor ($k$). It is not unusual to have a column's capacity differ by 2–3 times for dissimilar compounds or retention.

Greater column capacity is obtained when the polarity of the stationary phase and compound are similar. Larger diameter and thicker film columns have greater capacities. Column capacity decreases as compound retention increases. At the same concentration, the early eluting peaks in a chromatogram have satisfactory shape while the later eluting peaks exhibit overloading characteristics. Table 2-3 lists approximate column capacities for common column dimensions.

A column is not harmed by injecting too much sample. The primary problem is the asymmetric peak. It is visually unappealing and it may not be integrated properly by the data system.

**Table 2-3** Approximate column capacities.

| Film thickness (µm) | Column diameter (mm) | | | | |
|---|---|---|---|---|---|
| | **0.10** | **0.18–0.20** | **0.25** | **0.32** | **0.53** |
| 0.10 | 10–15 | 20–35 | 25–50 | 35–75 | 50–100 |
| 0.25 | 25–30 | 35–75 | 50–100 | 75–125 | 100–250 |
| 0.50 | 40–60 | 75–150 | 100–200 | 125–250 | 250–500 |
| 1.00 | | 150–250 | 200–300 | 250–500 | 500–1000 |
| 1.50 | | | 250–375 | 300–625 | 625–1250 |
| 3.00 | | | 400–600 | 500–800 | 1000–2000 |
| 5.00 | | | 1000–1500 | 1200–2000 | 2000–3000 |

For similar polarity stationary phases and compounds.
Capacities reported as ng per compound.

# 3
# Capillary GC Columns: Tubing

## 3.1
## Fused Silica Capillary Columns

Capillary columns can be thought of having three distinct "layers" or parts. They are the fused silica tube, the protective coating and the stationary phase (Figure 3-1). It is assumed that "fused silica tubing" refers to tubing with the protective coating. The stationary phase is responsible for compound retention and separation (see Chapter 4). Fused silica tubing without stationary phase is available.

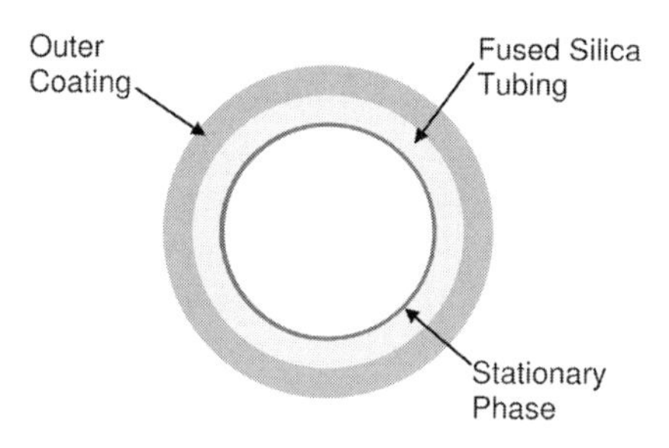


**Figure 3-1** Cross-sectional view of a capillary GC column (not to scale).

## 3.2
## Fused Silica Tubing

The only interaction occurring inside a capillary column should be between the stationary phase and the sample compounds. This requires an inert inner surface that does not interact with the compounds. Tubing that is not inert is called active, thus it exhibits activity. Fused silica is the material of choice for capillary column tubing since its purity is unsurpassed by any other suitable glasses. Fused silica is a synthetic, quartz-like glass containing typically less than 1 ppm of metallic oxide impurities. A very low concentration of these oxides is required to minimize the reactivity of the fused silica surface.

Untreated fused silica is very active due to the presence of silanol groups. Silanol groups interact with compounds with hydrogen bonding capabilities (Figure 3-2).

*The Troubleshooting and Maintenance Guide for Gas Chromatographers, Fourth Edition.* Dean Rood

ISBN: 978-3-527-31373-0

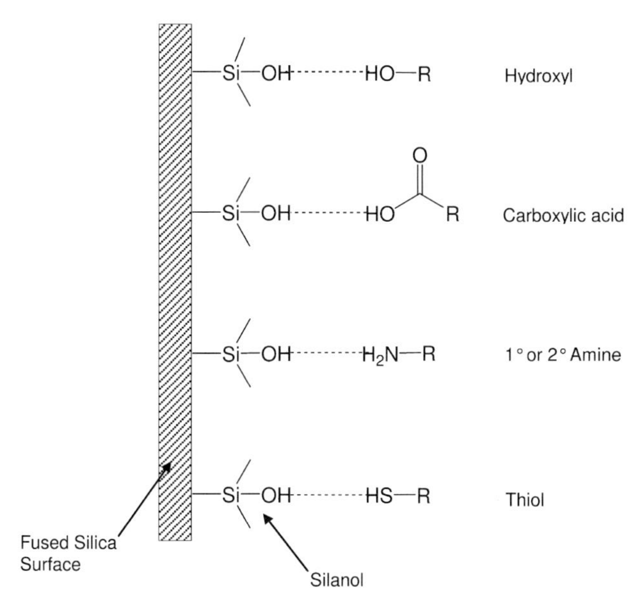


**Figure 3-2** Silanol and active groups.

In general, these types of compounds have thiol (-SH), hydroxyl (-OH), amine (-NH) or carboxylic acid (-COOH) functional groups. Interaction with an active surface results in tailing or reduced size peaks. Fused silica tubing is usually deactivated via surface modification using acid leaching and silylating reagents. The deactivation process eliminates the silanols by converting them into relatively nonpolar and nonreactive groups. Even after deactivation processes, some surface silanols are still present and contribute to the residual activity of columns. Some compounds always exhibit peak tailing due to their chemical structure and residual silanol groups.

Surface deactivation processes are also intended to create a suitable surface for coating of the stationary phase. The structure of the stationary phase may also dictate which deactivation process is used for a column. An incompatible surface results in an uneven coating of the stationary phase, thus a lower number of theoretical plates.

Columns lose some of their inertness with continued use. Slow loss of the deactivating groups with conversion back to the silanols occurs. Also, normal stationary phase degradation with column use creates active sites in the stationary phase. These processes are not easily reversed. There are column treatment solutions that are available, but they have little to no permanent effect.

## 3.3 Outer Coating

Fused silica tubing is fairly fragile and requires some type of protective coating. The most common coating material is polyimide. This polymer coating serves a dual purpose. It fills flaws in the tubing which prevents the further growth of any defects, and it acts as a barrier to prevent corrosion and damage to the fused silica tubing. Polyimide coated columns are very strong and do not require delicate handling. However, any handling or process causing scratches or abrasions in the polyimide coating results in a weak spot with possible breakage at a later time.

Standard polyimide coating is stable for only several days above 360 °C. Longer exposure times or higher temperatures makes the polyimide begin to discolor, flake and/or peel away from the tubing. The tubing may break or become brittle at these locations. The polyimide coating has excellent stability up to 360 °C and down to –60 °C. Below –60 °C, polyimide coated fused silica tubing may be brittle and easy to break upon application of stress. Recently, tubing with polyimide coatings stable to 400 °C has become available. This tubing is usually found on high temperature columns intended to be regularly used at 350–400 °C.

The polyimide coating is responsible for the brown color of fused silica tubing. The intensity of the brown color is influenced by slight batch-to-batch differences and the temperature of the polyimide coating process. The color has no effect on the performance of a capillary column. The coating is on the outside of the tubing while the separations occur inside the tubing. Some slight tubing strength or flexibility differences may be experienced with different color columns; however, there appears to be no pattern to the color and strength relationship. Column colors may range from very dark brown to a very light yellow. The color may darken with use especially if the column is routinely held at higher temperature.

Aluminum coated ("clad") fused silica tubing is also used. This tubing has a much higher upper temperature limit (~450 °C) than polyimide coated tubing. One drawback to aluminum clad tubing is its brittleness when exposed to continuous changes in temperature. It is satisfactory for isothermal temperature work; however, the tubing becomes weak upon cycling of the oven temperature during temperature programming situations. This has precluded widespread use of aluminum as a coating material.

## 3.4 Other Tubing Materials

Stainless steel clad columns are available. This tubing is probably better described as stainless steel tubing with a thin, interior coating of fused silica or a similar type of material. This tubing exhibits slightly higher activity than polyimide coated, fused silica tubing. This limits its use for compounds with little to moderate activity. Cracks in the fused silica can develop exposing the compounds to bare metal surfaces. This usually occurs if the tubing is bent at a sharp angle or kinked.

The exposed metal surfaces are very active. An increase in activity problems may occur with continued use of the column. The metal coated tubing is not a flexible as polyimide coated tubing which makes installation slightly more difficult. Also, the metal coated tubing is more difficult to properly cut. The primary advantages of this tubing are its high temperature limits and ruggedness. This tubing is the not same as the undeactivated stainless steel tubing used to make capillary columns in the 1960's.

## 3.5 Polyimide Fused Silica Tubing Bending Stress

Due to the extreme lengths of capillary columns, the tubing is wound onto a round cage or basket with diameters of 5–8 inches (~13–20 cm). Each column manufacturer uses a slightly different style of column cage, but the overall design is the same and the tubing is bent in order to be wound onto the cage. Bending fused silica tubing produces localized tension or bending stress which increases as the diameter of the column cage decreases. The probability of tubing breakage increases as the bending stress increases.

Table 3-1 lists the minimum recommended cage or bending diameters for fused silica columns. Bending tubing beyond the recommended minimums will significantly increase of the probability of premature tubing breakage. Column manufacturing processes often weaken the tubing, thus the values in Table 3-1 are below the published values for unprocessed fused silica tubing coated with polyimide.

**Table 3-1** Minimum bending diameters for polyimide fused silica tubing.

| Tubing i.d. (mm) | Tubing o.d. (mm)[1)] | Minimum bending diameter |
|---|---|---|
| 0.05–0.25 | 0.34–0.40 | 5 cm (2 in) |
| 0.32 | 0.40–0.48 | 7.5 cm (3 in) |
| 0.45–0.53 | 0.65–0.75 | 13 cm (5 in) |

1) Typical o.d. range for the respective i.d.'s.

# 4
# Capillary GC Columns: Stationary Phases

## 4.1
## Stationary Phases

Capillary column stationary phases are polymers deposited on the inner walls of the tubing in a thin, uniform film. The polymers must withstand high temperatures for prolonged periods of time, be non-reactive, and evenly coat the surface of the tubing. The uniformity, thickness and chemical nature of the stationary phase are very important to the overall behavior and performance of a column.

The structure of the stationary phase influences the *separation* of the compounds. Column dimensions primarily affects the *resolution*. A column with very high resolution is not useful if the stationary phase can not separate the desired compounds. Capillary columns require relatively few stationary phases to achieve the separations necessary for complex samples. For this reason, only 12–15 distinct stationary phases are available for capillary columns.

When the separation of two compounds does not occur with one stationary phase, it is very likely that separation can be achieved with a different stationary phase. A stationary phase is selected primarily based on whether it provides adequate compound separation. There are other considerations and limitations involved with stationary phase selection; however, these are relatively minor when compared to obtaining the required separations.

Unfortunately, there are no simple tables or foolproof formulas that can be used to easily select the proper stationary phase. On the other hand, there are some relatively simple concepts that can be successfully used in most cases. A basic understanding of the structure of common stationary phases and the properties of various compound functional groups is all that is necessary.

*The Troubleshooting and Maintenance Guide for Gas Chromatographers, Fourth Edition.* Dean Rood

ISBN: 978-3-527-31373-0

## 4.2 Types of Stationary Phases

### 4.2.1 Polysiloxanes or Silicones

The predominant polymers used as stationary phases are the substituted polysiloxanes. These stationary phases are sometimes referred to by their common name of silicones. Polysiloxanes are considered to be the stationary phases most resistant to abuse and they have superior lifetimes compared to most other types of stationary phases. Polysiloxanes are distinguished by their alternating silicon and oxygen linear backbone and two functional groups attached to each silicon (Figure 4-1). It is the structure of the functional groups and amount of each group that distinguishes each stationary phase.

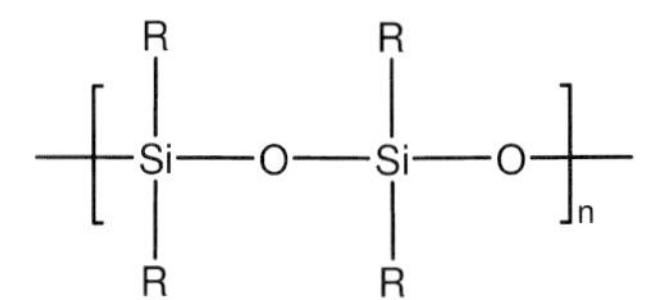


**Figure 4-1** Base siloxane backbone unit.

There are only a few structures used as functional groups. They are limited to ones that can be attached to the polysiloxane backbone without significantly impairing the thermal stability of the polymer. Also, each group needs to change the stationary phase in a manner to make it have unique separation characteristics. The primary groups used are methyl, phenyl, cyanopropyl and trifluoropropyl. These groups are used in various amounts and combinations to impart specific separation characteristics to each stationary phase. The total numbers of each group is precisely regulated to maintain the bulk properties of the polymer. The functional groups are usually not added to the backbone in a irregular pattern, but in a semi-ordered fashion. The base structures of the most common stationary phases are illustrated in Figure 4-2.

Stationary phase descriptions provide information about the type and amount of substitution on the polysiloxane backbone. Manufacturer's brand names often have no real meaning or significance and are not a reliable method to obtain structural information about a stationary phase. There are plenty of examples where similar brand name columns are not equivalent in structure, and hence separation behavior.

Polysiloxane stationary phases use a percentage substitution system to designate stationary phase structures. For example, a 5% phenylmethylpolysiloxane is comprised of a phenyl group attached to 5%, by number, of the sites on the backbone silicon atoms; the remaining 95% of the silicon sites have methyl groups. Sometimes a stationary phase is named using a shorthand system. The previous example could be called a 5% phenyl phase. When not stated, the remaining group is methyl and at a percentage to make the total 100%.

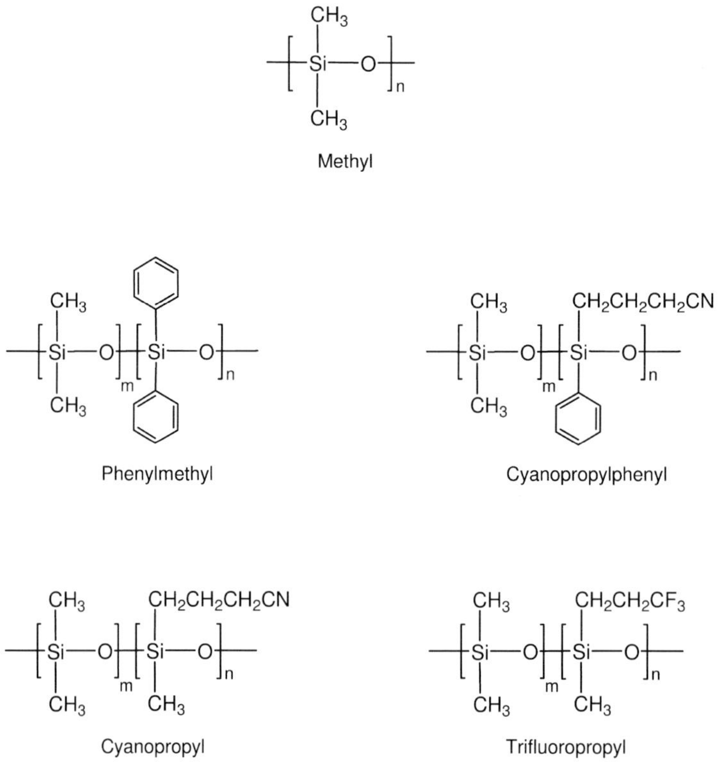

Methyl

Phenylmethyl

Cyanopropylphenyl

Cyanopropyl

Trifluoropropyl

Figure 4-2 Common substituted polysiloxanes.

The naming of cyanopropylphenyl polysiloxanes can be confusing. The cyanopropyl and phenyl groups are separate, but attached to the same silicon atom; the cyanopropyl group is not attached to the phenyl ring. For example, a 14% cyanopropylphenylmethyl polysiloxane has a cyanopropyl attached to 7% of the silicon sites, a phenyl on 7% of the sites (on the same silicon atoms as the cyanopropyl groups) and a methyl on 86% of the sites. Realizing there are only four major functional groups (i.e., methyl, phenyl, cyanopropyl and trifluoropropyl) helps to clarify most stationary phase naming confusions.

Commonly, about 1% of the silicon backbone atoms are substituted with a vinyl group which is used for bonding and cross-linking purposes (Section 4.3.1). In most cases, the vinyl content is not stated. The carbon-carbon bonds comprising the bonding and cross-linking groups do not significantly influence the separation characteristics of the stationary phase, thus they are often ignored.

### 4.2.2
### Arylene-Modified Polysiloxanes

Aryl-poly or arylene stationary phases are similar to standard polysiloxanes except the polymer backbone contains phenyl groups (Figure 4-3). The actual location and amount of phenyl substitution is usually considered proprietary by column manufacturers. Arylene stationary phases exhibit lower column bleed (Section 4.7) and higher temperature limits (Section 4.6) than their polysiloxane counterparts.

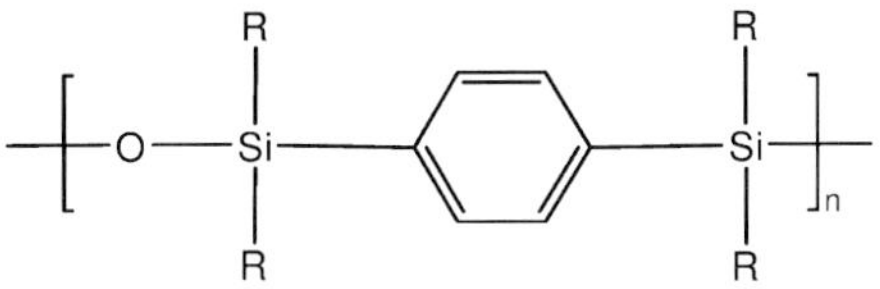

**Figure 4-3** Arylene stationary phase.

Most low bleed columns contain arylene stationary phases; however, changes to the manufacturing process or column testing criteria may be used by some manufacturers to obtained lower bleed columns. Most arylene stationary phases are designed to be equivalent to a familiar stationary phase such as a 5% phenylmethylpolysiloxane (e.g., DB-5ms and DB-5). The separation characteristics are very similar, but not exactly the same. The separation differences are usually very small, thus insignificant. A few arylene or low bleed columns may contain an unique stationary phase without an equivalent polysiloxane counterpart (e.g., DB-XLB).

### 4.2.3
### Polyethylene Glycols

A variety of different polyethylene glycols (PEG) are commonly used as capillary GC column stationary phases (Figure 4-4). Polyethylene glycol stationary phases differ primarily by the molecular weight or chain length of the polymer chains. PEG stationary phases have some unique separation characteristics not exhibited by polysiloxanes. The major disadvantage of PEG stationary phases is their high sensitivity towards oxygen, especially at high temperatures. The presence of oxygen in the carrier gas causes rapid destruction of most capillary column stationary phases; polyethylene glycols are among the stationary phases most susceptible to this problem. PEG stationary phases have narrow temperature ranges where they can be safely used without damage or performance problems. The liabilities are

$$HO-\left[-CH_2OCH_2O-\right]_n-H$$

**Figure 4-4** Polyethylene glycol stationary phase.

tolerated since PEG stationary phases have been widely used for a long time, and there is a large database of separation data available in the published literature.

When analyzing samples that are acidic or alkaline, a PEG stationary phase may begin to mimic some of the pH characteristics of the sample. For example, if a large number of amine samples have been injected into the column, it may become slightly alkaline. A subsequent injection of an acidic compound may result in a tailing peak or no peak at all. This pH shift is reversible by solvent rinsing the column (only if it is a bonded and cross-linked phase) with a small amount of water, followed by acetone then hexane (Section 11.7).

Acidic or alkaline PEG columns are available. They are used when minimally tailing peaks are required for organic acids or bases that are separated with a PEG column. Obviously, an acidic column excludes the use of the column for alkaline compounds and vice versa. Unfortunately, many of these columns are not very stable and require frequent replacement.

### 4.2.4
### Porous Layer Stationary Phases

An interesting type of stationary phase is the porous polymers. Capillary columns with these stationary phases have a layer of small, porous particles coated onto the inner wall of the fused silica tubing. The particles are usually held in place by some type of chemical binder. These types of columns are called porous layer open tubular or PLOT columns. Stationary phases of aluminum oxide (Alumina), molecular sieves and a series of porous polymers (Poraplot-like) are currently available.

For PLOT columns, gas-solid adsorption is the primary separation mechanism involved rather than gas-liquid partitioning as for polysiloxanes and PEG. The stationary phases used in PLOT columns are more retentive than liquid stationary phases. This makes PLOT columns well suited for the analysis of light hydrocarbons, sulfur gases, permanent gases or other very volatile compounds. Polysiloxanes require column temperatures well below 35 °C to separate these types of compounds, thus necessitates the use of liquid nitrogen to cool the GC oven. This adds expense, complexity and time to the analysis. Many PLOT columns can separate these very volatile compounds at column temperatures of 35 °C or above.

Since PLOT columns are so retentive, they are not suited for many types of compounds. Unless the compound is a gas at room temperature or a very low boiling liquid, the compound requires a very long time to exit the column; some may not exit the column at all. Peak width increases very rapidly with retention with PLOT columns. Beyond a $k$ of about 10, the peak widths are very wide. PLOT columns are less efficient than columns with liquid stationary phases and are usually limited to wider diameter fused silica tubing. Depending on the stationary phase and conditions, some stationary phase particles can become dislodged from the column and possibly foul the GC detector.

## 4.3
## Characteristics of Stationary Phases

### 4.3.1
### Bonded and Cross-linked Stationary Phases

Long chain length polymers are used for most stationary phases. To improve the stability and longevity of the stationary phases, the individual chains are interconnected using a two carbon linking unit using a process commonly called cross-linking. Upon linking, the stationary phase becomes like a very large, branched polymeric molecule. Polymers are also attached to the surface of the fused silica tubing using the same type of two carbon unit linkage. This anchors the stationary phase to the tubing and aids in prevents the stationary phase from leaving the column. Stationary phases stabilized in such a manner are commonly referred to be as being bonded.

Columns with non-bonded and cross-linked stationary phases require substantial maintenance and are subject to some limitations in use. Non-bonded stationary are subject to damage or disruption upon the injection of some solvents. This usually caused poor peak shape and loss of column efficiency. Organic solvents and water do not damage bonded and cross-linked stationary phases. Peak shape problems may occur with some compound, stationary phase and solvent combinations, but stationary phase damage does not occur. An added benefit of bonded and cross-linked stationary phases is that the column can be washed with solvents to remove contaminants. This substantially prolongs the useable life of the column and minimizes the potential damage caused by dirty samples.

Whenever possible, a column with a bonded and cross-linked stationary phase should be used. The long term stability, reproducibility, and performance are better. A few stationary phases are not available in bonded form. In these cases, greater attention to column care and maintenance is required to maximize column life and performance. The use of a retention gap (Section 11.8) is strongly recommended to minimize the amount of problems encountered with non-bonded stationary phases.

### 4.3.2
### Stationary Phase Polarity

Stationary phase polarity is a bulk property of the polymer comprising the stationary phase. Stationary phase polarity is directly related to the amount and polarity of each functional group. The polarity of a stationary phase increases as the polarity of the substituted groups and their relative amounts increase. Table 4-1 lists the most common capillary column stationary phases in the order of increasing polarity.

Stationary phase polarity influences compound separation; however, stationary phase selectivity (Section 4.3.3) is a more important factor when considering separations. Stationary phase polarity should be considered when temperature

**Table 4-1** Stationary phase polarities.

| In order of increasing polarity | |
|---|---|
| 100% Methyl | 50% Phenyl-methyl |
| 5% Phenyl-methyl | 65% Phenyl-methyl |
| 6% Cyanopropylphenyl-methyl | 50% Trifluoropropyl-methyl |
| 20% Phenyl-methyl | Polyethylene glycol (PEG) |
| 35% Phenyl-methyl | 50% Cyanopropyl-methyl |
| 14% Cyanopropylphenyl-methyl | 80% Cyanopropyl-methyl |
| | 100% Cyanopropyl-methyl |

Based on Kovats indices.

ranges, column lifetime, capacity and bleed are important. It is common for more than one stationary phase to provide the desired separations. In these cases, use the most non-polar stationary phase. Non-polar stationary phases are superior to polar phases in several areas. Non-polar stationary phase columns can be used over a wider temperature range. For very low or high temperature analyses, stationary phase selection is primarily limited to the non-polar ones. Polar stationary phases are more fragile than non-polar phases. Under equivalent conditions, a non-polar stationary phase column has a longer lifetime. Polar stationary phases generate a higher background signal (bleed) than non-polar stationary phases. This makes non-polar stationary phase columns better for low level analysis of compound that elute at high temperatures. Greater capacity is obtained for stationary phases close in polarity to the compounds. For example, a column with a polar stationary phase has greater capacity for polar compounds than non-polar compounds.

Using a more polar phase never guarantees the separation of compounds that co-elute on a less polar phase. Polar columns are often tried as a means to improve a separation in the mistaken belief that increased polarity translates into better separation. This approach is successful on occasion, but due to other factors not directly related to the change in stationary phase polarity.

### 4.3.3 Stationary Phase Selectivity

For two compounds to be separated, each must spend a different amount of time in the column. This is accomplished when one compound spends more time in the stationary phase than the other. The time spent in the stationary phase depends on the interaction of each compound with the stationary phase. These interactions affect retention, and thus also compound separation.

The interactions are dependent on the structure of the compounds and the stationary phase. Each stationary phase has a particular set of interactions that it can undergo with the compounds. If a compound has one of the characteristics required to participate in the interactions, the separation is influenced by those interactions. If a compound does not have a characteristic required for a particular interaction, that interaction does not occur and does not have an influence on the separation. This behavior is called stationary phase selectivity. The stationary phase selectively interacts with some compounds at the exclusion of others.

## 4.4 Stationary Phase Interactions

There are numerous ways to describe or explain stationary phase interactions. *In order to discuss selectivity on a practical level, some over-simplifications and generalization have to be made.* Instead of describing the actual interactions on a molecular level, retention and separation behavior can be described using common physical and chemical properties. The actual and complex molecular interactions can be simplified and reduced into three interactions – dispersion, dipole and hydrogen bonding (refer to the paper in the Reference Section for more detailed information on these interactions). These simplifications are valid in most cases and can be used to better understand the complex behaviors of stationary phases and compounds.

### 4.4.1 Dispersion Interaction

Dispersion is the most basic and universal of the stationary phase interactions. As shown in Table 4-2, the dispersion interaction is high for all stationary phases, thus the primary separation mechanism for nearly all compounds. The dispersion interaction is related to the intermolecular attraction between the compound and stationary phase. The polarizability of the compound plays a major role in this interaction. Larger molecules are more polarizable than smaller molecules of similar functionality.

**Table 4-2** Dispersion interaction.

| Functional group or phase | Dispersion interaction strength |
|---|---|
| Methyl | Strong |
| Phenyl | Very strong |
| Cyanopropyl | Strong |
| Trifluoropropyl | Strong |
| PEG | Strong |

Interaction strengths: none, weak, moderate, strong, very strong.

An erroneous, but common simplification for dispersion-only stationary phases is to relate retention to compound boiling points. That is, compounds with higher boiling points are more strongly retained. This can be a useful simplification in some cases (e.g., compounds in a homologous series); however, there are many examples that do not follow this simplification (e.g., a collection of compounds with a mixture of functional groups and structures). A better simplification is to relate retention to compound vapor pressures. Compounds with lower vapor pressures at the column temperature are more strongly retained on dispersion-only stationary phases. The vapor pressure simplification still involves some error, but it is much smaller than using compound boiling points. Unfortunately, compound vapor pressures are not as readily available in references sources as compound boiling points.

The solubility of the compound in the stationary phase is one aspect of dispersion. Higher compound solubility in the stationary phase results in greater retention. Compound solubilities in stationary phases can be difficult to estimate. Polar compounds are more soluble (i.e., more retained) in a polar stationary phase than a non-polar stationary phase; non-polar compounds are more soluble in a non-polar stationary phase than a polar stationary phase. Difficulties arise when trying to determine the retention compounds of diverse polarities and structures. For example, the relative retention of a large non-polar compound and a small polar compound on a polar stationary phase can be difficult to predict. The small size of the polar compound results in low retention while its polar character results in high retention on the polar stationary phase; the large size of the non-polar compound results in high retention while its non-polar character results in low retention on the polar stationary phase. It is not easy to determine whether compound size or polarity has the greatest influence on retention with this stationary phase.

### 4.4.2 Dipole Interaction

The dipole interaction is related to the interaction between the stationary phase and compounds with dipole moments. Cyanopropyl substituted polysiloxanes and PEG stationary phases have particularly strong dipole interactions (Table 4-3). If a compound does not have a dipole moment (permanent or induced), the dipole interaction does not have an influence on retention. Also, if a compound has a dipole, but the stationary phase does not exhibit a dipole interaction, a dipole interaction does not occur.

A compound has an appreciable dipole if there is uneven charge distribution due to the structure of the compound. This does not mean that the compound has a charged group. As long as part of the molecule is slightly positive and another is slightly negative, the compound has a dipole moment. There are some generalizations that aid in determining whether compounds have moderate to large dipole moments. Compounds with heteroatoms (O, N, S, P) or halogens (Cl, F, Br, I) usually possess a dipole moment. Also, compounds with hydroxyl (-OH) and amine (-NH) functional groups usually have large dipoles. Compounds with

**Table 4-3** Dipole interaction.

| Functional group or phase | Dipole interaction strength |
|---|---|
| Methyl | None |
| Phenyl | None |
| Cyanopropyl | Very Strong |
| Trifluoropropyl | Moderate |
| PEG | Strong |

Interaction strengths: none, weak, moderate, strong, very strong.

double bonds, ethers, ketones, and many esters usually have moderate to small dipole moments. Saturated hydrocarbons and compounds that are completely symmetrical (even if they have heteroatoms or halogens) usually have very little to zero dipoles. For compounds with complex structures or multiple functional groups, it can sometimes be difficult to determine if there is a significant dipole moment. Dipole moments are more dependent on the overall structure of the compounds and not on the functional groups themselves. There is a limited amount of dipole information available in the literature, but it is usually limited to common and widely used compounds.

It is the difference in the dipole moments between the compounds rather than the size of their dipoles that is most important. Regardless of the strength of the dipoles, if they are the same for two compounds, their dipole interactions are the same. If two compounds have different dipoles, the different dipole interactions may lead to different retention (ignoring the other interactions).

If the dipole difference between two compounds is small, a stationary phase with a large dipole interaction is needed. A large amount of the dipole interaction characteristic is necessary to distinguish the small difference between the two compound's dipole moments. For example, a PEG stationary phase or one with a high cyanopropyl content is required to separate compounds with small dipole differences that do not separate on other types of stationary phases.

### 4.4.3 Hydrogen Bonding Interaction

The hydrogen bonding interaction is related to the ability of the stationary phase and compound to hydrogen bond. It is not overly important which is the proton acceptor or donor as long as the interaction occurs. Table 4-4 shows that the strongest hydrogen bonding interaction occurs for the cyanopropyl and PEG stationary phases.

In general, a compound exhibits strong hydrogen bonding if it contains a hydroxyl (-OH) or amine (-NH) group. Other functional groups such as esters, ethers and ketones have moderate to small hydrogen bond strengths. Alkyl

**Table 4-4** Hydrogen bonding interaction.

| Functional group or phase | Hydrogen bonding interaction strength |
|---|---|
| Methyl | None |
| Phenyl | Weak |
| Cyanopropyl | Moderate |
| Trifluoropropyl | Weak |
| PEG | Moderate |

Interaction strengths: none, weak, moderate, strong, very strong.

groups (hydrocarbons) and halogens have weak to no hydrogen bonding strength. Hydrogen bonding is more dependent on the actual functional groups and not on the overall structure of the compound. There is little information in the literature on hydrogen bonding strengths.

It is the difference in the hydrogen bonding strength of the compounds and not the absolute strength of the bonding that determines the result of the interaction. Ignoring all other interactions, if two compounds have different hydrogen bonding strengths with the stationary phase, differences in retention may be observed. The smaller the difference in the hydrogen bonding strengths, the greater the hydrogen bonding interaction needed by the stationary phase. A PEG stationary phase or one with high cyanopropyl content is required to separate compounds with small hydrogen bonding differences that do not separate on other types of stationary phases.

### 4.4.4
### When There are Multiple Interactions

Many polysiloxane stationary phases exhibit multiple interactions due to their structure (Table 4-5). The influence of each interaction on the separations is difficult to predict. One interaction may be stronger than another and each interaction affects each compound to a different extent. The amount of each functional group influences the overall interaction properties of the stationary phase. For example, both cyanopropyl and PEG stationary phases have strong dipole and hydrogen bonding interactions. PEG stationary phases are usually 100% polyethylene glycol while the cyanopropyl content is usually less than 100% for most polysiloxanes. The dipole and hydrogen bonding interactions are stronger for the PEG stationary phase and have a greater impact on the separations.

Another problem with predictions about separations is the conflicting effects of two different interactions. One interaction may move the compound peaks further apart while the other has the effect of moving them closer together. For example, two compounds may separate due to the dispersive interaction, but the dipole interaction may alter the relative retention enough to cause co-elution.

**Table 4-5** Stationary phase interactions.

| Function group or phase | Dispersion | Dipole | Hydrogen bonding |
|---|---|---|---|
| Methyl | Strong | None | None |
| Phenyl | Very strong | None | Weak |
| Cyanopropyl | Strong | Very strong | Moderate |
| Trifluoropropyl | Strong | Moderate | Weak |
| PEG | Strong | Strong | Moderate |

Interaction strengths: none, weak, moderate, strong, very strong.

Using the various interactions to make separation predictions without any reference data is rarely successful. Polarity and interaction information is best used to determine whether a retention or separation shift is expected upon changing to a different stationary phase. If there is a separation problem that can not be solved by changing analysis conditions, changing to a different stationary phase is usually the next step. Success in finding a new stationary phase is improved by examining the compound and stationary phase interactions. At least certain stationary phases can be eliminated based on the lack of the necessary interactions or absence of significant change in the interactions. In some cases, there may be more than one stationary phase that provides adequate separation.

## 4.5 Stationary Phase Equivalencies

Most high quality column manufacturers control the synthesis of their stationary phases from start to finish, and they deactivate the fused silica tubing using proprietary processes. This control results in high quality and reproducible columns, but ones that may be slightly different. A column from one manufacturer may not be exactly the same as an equivalent stationary phase and size column from another supplier. In most cases, the differences are insignificant. Performance differences are more pronounced with polar stationary phases. The most common differences are the degree of inertness and slight retention variations. Quality differences are evident with efficiency, bleed and reproducibility being the main distinguishing characteristics.

Particular care must be taken when comparing specialty phases. These are columns designed for specific analyses and the stationary phase structures are usually proprietary. The stationary phases may be different for columns with apparently similar brand names (e.g., DB-608 and SPB-608). This can make specialty column comparisons difficult since the stationary phases may be different, but their structures are unknown. The separations obtained with these columns should be the guide when determining the equivalencies of specialty columns. Each column may have significantly different separations, making one column a better choice for a specific analysis or conditions.

## 4.6 Column Temperature Limits

All stationary phases have a temperature range over which they are stable and functional for extended periods. This range usually defines the temperature limits for a capillary column. Exceeding the temperature limits may lead to permanent damage to the column and loss of performance.

Below the lower temperature limit, the stationary phase loses its chromatographic properties. If the column is below its lower temperature limit, the peaks become broad and rounded. The problems becomes worse the further the column temperature is below its limit. The peaks regain their narrow and sharp form when the column is heated above the lower temperature limit. Permanent damage does not result if the column is exposed to temperatures below its lower limit. Cryocooling a column to focus a sample at the front of the column does not damage a column. Separation characteristics return as soon as the column heats up beyond its lower limit. At temperature below –60 °C, some brittleness of the fused silica tubing may be experienced.

The upper temperature limit is usually given as two numbers (e.g., 325/350 °C). In some cases, there may only be one upper limit. The first temperature is called the isothermal limit. This is the highest temperature that the column can be exposed to for indefinite periods of time. The second temperature is called the temperature program limit. The column can be left at this temperature for 10–15 minutes.

Exceeding the upper temperature limits results in damage to the stationary phase. Slight overexposure of the column to excessive temperatures does not result in the instant destruction of the column. In most cases, a small amount of column lifetime is sacrificed. The amount of lifetime loss is dependent on the temperature, the duration at the elevated temperature, the stationary phase, and the amount of oxygen present in the carrier gas. Greater damage occurs with longer exposure times, larger temperature excesses, higher oxygen concentrations, and a lack of carrier gas flow. Polar stationary phases are more susceptible to damage and degrade at a faster rate when exposed to excessively high temperatures.

Thermal damage is evident as excessive column bleed and peak tailing for active compounds. Loss of retention and efficiency (resolution) may be evident for severe cases of thermal damage. Temperature limits do not have any extra safety factors built into the values. Maintaining the column below its upper temperature limits should be practiced to ensure reasonable column life and performance.

Column temperature limits may differ slightly between manufacturers. Depending on column lifetime and especially bleed criteria, the limits may be set 10–20 °C lower or higher. Be careful to distinguish between a high temperature version of a stationary phase and the regular version.

## 4.7 Column Bleed

### 4.7.1 What is Column Bleed?

Column bleed is one of the most misunderstood aspects of capillary columns. Column bleed is evident as a rising or elevated baseline at temperatures near the upper temperature limit of the column. The rising baseline is caused by the elution of the degradation products of a column's stationary phase. This degradation is normal and increases at elevated temperatures. The degradation products are always present and are not necessarily a sign of a damaged column. Every column regardless of the source or quality exhibits some stationary phase bleed.

The primarily degradation process is breakage of the polymer chain at the Si-O bond. The amount of bond cleavage increases at higher temperatures. If the breakage occurs near the end of a polymer chain, the resulting small fragment is volatile enough to elute from the column. It is the continuous elution of these fragments that is responsible for column bleed. The sharp rise in the baseline is caused by the increasing amount of degradation species as the temperature increases. Oxygen acts to increase the amount of bond breakage, thus accelerating the rate and amount of stationary phase degradation. In general, polar stationary phases are less stable, thus their bleed levels are higher. Temperature limits are adjusted to partially accommodate for this behavior

Stationary phase degradation is temperature dependent. The rate and amount of degradation product formation is constant under isothermal column conditions. This means that the baseline should be flat at constant temperature. Any rise or deviation in the baseline during constant temperature situations is not due to column bleed, but to another source. In most cases, the baseline deviation is caused by system contamination.

Stationary phase degradation increases at higher column temperatures. If a column is temperature programmed, the degradation increases as the temperature program progresses. At lower temperature ranges, the baseline rise is small or not evident since the amount of stationary phase degradation is negligible. As the column temperature approaches the column's upper temperature limit, the amount of degradation becomes much greater. The degradation starts to rapidly increase at 30–40 °C below the isothermal temperature limit of the column, thus a sharply rising baseline is seen as the temperature program passes through this temperature region.

The appearance of column bleed is best illustrated by a bleed profile (Figure 4-5). A bleed profile is generated by running a temperature program without an injection (i.e., blank run). The temperature program should follow these guidelines: initial temperature of 50–100 °C, ramp rate of 10–20 °C/min, final temperature equal to the isothermal temperature limit of the column, and a final hold time of 15–20 minutes. The data system should be set so that the size of the baseline rise is 25–50% of full scale.

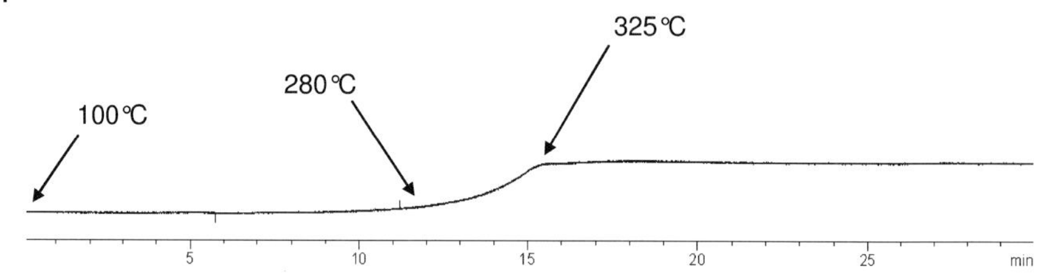


**Figure 4-5** Typical bleed profile.
*Column:* DB-5ms, 30 m × 0.25 mm, 0.25 μm
*Oven:* 100–325 °C at 15°/min, 325 °C for 15 min

The features of a bleed profile are: (1) No to little rise in the baseline at the lower temperature regions; (2) sharp rise starting at 30–40 °C below the upper temperature limit; (3) flat baseline during isothermal temperature conditions. A small rise and baseline drop may be evident for a few minutes after a column reaches and is held at its upper temperature limit.

It is important to notice the absence of significant peaks in the bleed profile. At very sensitive settings, a few very small peaks may be evident; however, they are contaminants in the GC and do not originate from the column. A column can not generate any peaks by itself. Introduction of a compound into the column followed by eventual elution from the column is required to obtain a peak. Bleed species originate along the entire length of column and elute from the column in a continuous stream. Bleed is a continuous process, thus a continuous signal is generated at the detector. This is the reason that column bleed is observed as part of the baseline signal instead of individual peaks. Stationary phase degradation can not be the source of peaks. Even a defect at a single point in the stationary phase does not result in peaks. Peaks are due to individual compounds originating from a contamination source somewhere in the GC system.

### 4.7.2 Measuring Column Bleed

Column bleed can be difficult to accurately measure. There are many contributors to the background signal with column bleed being only one of them. There are several methods used to measure column bleed. The detector signal at the column's temperature limit can be used. This signal level is obtained from the data system/ integrator or detector output signal on the GC. Other sources contribute to the magnitude of this signal; thus the contribution from the column is difficult to determine. A better technique is to compare the signal at the upper temperature limit to the one obtained at a lower temperature (usually between 50–100 °C). Most of the contributions from other sources are constant at the two temperatures; thus the difference in the signals is primarily due to the amount of column bleed. This technique is used by most column manufacturers in their quality control tests.

Collecting a bleed profile is another method to measure column bleed. If the same GC and data system settings are used, bleed profiles can be directly compared. While this method does not provide absolute numbers, column bleed changes or differences are readily visible upon comparison.

### 4.7.3
### Sensitivity Considerations

The size of the baseline rise as drawn on integrator paper or data system screen can be deceptive. The actual size of the baseline rise is dependent on the attenuation or scale setting for the recording device. For low level analyses, the chromatograms are expanded (small attenuation values) so that the small peaks are visible. At these settings, any rise in the baseline appears to be quite large. When the chromatograms are reduced in scale, the rise appears much smaller or is not visible at all. The amount of column bleed is the same, but the rise is amplified in the former situation.

Column bleed is a greater concern for low level analyses. The very small baseline rise present at lower temperatures becomes more visible and has a greater impact on peak integration. The large baseline rise at high temperatures goes off scale and any peaks in this region are not visible. The chromatogram has to be redrawn at smaller scale to keep the baseline on scale; however, the peaks of interest may become too small to be visible. Column bleed becomes less of a concern at higher sample concentrations. Even large increases in column bleed are not readily visible and do not interfere with the peaks.

### 4.7.4
### Detector Considerations

Due to the extreme sensitivity of some detectors for some stationary phase degradation products, some stationary phases appear to have substantially greater bleed than others. For example, the baseline rise at higher temperatures for a cyanopropyl substituted polysiloxane is much greater with a nitrogen phosphorus detector (NPD) than with a flame ionization detector (FID). The NPD is more sensitive to the presence of nitrogen than a FID. The amount of column bleed is the same for the two detectors, but the NPD is more sensitive to the bleed species and produces a much larger baseline rise at the same recorder settings. If the larger baseline rise does not create any problems, there is no damage to the column or GC.

From a chromatographic perspective, mass spectrometers are not significantly different from other, less complex GC detectors. A mass spectrum of the baseline contains any of the stationary phase degradation products eluting from the column when the spectrum was collected. The spectrum can be obtained anytime during the chromatographic run, but the bleed is most intense at the upper temperature limit of the column. A mass spectrum of the baseline is not the spectrum of a single column bleed compound, but of the sum of the stationary phase degradation products. The major fragments in the mass spectrum usually correspond to the more stable degradation products and are often the same as the mass fragments of the larger, less stable degradation products. The larger mass species are usually only visible for higher bleed columns. The mass spectra of the most common stationary phases can be found in Appendix D.

A common error is to attribute a rising baseline to excessive column bleed because siloxane materials are identified by the mass spectrometer's library program. Some types of contaminants are often misidentified as siloxanes. Silicone oils used in flow controllers and pressure regulators, and silicone septum bleed species often produce mass spectra very similar to those of many common stationary phases.

Adding to the problem is the presence of the major stationary phase degradation products at all times. In cases where the stationary phase bleed ions are more abundant than the compound ions (e.g., higher column temperatures, older column, low compound concentrations), the mass spectrum is reasonably close enough to authentic column bleed that the library search labels a peak or baseline areas as a polysiloxane. From this match, a conclusion of excessive column bleed is made. Obtaining reference mass spectra for a new column in a verified clean GC/MS system is recommended to minimize the possibility of misidentifying column bleed.

### 4.7.5 Minimizing Column Bleed

Complete elimination of column bleed is not possible. Column bleed continues to increase with column use. The increase in bleed is very small for a new column then it starts to elevate at a much faster rate. Eventually, the bleed increases to a level that renders the column unusable. The onset of excessively high bleed is dependent on the stationary phase polarity, temperature of column usage and the oxygen concentration in the carrier gas.

There are two conditions that increase or accelerate column bleed – high temperature and oxygen. Stationary phase degradation increases with temperature. Using columns at least 20–30 °C below their upper temperature limit avoids the temperature region where column bleed is the highest. Oxygen increases the rate of stationary phase degradation. Maintaining a leak free system and using high quality (low oxygen concentration) carrier gas slows down the onset of higher bleed.

## 4.8 Selecting Stationary Phases

Selecting the best stationary phase is a mixture of science, experience and a bit of guess work. A variety of factors need to be considered before deciding which stationary phase is best. The first step is to consult the literature. It is very likely that chromatograms or analysis information similar to the analysis of interest can be found in a published paper. Stationary phase selection is more certain and easier even with partial information from previous work. GC and column manufacturers are also good sources of applications information. Chromatograms are usually employed to help market and sell their products, and many manufacturers offer technical support services at no cost.

Selecting a stationary phase without any previous information is an inexact science. It is usually easier to determine which stationary phases will not work. This narrows the choices to the ones most likely to work. Before selecting a stationary phase, details about the compounds and their concentrations, available GC hardware, analysis time considerations, and any other important analysis factors need to be determined. These may influence stationary phase (and column size) selection.

It is fairly common for more than one stationary phase to separate all of the sample compounds. There are plenty of other factors that may influence stationary phase selection. When selecting a stationary phase, the following factors should be considered:

1. *Previous separation information:* Examine the compound separations on other stationary phases. These separations are often very helpful in selecting another stationary phase. Literature references and application notes are great sources of this type of information. If other columns are available in the laboratory, try the analysis on these columns providing the column dimensions are suitable. One of these columns may provide satisfactory separation.
2. *Selectivity:* Determine the potential compound/stationary phase interactions. If the compounds have different dipoles or can hydrogen bond, consider a selective stationary phase with these interactions. If the compounds are hydrocarbons or are part of a homologous series, a selective stationary phase is probably not necessary.
3. *Polarity:* Use the most non-polar stationary phase that provides the required separations. Non-polar stationary phases are more versatile, have longer lifetimes and generally exhibit better efficiency.
4. *Temperature limits:* Compounds with high boiling points or molecular weights require high column temperatures to avoid extremely long retention times. The lower temperature limits of polar stationary phases restricts their use to low and moderate boiling point compounds. When multiple columns are in the same GC oven, a lower temperature limit of one column may restrict the temperature range of the other.
5. *Compound activity:* Columns with non-polar stationary phases usually are the most inert. Greater peak tailing or adsorption may be experienced with polar stationary phases.
6. *Analysis time:* Some stationary phases provide satisfactory separations in less time.
7. *Capacity:* Stationary phases similar in polarity to the compounds have greater capacity for those compounds. If one or more of the compounds are present

at very high levels, the overloaded peak may interfere with some of the other peaks.

8. *Bleed:* In general, non-polar stationary phases have lower bleed. Compounds with high boiling points or molecular weights elute in the higher temperature regions where column bleed is more severe. Non-polar stationary phases not only bleed less, but the onset of higher bleed occurs at higher temperatures.

9. *Selective detectors:* Avoid stationary phases containing the species or functional group that generates a strong response with a selective detector. Extreme baseline rises and higher noise levels usually occur.

10. *Versatility:* For multiple analyses, different stationary phases may be required for optimal separation. In some cases, several analyses can be performed with a single stationary phase providing some sacrifices are acceptable. This reduces the number of columns needed which may reduce complexity and cost.

11. *Critical separations:* The separation of some peaks may be more important than others. Greater separation of these peaks with less separation of others may be desired. This helps to compensate for errors, concentration induced retention shifts and variable matrix interferences that can reduce the separation of the critical peaks.

# 5
# Capillary GC Columns: Dimensions

## 5.1
## Introduction

Capillary columns are available in a wide variety of sizes. Combined with the different stationary phases, there are hundreds of different possible columns. The large number of column possibilities seems to make column selection difficult. A knowledge of the effect of each column dimension on the chromatography greatly facilitates column selection and use. Column length and diameter mainly affect peak resolution while film thickness mainly affects peak retention. The stationary phase is responsible for the separation of the peaks. If better separation is needed, changing a column dimension rarely improves the situation; changing the stationary phase or column temperature is required. If better resolution is needed, changing one or more of the column dimension is required.

## 5.2
## Column Length

The most frequently used capillary column lengths are between 12 and 30 meters. Columns as long as 150 meters and short as 5 meters are commercially available. Differences of 10–30 meters between commercially available lengths are common. Fortunately, column lengths are easily shortened by cutting the tubing and lengthened by joining two pieces with an union. Most column manufacturers will supply columns in lengths other than their regular offerings on a special request basis. Using the standard column lengths is easier and makes the most sense (and usually less expensive) since column length is not a critical column dimension in most cases.

### 5.2.1
### Column Length and Efficiency/Resolution

The number of total theoretical plates is directly proportional to column length (Table 5-1). For example, doubling column length doubles the total number of theoretical plates or halving the length reduces the number of plates by half.

*The Troubleshooting and Maintenance Guide for Gas Chromatographers, Fourth Edition.* Dean Rood

ISBN: 978-3-527-31373-0

**Table 5-1** Changes in efficiency and resolution with column length.

| Length (m) | *N* | *R* |
|---|---|---|
| 5 | 23,765 | 0.41 |
| 12 | 57,035 | 0.63 |
| 15 | 71,295 | 0.71 |
| 25 | 118,825 | 0.91 |
| 30 | 142,590 | 1.00 |
| 50 | 237,650 | 1.29 |
| 60 | 285,180 | 1.41 |
| 75 | 356,475 | 1.58 |

Theoretical maximum efficiency for 0.25 mm i.d. column; $k = 5$.
Resolution normalized to the 30 meter column.

Increasing the number of theoretical plates results in better resolution, but the resolution increase is not directly proportional to the increase in the number of theoretical plates.

Resolution is proportional to the square root of the number of theoretical plates. Since the number of theoretical plates and column length are directly proportional, resolution is proportional to the square root of column length (Table 5-1). For example, doubling column length increases resolution by 41% (not 100%) or halving the length reduces resolution by 29% (not 50%). In practice, resolution changes are smaller than theoretically expected. For isothermal temperature conditions, the resolution changes are close to the theoretical values (Figure 5-1). For temperature program conditions, the actual resolution changes are often smaller than the theoretical values for low to moderate $k$ solutes ($k \sim 7$ or lower) and resolution can change in the opposite direction than expected for higher $k$ solutes ($k \sim 7$ or higher) (Figure 5-2). Adjustments to the later portion of the temperature program is often necessary to maintain or gain resolution when changing to a longer column. In most cases, decreasing the ramp rate makes the largest improvement.

The square root relationship means column length has to increase by four times to double the resolution by changing column length alone. For shorter columns, this is not much of a problem. A 15 meter column can be replaced with one 60 meters long since 60 meter columns are readily available. For longer columns, this is a problem. A 30 meter column has to be increased to 120 meters to double the resolution; 60 meters is often the longest standard length columns available.

The square root relationship of length and resolution is beneficial when shortening column lengths. A 30 meter column has to decrease to 7.5 meters before the resolution is halved. A substantial length of column can be removed without negatively affecting resolution. For example, removing 1 meter from a

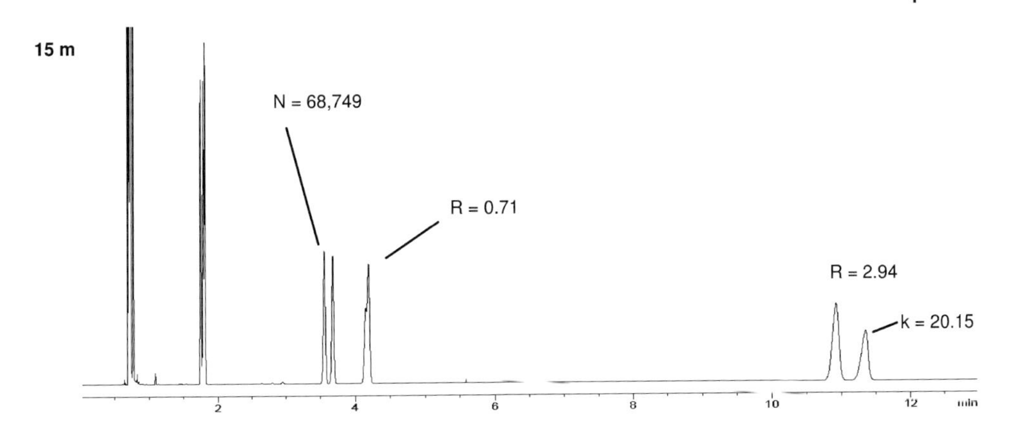


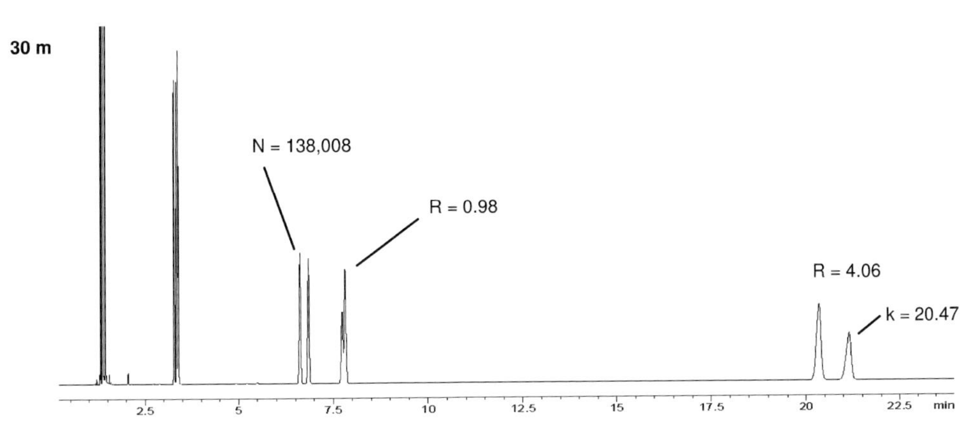


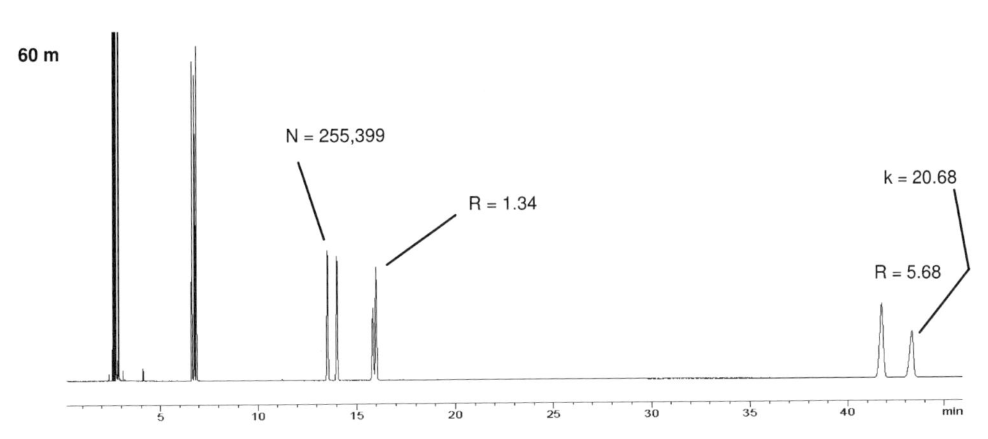


**Figure 5-1** Changes in efficiency, resolution and retention with column length – *isothermal*.
*Column:* 0.25 mm i.d., 0.25 μm
*Carrier:* Hydrogen at 50 cm/sec
*Oven:* 65 °C

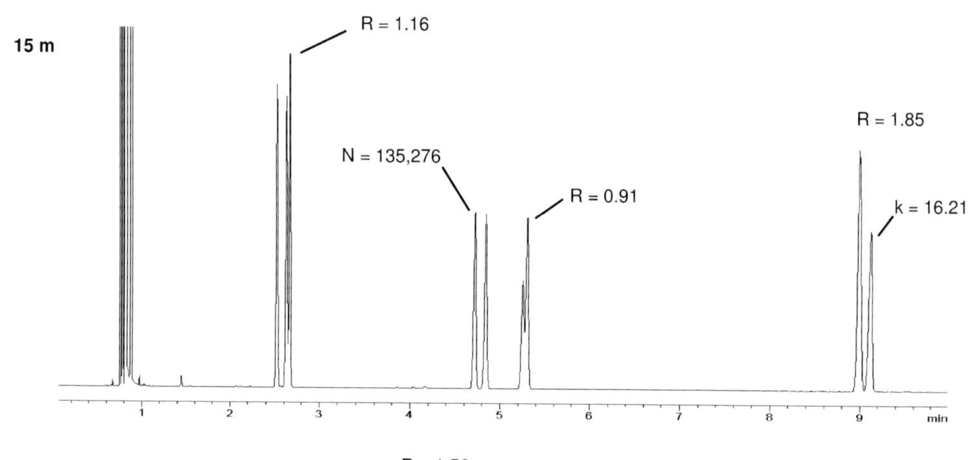


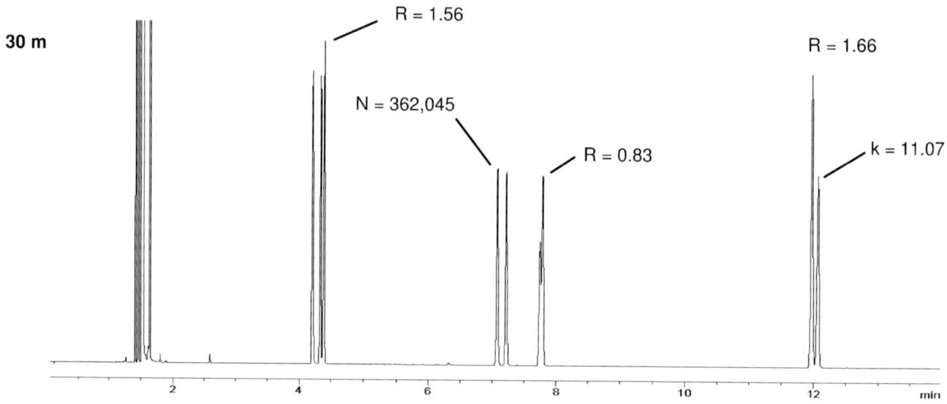


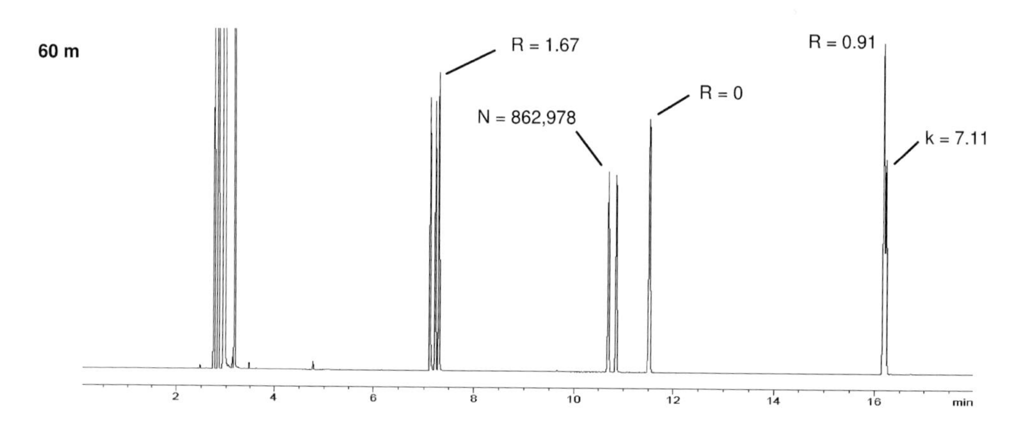


**Figure 5-2** Changes in efficiency, resolution and retention with column length – *temperature program*.
*Column:* 0.25 mm i.d., 0.25 µm
*Carrier:* Hydrogen at 50 cm/sec
*Oven:* 50 °C for 1 min, 50–140 °C at 5°/min

30 meter column reduces resolution by less than 2%. Trimming short pieces off a column such as during column installation has virtually no impact on resolution.

Changes in column length have less of an influence on resolution as column lengths become longer. Adding 5 meters to a 60 meter column results in a theoretical 4% increase in resolution; however, adding 5 meters to a 10 meter column increases resolution by up to 22%. Removing 5 meters from a 60 meter column results in a theoretical 4% decrease in resolution; however, removing 5 meters from a 10 meter column decreases resolution by as much as 29%.

### 5.2.2
### Column Length and Retention

Retention as measured by the retention factor ($k$) is not affected by a change in column length for isothermal conditions; however, retention time is affected by a change in the column length. For isothermal temperature conditions, the change in the retention time is roughly proportional to the change in column length. For example, doubling column length almost doubles the retention times or halving the length reduces retention time by almost half (Figurc 5-1). For temperature program conditions, the change in retention time is less than for isothermal conditions. In most cases, the change in retention time is between 25 and 75% of the isothermal (Figure 5-2). The change in retention factor ($k$) for temperature program conditions depends on the program parameters and $k$ can actually decrease with longer columns. The chromatograms in Figure 5-2 show this specific type of retention behavior.

### 5.2.3
### Column Length and Pressure

The column head pressure is directly proportional to column length. For example, doubling the column length doubles the head pressure required to maintain a particular carrier gas average linear velocity (or flow rate) and halving the column length reduces the required head pressure by half.

Most GC system work better when the head pressure is greater than 1 psig. Controlling and consistently setting the average linear velocity is very difficult when the pressure gauge reads below 1 psig even for electronically controlled pressure systems. Large diameter (e.g., 0.53 mm i.d.) columns require low head pressures to deliver the proper average linear velocity. Even with helium as the carrier gas, 15 meter × 0.53 mm i.d. columns barely have enough back pressure to generate a head pressure greater than 1 psig.

Narrow diameter columns have the opposite problem of those with wide diameters. Long lengths of narrow diameter columns require high head pressures. For example, a 60 m × 0.25 mm i.d. column at 100 °C requires about 34 psig (235 kPa) of helium to deliver an average linear velocity of 30 cm/sec. The pressure gauges on many GC's have an upper range of 30 psig (210 kPa). The pressure gauge has to be changed to a higher range one to accommodate the long, narrow column.

Mass spectrometers are under vacuum and the column exit is in this vacuum. The vacuum helps to pull the carrier gas through the column. Significantly lower head pressures are required to maintain suitable carrier gas velocities in GC/MS systems. Even 15 m × 0.25 mm i.d. columns using helium as a carrier gas operate around 1 psig. Usually longer columns are necessary with GC/MS systems especially if wider diameter columns are being used.

### 5.2.4 Column Length and Bleed

Column bleed increases as column length increases. The bleed increase is usually small. In theory, the bleed change should be roughly proportional to the change in the column length. In practice, the column-to-column bleed variation between columns of the same description is close to the difference in bleed between columns of different lengths. Rarely is the increased column bleed with longer columns ever a factor when selecting column length.

### 5.2.5 Column Length and Cost

For a particular stationary phase and diameter, column cost is primarily dependent on column length. Cost and column length are almost proportional. Using excessively long columns can be a costly error. Since resolution is dependent on the square root of column length, increasing resolution by increasing column length is an expensive method. Along with the additional column cost, the longer run times is not a cost effective situation in many laboratories.

The cost of two 30 meters columns is 10–20% greater than one 60 meter column of the same description. Two 30 meter columns require twice the packaging, handling and testing than one 60 meter column. It is tempting to cut a longer column into shorter lengths to save some money. In most cases, there are no significant differences between the individual lengths from a single column. Differences between the individual pieces start to appear with increases in stationary phase polarity, length and stationary phase film thickness, and a decrease in diameter of the original column. The possibility of one of the pieces being different increases as the number of pieces cut from the original column goes up.

### 5.2.6 Selecting Column Length

Most analyses can be easily done on 12–30 meter columns. Columns longer than 30 meters should only be used when all other reasonable methods of improving resolution have been exhausted or are unavailable. Usually, the increase in analysis time and cost are greater than the return in resolution improvement. Only very complex samples or the most difficult to resolve compounds should be analyzed using a 60 meter column. Many analyses are performed with columns that are too long. Unfortunately, columns are available in fairly large increments of length. One length is too short while the next longer length is too long. Often the longer column has to be used with its excessive resolution and run times. Increasing the column temperature often eliminates some of the excess analysis time while still obtaining an increase in resolution. Occasionally, a longer column may be needed to increase carrier gas head pressures in GC/MS systems. Sometimes, reducing the column diameter instead of increasing column length is better method to increase the head pressure.

## 5.3 Column Diameter

The most frequently used capillary column inner diameters are between 0.25 and 0.53 mm. Columns with diameters as small as 0.05 mm are available, but they are rarely used for GC applications. The common name of Megabore$^{TM}$ is often applied to 0.53 mm i.d. columns. The differences between the available diameters seem small, but the relative differences are quite large (30–70%).

### 5.3.1 Column Diameter and Efficiency/Resolution

The number of theoretical plates is inversely proportional to column diameter. In other words, theoretical plates per meter increases as column diameter decreases (Table 5-2). Decreasing column diameter is one of the easiest methods to increase

**Table 5-2** Changes in efficiency and resolution with column diameter.

| Diameter (mm) | *N* | *R* |
|---|---|---|
| 0.10 | 375,000 | 1.62 |
| 0.18 | 198,025 | 1.18 |
| 0.20 | 178,220 | 1.12 |
| 0.25 | 142,575 | 1.00 |
| 0.32 | 111,390 | 0.88 |
| 0.45 | 79,210 | 0.75 |
| 0.53 | 67,255 | 0.69 |

Theoretical maximum efficiency for 30 meter column; $k = 5$.
Resolution normalized to the 0.25 mm i.d. column.

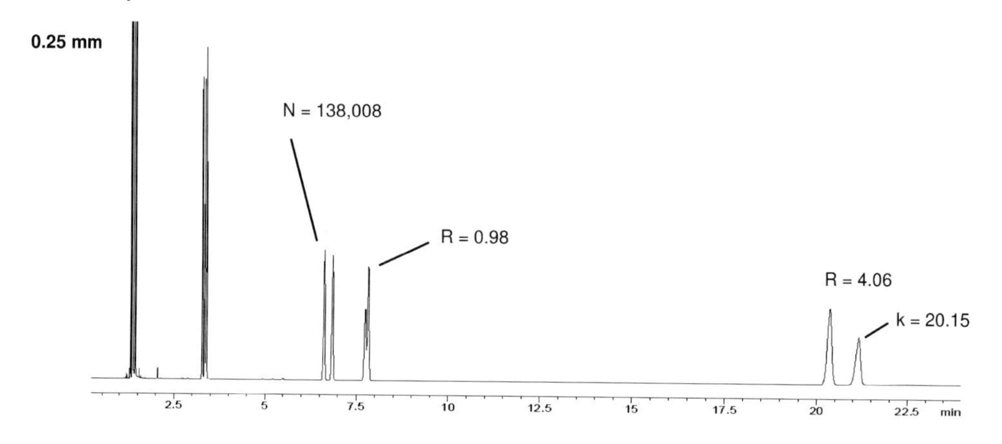


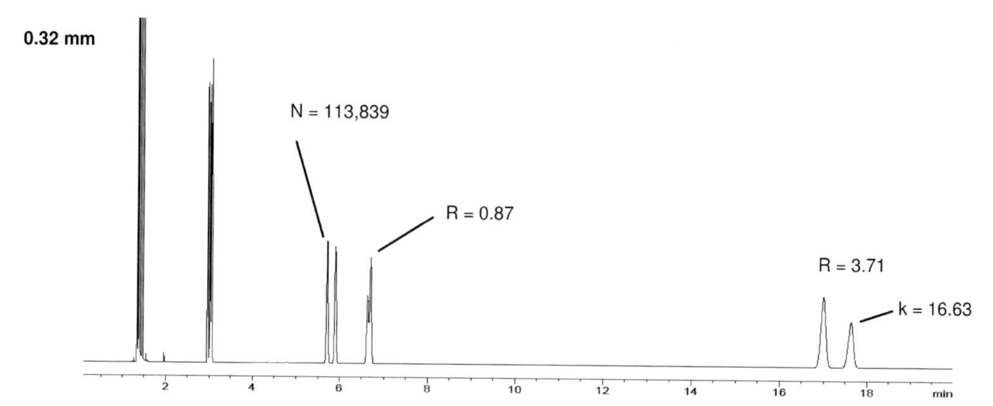


**Figure 5-3** Changes in efficiency, resolution and retention with column diameter – *isothermal*.
*Column:* 30 m, 0.25 μm
*Carrier:* Hydrogen at 50 cm/sec
*Oven:* 65 °C

efficiency; however, very small diameter columns are not practical in the standard laboratory environment. The smallest diameter column easily used for most GC applications is about 0.18 mm i.d. Increasing the number of theoretical plates results in better resolution, but the resolution increase is not directly proportional to the increase in the number of theoretical plates.

Resolution is proportional to the square root of the number of theoretical plates. Since the number of theoretical plates and column diameter are directly proportional, resolution is proportional to the square root of column diameter (Table 5-2). For example, halving column diameter increases resolution by 41% (not 100%) or doubling the diameter reduces resolution by 29% (not 50%). In practice, resolution changes are smaller than theoretically expected. For isothermal temperature conditions, the resolution changes are close to the theoretical values (Figure 5-3). For temperature program conditions, the actual resolution changes are often smaller than the theoretical values for low to moderate $k$ solutes ($k \sim 7$

or lower) and resolution can change in the opposite direction than expected for higher $k$ solutes ($k \sim 7$ or higher) (Figure 5-4).

The square root relationship means that column diameter has to decrease by four times to double the resolution by changing column diameter alone. For larger diameter columns, this is not much of a problem. Changing a 0.53 mm i.d. column to a 0.25 mm i.d. column increases the number of theoretical plates by over two fold with a resolution increase of approximately 45%. Attempting the same improvement with a smaller diameter column is difficult because very narrow diameter columns are required to obtain the same results. A 0.25 mm i.d. column has to be replaced with a 0.12 mm i.d. column to obtain similar improvements. Columns of this small diameter are difficult to use and are not recommended for routine GC applications.

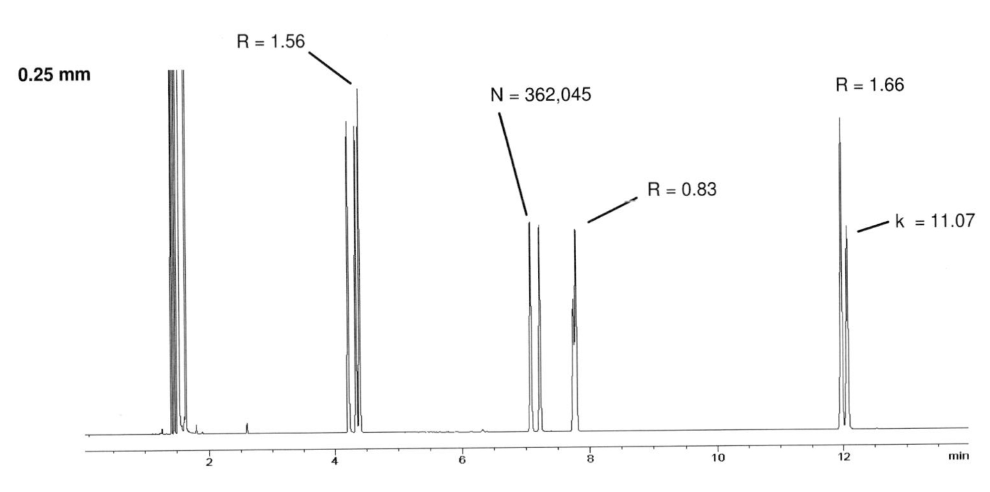


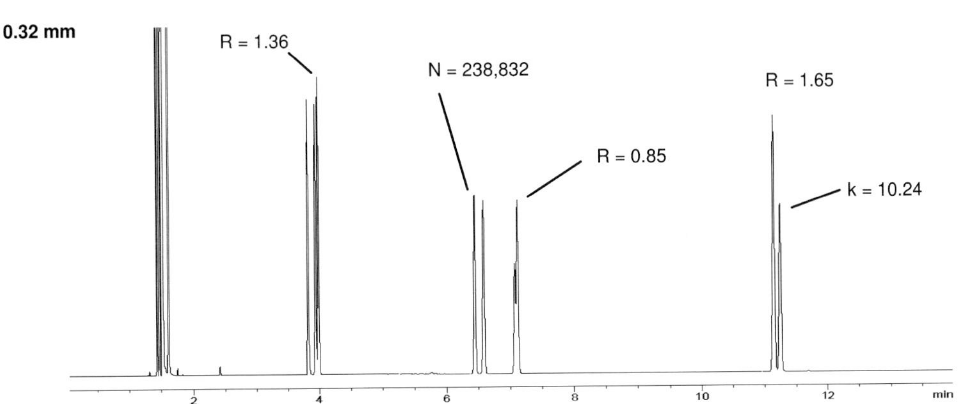


**Figure 5-4** Changes in efficiency, resolution and retention with column diameter – *temperature program.*

*Column:* 30 m, 0.25 μm
*Carrier:* Hydrogen at 50 cm/sec
*Oven:* 50 °C for 1 min, 50–120 °C at 5°/min

### 5.3.2
### Column Diameter and Retention

For isothermal conditions, the retention factor ($k$) is inversely proportional to column diameter. Retention times are similarity affected, but not in the same linear manner as $k$. For example, halving column diameter doubles retention or doubling the diameter reduces retention by half, but the retention time changes are not quite directly proportional. For temperature program conditions, the change in retention is less than for isothermal conditions. In most cases, the amount of retention change is between 25 and 75% of the isothermal values (Figure 5-4). Rarely is the difference in retention between columns of the various diameters ever a factor when selecting column diameter.

### 5.3.3
### Column Diameter and Pressure

The column head pressure varies with the diameter roughly according to an inverse square relationship (Table 5-3). For example, reducing column diameter by half results in a head pressure increase of over 4 times to maintain a specific carrier gas average linear velocity. For this reason, smaller diameter columns are usually offered in shorter lengths to compensate for the higher head pressure requirements.

The standard pressure gauges on many GC's have an upper range of 30 psig (210 kPa). The pressure gauge may have to be changed to one with a higher range to accommodate long, narrow columns. When using helium as the carrier gas,

**Table 5-3** Changes in column head pressure with diameter.

| Diameter (mm) | Pressure (psig) | | |
|---|---|---|---|
| | $H_2$ | He | $N_2$ |
| 0.05 | 404 | 497 | 222 |
| 0.10 | 93.3 | 114 | 49.7 |
| 0.18 | 25.9 | 32.3 | 14.0 |
| 0.20 | 20.7 | 25.7 | 11.3 |
| 0.25 | 12.9 | 16.0 | 7.1 |
| 0.32 | 7.8 | 9.5 | 4.3 |
| 0.45 | 3.9 | 4.8 | 2.2 |
| 0.53 | 2.8 | 3.4 | 1.6 |

$H_2$ at 50 cm/sec; He at 30 cm/sec; $N_2$ at 15 cm/sec.
Column length: 30 m.
Column temperature: 100 °C.

changing to hydrogen reduces head pressures by 20–25%. Using hydrogen makes the use of narrow diameter columns easier. Reducing column head pressures has an additional benefit in that higher pressure require more frequent septum changes and increases the probability of carrier gas leaks.

### 5.3.4 Column Diameter and Bleed

Column bleed increases as column diameter increases. The increased surface area of the larger diameter column requires a greater mass of stationary phase to maintain the same film thickness. The greater mass of stationary phase results in a greater amount of degradation products eluting from the column. In practice, the bleed increase with wider diameter column is usually small. The column-to-column variations in bleed for a given diameter are usually greater than the differences between the bleed levels of the different diameters. The increased bleed with larger diameter columns is rarely a factor when selecting column diameters.

### 5.3.5 Column Diameter and Capacity

Column capacity increases as column diameter increases in a fairly proportional manner (Table 5-4). The increased surface area of the larger diameter column contains a greater mass of stationary phase to maintain the same film thickness. Since there is more stationary phase in a given length of column, a greater amount of a compound can be dissolved in the stationary phase before overloading occurs. A full list of capacities for a variety of column diameters and film thicknesses can be found in Section 2.11.

Column capacity is also dependent on the polarity of the stationary phase and the compounds. Close polarities of compounds and stationary phases result in higher capacities. For example, a polar stationary phase has a greater capacity for a polar compound than a non-polar stationary phase. This makes it possible for a smaller diameter, polar stationary phase column to have higher capacity for a polar compound than a larger diameter, non-polar stationary phase column.

**Table 5-4** Capacity and column diameter.

| Diameter (mm) | Capacity range (ng) |
|---|---|
| 0.18–0.20 | 25–50 |
| 0.25 | 50–100 |
| 0.32 | 50–125 |
| 0.53 | 100–250 |

For similar polarity stationary phases and compounds.
Film thickness: 0.25 μm.

**Table 5-5** Carrier gas flow rates and column diameter.

| Diameter (mm) | Flow rate at columns exit (mL/min) | | |
|---|---|---|---|
| | $H_2$ | He | $N_2$ |
| 0.18 | 1.65 | 1.19 | 0.34 |
| 0.20 | 1.72 | 1.21 | 0.37 |
| 0.25 | 2.07 | 1.39 | 0.49 |
| 0.32 | 2.76 | 1.78 | 0.71 |
| 0.45 | 4.61 | 2.88 | 1.28 |
| 0.53 | 6.08 | 3.76 | 1.72 |

$H_2$ at 50 cm/sec; He at 30 cm/sec; $N_2$ at 15 cm/sec.
Column temperature: 100 °C.

### 5.3.6 Column Diameter and Carrier Gas Volume

Smaller diameter columns use a lower volume of carrier gas per unit time (mL/min) to operate at their most efficient level. This may not seem significant, but there are cases were carrier gas volume is important. Many bench top GC/MS systems have a maximum pumping capacity of 1–2 mL/min of carrier gas. The carrier gas volumes easily exceed this range for columns with inner diameters of 0.32 mm or greater (Table 5-5). Decreased sensitivity may occur at higher carrier gas volumes. For portable GC systems or in locations where high purity gases are difficult or expensive to obtain, the low carrier gas consumption of smaller diameter columns is beneficial.

Some types of external sampling devices (e.g., headspace, purge and trap) require high volume gas flows to efficiently transfer the sample into the GC. The flow rates used are significant beyond the values suitable for smaller diameter columns. Either splitting the gas flow prior to the column (i.e., discarding a large amount of the gas volume and sample) or cryofocusing is required to obtain acceptable results. For this reason, 0.53 mm i.d. columns are usually used with external samplers since they can tolerate high carrier gas flow rates.

### 5.3.7 Column Diameter and Injector Efficiency

Most capillary injectors work better at high carrier gas flow rates. The higher volume of carrier gas sweeps the vaporized sample into the column much faster. Larger diameter columns have higher flow rates at the same linear velocity as a smaller diameter column. In cases where the sample is not significantly trapped or focused at the front of the column, the faster transfer rate results in a shorter sample band. Better efficiency is obtained when the sample band is shorter.

For splitless injections, better efficiency for low $k$ peaks is obtained with a 0.32 mm i.d. column than a 0.25 mm i.d. The faster transfer of the sample and subsequent short band width overcomes the loss of efficiency of the larger diameter column. The gain in efficiency (N/m) decreases as retention ($k$) increases until better efficiency is obtained with the 0.25 mm i.d. column.

Split injectors rapidly transfer the sample into the column, and splitless and on-column injectors utilize a focusing effect to maintain reasonable injector efficiencies. Direct injectors usually do not utilize any type of focusing effect, thus high carrier gas flow rate are needed to rapidly transfer the sample into the column. For this reason, only 0.45–0.53 mm i.d. columns are suitable for use in most direct injectors. Smaller diameter columns require much lower carrier gas flow rates, thus slow sample transfer occurs with the resulting poor injector efficiency.

### 5.3.8 Column Diameter and Breakage

The largest internal diameter fused silica column regularly available is 0.53 mm. Larger diameter tubing proves to be too fragile and excessive breakage occurs in normal use. Smaller diameter columns are stronger and break less often. It is unusual for 0.53 mm i.d. tubing to be available on a cage that is substantial less than 7 inches (18 cm) in diameter. A smaller diameter cage places to much stress on the large bore tubing and excessive breakage occurs. Tubing with diameters of 0.25 mm i.d. or less has been wound in diameters of 1–2 inches without any breakage problems. Table 3-1 lists the minimum bending diameters for the most common diameters of fused silica tubing.

### 5.3.9 Column Diameter and Cost

As a rule, tubing costs increase as diameter increases. For this reason, larger diameter columns are slightly higher in price than a smaller diameter column of the same length. The price for 0.05 mm and 0.10 mm i.d. columns increases due to the higher tubing cost and manufacturing difficulties. Columns with these diameters are normally not used for GC work, thus their higher price should not be an issue.

### 5.3.10 Selecting Column Diameter

The most common column diameters for use with split and splitless injectors are 0.25 and 0.32 mm. Column diameters of 0.25 mm are used when higher efficiency and lower carrier flow rates are needed. Smaller diameter columns are best suited for complex samples, difficult to resolve compounds and GC/MS systems. The higher efficiency of the 0.18 and 0.20 mm i.d. columns may be beneficial; however, these small diameter columns are often only available in shorter lengths to reduce carrier gas head pressures. The shorter lengths may negate the efficiency improvements with the smaller diameter column. Columns diameters of 0.32 mm are used when higher capacity and high carrier gas flow rates are needed. These column diameters are often used with on-column and splitless injectors. The loss in efficiency is often compensated by the gains in capacity and injector efficiency. Columns with 0.53 mm i.d. diameters are intended for use with direct injectors. Nothing prevents 0.53 mm i.d. columns from being used in split, splitless or on-column injectors, but the loss in efficiency is rarely worth any of the potential benefits. The high capacity and carrier gas flow rate of 0.53 mm i.d. columns are probably the only valid reasons to use such a large diameter column without a direct injector.

## 5.4 Column Film Thickness

The most frequently used film thicknesses are between 0.25 and 1.0 μm. Columns with films as thin as 0.1 μm and thick as 10 μm are commercially available. Film thicknesses outside the range of 0.1 to 5 μm present a large series of different problems which limits their functionality. The differences between the available film thicknesses are fairly large (200–400%). Film thickness is not manipulated or an option with many PLOT columns.

### 5.4.1 Column Film Thickness and Retention

For isothermal conditions, retention ($k$) is directly proportional to film thickness. Retention times are similarly affected, but not in the same linear manner as $k$. For example, increasing the film thickness by four times increases retention by four times or reducing the film thickness by four times decreases retention by four times, but the retention time changes are not quite directly proportional (Figure 5-5). For temperature program conditions, the change in retention is less than for isothermal conditions. In most cases, the amount of retention change is between 25 and 75% of the isothermal values (Figure 5-6). When increasing retention by increasing film thickness, one potential drawback is the loss of resolution for high $k$ solutes ($k \sim 7$ or higher). This is evident in the chromatograms shown in Figure 5-6.

One question that arises is why use film thickness to change retention when adjusting column temperature seems to be easier and more flexible? Some analyses require temperature conditions where the GC or column is at the extreme of its temperature range. Adjusting retention using column temperature may not be practical or possible in these cases. Most GC's are limited to a lower temperature of about 35 °C without the use of cryogenic cooling. Limitations occur when greater retention is desired to improve peak separation or to move a peak away

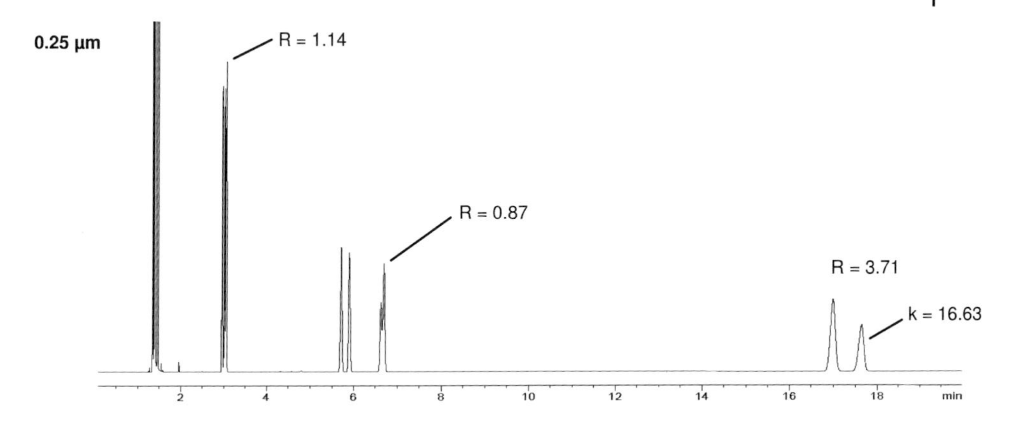


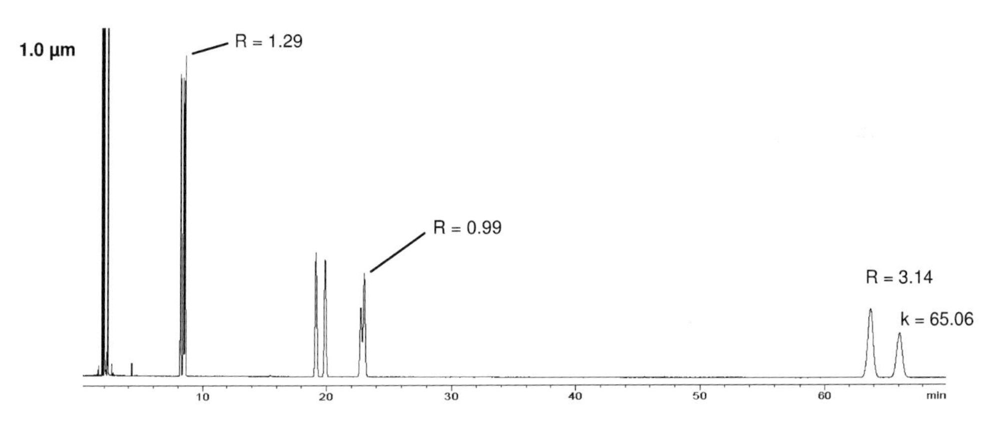


**Figure 5-5** Changes in retention and resolution with column film thickness – *isothermal*.
*Column:* 30 m × 0.32 mm i.d.
*Carrier:* Hydrogen at 50 cm/sec
*Oven:* 65 °C

from the solvent front, but the GC is already at its lowest temperature. Changing to a thicker film column increases retention without the need to lower the column temperature. In most cases, a slightly higher column temperature than originally used with the thinner film column can be used. A higher initial temperature in a temperature program reduces the time required to cool down the GC after the program has finished. Limitations also occur when less retention is desired, but the column is already at or near its upper temperature limit. Changing to a thinner film column decreases retention without the need to increase the column temperature. In some cases, a lower temperature than originally used with the thicker film column can be used. The lower column temperature usually increases column lifetime and reduces bleed and may benefit any compounds that decompose in the column at high temperatures. The lower column temperature does not subject the compound to the higher temperatures where greater degradation may occur.

0.25 µm

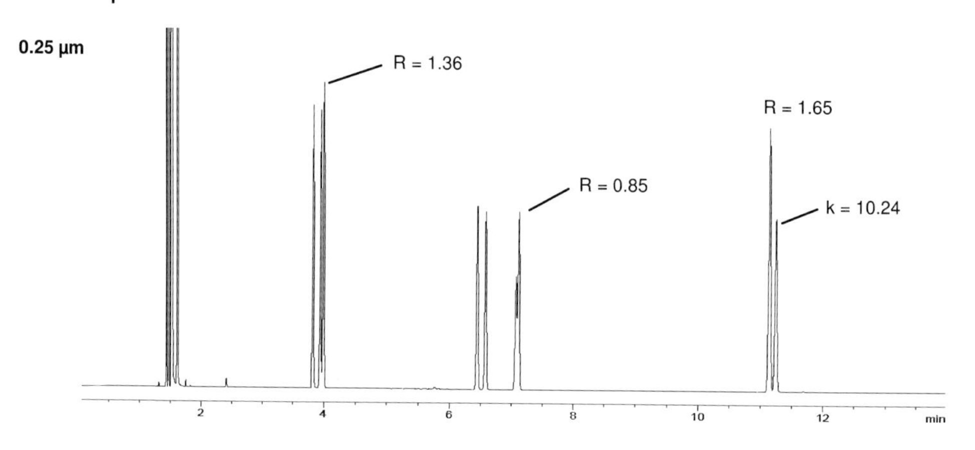


1.0 µm

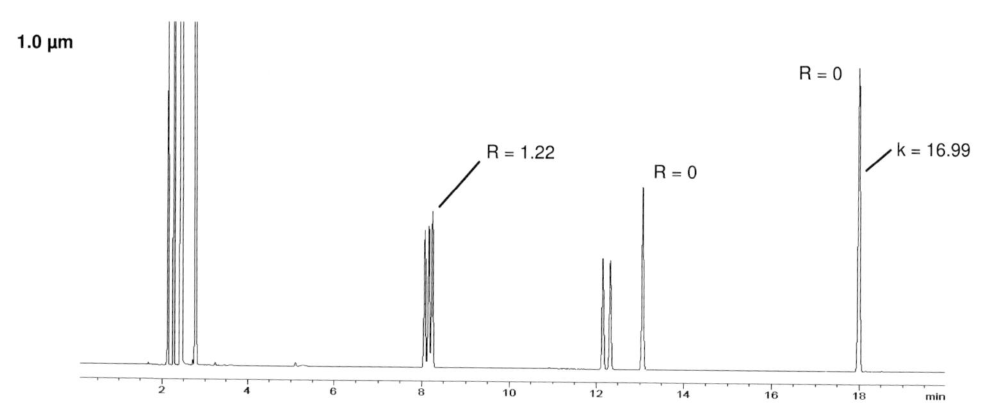


**Figure 5-6** Changes in retention and resolution with column film thickness – *temperature program.*

*Column:* 30 m × 0.32 mm i.d.
*Carrier:* Hydrogen at 50 cm/sec
*Oven:* 50 °C for 1 min, 50–150 °C at 5°/min

Column temperature and retention are not directly proportional – they are related by an inverse natural log (ln) function. This means that when column film thickness is changed, a much smaller relative temperature change is needed to maintain the same retention as before. For isothermal conditions, the change in column temperature is related to the inverse natural log of the change in the film thickness. For example, if the compound peak has a retention time of 10 minutes at 100 °C with a 0.25 µm film column, about 140 °C is needed for a 1 µm film column to keep the peak at 10 minutes. For temperature program conditions, it is more difficult to predict the magnitude of the retention changes. The inverse natural log relationship between retention and column temperature still applies, but changes in the initial and final temperatures, the ramp rate and the hold times all affect the retention shift. The initial temperature and its hold time have the greatest impact on the earlier eluting peaks; the final temperature and ramp rate have the greatest impact on the later eluting peaks. Peaks eluting in the middle part of the chromatogram are influenced to different degrees by all of the program conditions.

### 5.4.2
### Column Film Thickness and Efficiency/Resolution

When properly calculated (i.e., using a peak with $k = 5–10$), the number of theoretical plates decreases as film thickness increases (Table 5-6). When incorrectly calculated (i.e., using a peak with $k < 5$), the number of theoretical plates increase as film thickness increases (as long as the peaks remain at $k < 5$ on the thicker film column). This behavior has a direct impact on using film thickness to improve peak resolution.

One method to improve resolution is to increase the retention of the peaks. This can be accomplished by lowering the column temperature or increasing the film thickness. When using film thickness to improve resolution via an increase in retention, there is a guideline to follow. If the peaks have $k$ values less than 5, increasing the column film thickness increases the resolution of peaks providing the peaks do not have $k$ values much larger than 10 on the thicker film column (Figure 5-7). If the peaks already have $k$ values greater than 10, increasing the column film thickness usually does not improve peak resolution and in some cases will make it worse (Figures 5-5 and 5-6).

The improved resolution of peaks with $k$'s less than 5 upon increasing film thickness is more of a guideline than a rule. The amount of resolution gain or loss is influenced by the stationary phase, column temperature and the compounds. The biggest problem with using film thickness to improve resolution is when the peaks in the chromatogram elute over a wide $k$ range. Improving resolution of the low $k$ peaks by increasing the film thickness often reduces the resolution of the high $k$ peaks. When this happens, increasing the ramp rate or programming to a higher final temperature sometimes reduces the extent of resolution loss of the high $k$ peaks.

**Table 5-6** Changes in efficiency with film thickness.

| Film thickness (μm) | *N* |
|---|---|
| 0.25 | 51,700 |
| 1.0 | 48,400 |
| 3.0 | 30,600 |
| 5.0 | 20,000 |

Column: 15 m × 0.32 mm i.d.
Carrier: Helium at 35 cm/sec.
Peak at $k = 7$.

0.25 µm

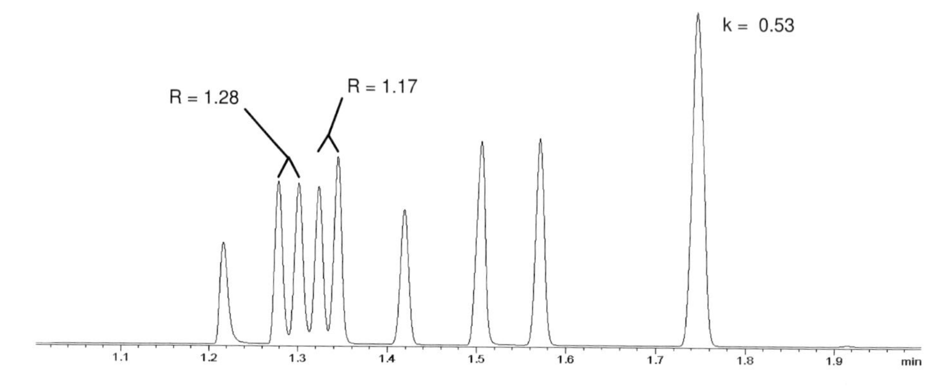


1.0 µm

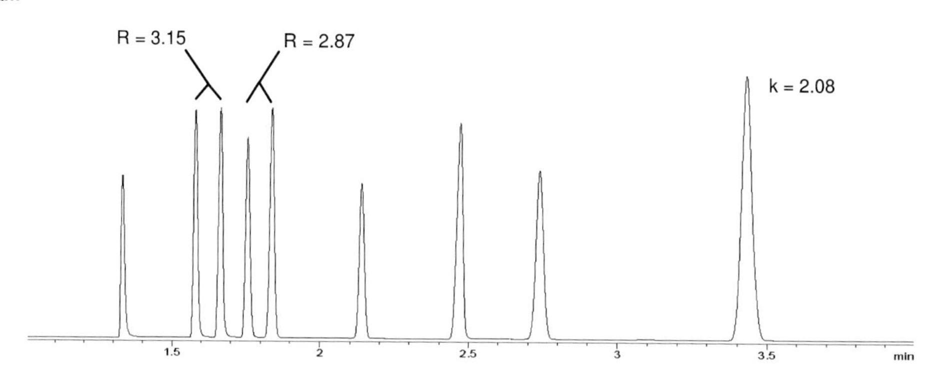


**Figure 5-7** Changes in resolution and retention with column film thickness for peaks with $k < 5$.
*Column:* 30 m × 0.32 mm i.d.
*Carrier:* Hydrogen at 45 cm/sec
*Oven:* 35 °C

### 5.4.3
### Column Film Thickness and Capacity

Column capacity increases as film thickness increases in a fairly proportional manner (Table 5-7). A greater amount of compound can be dissolved in the greater mass of stationary phase in a given length of column. When attempting to increase capacity, some of the increased retention (and possible resolution change) with the thicker film column can be overcome if the column temperature is raised. Since capacity increases as $k$ decreases, additional compound capacity is obtained with the higher column temperature. A full list of capacities for a variety of column diameters and film thicknesses can be found in Section 2.11.

Column capacity is also dependent on the polarity of the stationary phase and the compounds. Close polarities of compounds and stationary phases result in

**Table 5-7** Capacity and column film thickness.

| Film thickness (μm) | Capacity range (ng) |
|---|---|
| 0.10–0.20 | 20–50 |
| 0.25 | 50–125 |
| 0.50 | 125–250 |
| 1.0 | 250–500 |
| 3.0 | 500–1000 |
| 5.0 | 1000–2000 |

For a similar polarity stationary phase and compound.
Column Diameter: 0.32 mm.

higher capacities. For example, a polar stationary phase has better capacity for a polar compound than for a non-polar one. This makes it possible for a thinner film, polar stationary phase column to have higher capacity for a polar compound than a thicker film, non-polar stationary phase column.

### 5.4.4
### Column Film Thickness and Bleed

Column bleed increases as column film thickness increases. The greater mass of stationary phase results in a greater amount of degradation products eluting from the column. Very thick film columns (3 μm) usually have a lower value for the upper temperature limit than a thinner film counterpart. The high bleed of these very thick film columns requires their upper temperature limits to be set at lower values. Non-polar stationary phases have lower bleed levels than more polar stationary phases. This makes it possible for a non-polar, thicker film column to have lower bleed than a polar, thinner film column. Due to the higher bleed of polar stationary phases, film thicknesses of 3 μm or greater are usually available only for non-polar stationary phases.

### 5.4.5
### Column Film Thickness and Inertness

The surface of the tubing is the site of activity in capillary columns. Interaction of the compounds with these sites is undesirable and needs to be avoided. The interaction results in reversible (peak tailing) or irreversible (loss of peak size) adsorption. Thicker films shield the compounds from these active sites and reduces the number of interactions. Peaks that tail on one column may show reduced adsorption on a corresponding thicker film column. Problems are encountered when the film thickness is reduced and activity problems occur with the thinner film column that were not experienced with the thicker film column. When attempting to improve inertness, some of the increased retention (and possible resolution change) with the thicker film column can be overcome if the column temperature is raised.

### 5.4.6 Selecting Column Film Thickness

Improving the resolution for low $k$ compound peaks is the primary reason for using a thick film column. Reducing retention of highly retained compound peaks when limited by column temperature is the primary reasons for a thin film column. For most analyses, a film thickness of 0.2–0.50 µm is suitable for 0.20–32 mm i.d. columns, and a film thickness of 1.0–1.5 µm is suitable for 0.53 mm i.d. columns. Improving capacity or inertness by increasing the film thickness is accompanied by increased retention and bleed. Some of the increase in retention and possible change in resolution can be overcome by using a higher column temperature; however, the higher bleed of the thicker film column at the higher temperature can be a problem.

## 5.5 Manipulating Multiple Column Dimensions

Changing any of the column dimensions often affects more than one performance parameter. Also, the same parameters are often affected by each column dimension. Sometimes, these interconnected relationships can be advantageous. An example best illustrates this concept. A 15 m × 0.25 mm i.d. column is being used, but a lower carrier gas flow (volume) into the detector is desired. This is best accomplished by reducing the column diameter. Decreasing column diameter also increases efficiency and retention. While the increase in efficiency is good, the increase in retention may not be so good. Realizing that retention and efficiency decrease with shorter columns, using a shorter column may be a satisfactory adjustment. A 12 m × 0.2 mm i.d. column is available – does this column provide all of the desired improvements without too much of a sacrifice? Assuming a worst case scenario (i.e., isothermal), the 0.2 mm i.d. column is 25% more retentive than the 0.25 mm i.d. column; the 12 meter column is 20% less retentive than the 15 meter column. Concerning retention, these two dimension changes nearly cancel each other. An average 0.2 mm i.d. column has N/m = 5350 and an average 0.25 mm i.d. column has N/m = 4275. The total number of theoretical plates for the 12 m × 0.2 mm i.d. column is 64,200 and 64,125 for the 15 m × 0.25 mm i.d. column. Again, the two dimension changes nearly cancel each other. Using a 12 m × 0.2 mm i.d. instead of a 15 m × 0.25 mm i.d. column provides the lower carrier gas flow into the detector while causing relatively insignificant changes in retention and efficiency. There is a higher carrier gas head pressure and lower capacity affect when changing to a smaller diameter column. As long as this is not a problem, the 12 m × 0.2 mm i.d. column can be successfully used to lower carrier gas flow.

**Table 5-8** Relationships between column behavior and dimensions.

| Increase[1] | Length | Diameter | Film thickness |
|---|---|---|---|
| Efficiency | Longer | Smaller | Thinner |
| Resolution | Longer | Smaller | Thicker ($k < 5$)<br>Thinner ($k > 5$) |
| Retention time | Longer | Smaller | Thicker |
| Bleed | Longer | Smaller | Thicker |
| Capacity | N/A | Larger | Thicker |
| Head pressure | Longer | Smaller | N/A |
| Inertness | Shorter | Smaller | Thicker |
| Carrier gas flow | N/A | Larger | N/A |
| Injector efficiency | N/A | Larger | N/A |

N/A = No significant affect.

[1] To *increase* each parameter not improve.

It is important to realize that changing a column dimension changes other performance parameters. Sometimes these parameters changes are not acceptable. It is equally important to realize is another column dimension can be altered to counteract some of the changes to the parameters. Sometimes the parameter changes can not be offset by changing another dimension or another change occurs that is not acceptable. All of the various performance parameters and how they are related to column dimensions have to be reconciled. Table 5-8 summarizes the major relationships between column behavior and column dimensions.

# 6
# Carrier Gas

## 6.1
## Carrier Gas and Capillary Columns

The influence of the carrier gas in capillary column chromatography is often underappreciated and neglected. The type of carrier gas and its velocity through the column have a significant impact on resolution and retention time. The proper selection of the type and velocity of the carrier gas is critical for full exploitation of the high efficiencies of capillary columns. Small variations or deviations in the carrier gas velocity may result in noticeable losses in efficiency.

## 6.2
## Linear Velocity versus Flow Rate

The linear velocity is the speed (cm/sec) of the carrier gas as it travels through the column. The flow rate is the volume of carrier gas passing through the column per unit time (mL/min). The linear velocity is more important and meaningful for capillary columns. The linear velocity affects efficiency and retention time. There are situations where the carrier gas flow rate is important; however, they involve injectors and detectors and do not have a direct impact on column efficiency.

Carrier gas linear velocity is not uniform throughout the column, hence the average linear velocity ($\overline{u}$) is used for measurement purposes. There is an optimal linear velocity ($\overline{u}_{opt}$) for each type of carrier gas where maximum column efficiency is obtained. There are small differences between the optimal average linear velocities for columns of different diameters, film thickness and length. Because they are usually small, these differences are often ignored without any negative consequences.

## 6.3
## Controlling the Linear Velocity and Flow Rate

The carrier gas average linear velocity or flow rate is dictated by pressure of the carrier gas at the front of the column. This pressure is commonly called the head

*The Troubleshooting and Maintenance Guide for Gas Chromatographers, Fourth Edition.* Dean Rood

ISBN: 978-3-527-31373-0

pressure. Increasing the head pressure results in an increase in the velocity or flow; decreasing the head pressure has the opposite effect.

Older or basic GC models use manual controls to regulate the head pressure. The head pressure is adjusted by manually turning a knob on the GC and reading the resulting head pressure value from an analog gauge. Since the head pressure is the only controllable parameter, the carrier gas linear velocity or flow rate is measured or determined by other means (Section 6.5).

Newer GCs use electronically controlled pneumatics. A keypad on the GC or the data system software is used to enter the desired head pressure which is then automatically set by GC. Unlike manual systems, linear velocities and flow rates can also be directly set using electronically controlled systems, thus an independent measure of the velocity or flow rate may not be necessary.

## 6.4 Van Deemter Curves

The relationship between average linear velocity and efficiency is described by the Van Deemter equation (Equation 6-1). If efficiency as $H_{min}$ is plotted against the average linear velocity for a given column and set of conditions, a Van Deemter curve is obtained (Figure 6-1).

$$H_{min} = A + \frac{B}{\bar{u}} + C\,\bar{u}$$

$H_{min}$ = height equivalent to a theoretical plate
$\bar{u}$ = average linear velocity
$A$ = multi-path flow term
$B$ = longitudinal diffusion term
$C$ = resistance to mass transfer term

**Equation 6-1** Van Deemter equation.

The optimal linear velocity ($\bar{u}_{opt}$) is where the curve is at its lowest point which corresponds to the smallest value of $H_{min}$ (i.e., point of greatest efficiency). Also important, Van Deemter curves show that using too high (right side of the curve) or low (left side of the curve) a linear velocity results in a loss of efficiency. However, this is not always a completely undesirable situation.

Using an average linear velocity greater than $\bar{u}_{opt}$ is often recommended. This value is called the optimal practical gas velocity (OPGV) and is 1.25–2 times $\bar{u}_{opt}$. The OPGV is where the maximum efficiency per unit time is obtained. The average liner velocity corresponding to $\bar{u}_{opt}$ provides the highest efficiency, but average linear velocities in the OPGV range sacrifice a small amount of efficiency for substantial reduction in retention time. The reduction in retention time (and hence run time) is usually worth the small sacrifice in efficiency.

Most GC analyses are run in the constant pressure mode, thus the carrier gas linear velocity does not remain constant throughout the course of a temperature

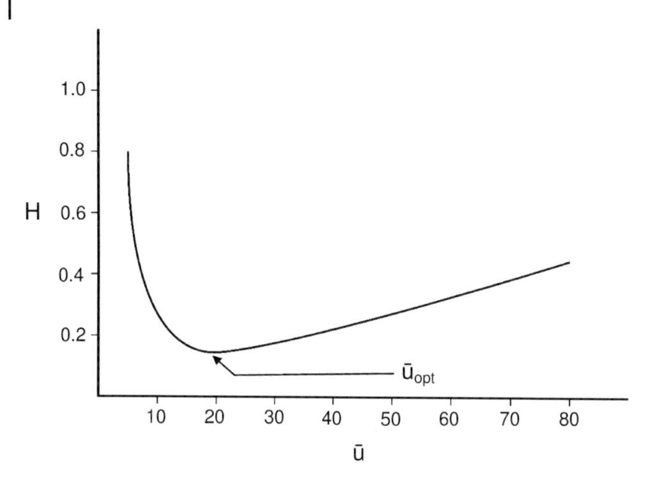


**Figure 6-1** Typical Van Deemter curve.

programmed run. As the column temperature rises, the viscosity of the carrier gas increases. As the carrier gas viscosity increases, there is a corresponding decrease in the carrier gas linear velocity since the head pressure remains constant (except when using certain direct injectors without a septum purge function). If the average linear velocity is set at $\bar{u}_{opt}$, the linear velocity decreases as the temperature program progresses to higher temperatures and the linear velocity proceeds to shift into a region of lower efficiency (left side of the Van Deemter curve). If the average linear velocity is set slightly above $\bar{u}_{opt}$, as the velocity decreases it shifts towards the minimum in the Van Deemter curve and into the region of higher efficiency. Most pressure programmable injectors can operate in the constant flow mode which changes the head pressure corresponding to the change in the column temperature. If this mode is used, average linear velocity remains at the set value and does not change during the run.

## 6.5
## Carrier Gas Measurements

### 6.5.1
### Average Linear Velocity ($\bar{u}$)

The average linear velocity is calculated using the retention time of a non-retained compound. Table 6-1 lists some recommended non-retained compounds. The non-retained compound's retention time is used in Equation 6-2 to calculate the average linear velocity. To obtain the proper units of cm/sec, the column length must be in centimeters and the retention time in seconds. For some very thick film columns and PLOT columns, using higher temperatures (50–75 °C) may be necessary to avoid retention of the compounds.

**Table 6-1** Suggested non-retained compounds.

| Detector | Non-retained compounds |
|---|---|
| FID | methane, butane[1] |
| ECD | methylene chloride[2],[3], dichlorodifluoromethane[2],[3] |
| NPD | acetonitrile[2],[4] |
| TCD, MS | methane, butane, argon, air |

1) A disposable lighter is a possible source of butane.
Place the syringe needle into the flame outlet of the lighter.
Depress the small button allowing the gas to escape.
Pull up 1–2 µL of the gas.
2) Headspace or diluted in solvent.
3) Use a column temperature of 40 °C or greater.
4) Use a column temperature of 90 °C or greater.

*Headspace:* Fill an autosampler vial with about 10 drops of solvent.
Tightly seal the vial with a cap. Shake the vial for several seconds.
Pierce the septum with the syringe needle, but do not insert the needle into the liquid.
Pull up 1–2 µL of the headspace above the liquid.

*Solvent dilution:* Add 25–50 µL of the appropriate solvent to 10 mL hexane
or iso-octane and thoroughly mix. Inject 0.1–0.2 µL.
Further dilution may be necessary depending on the sensitivity of the detector.

$$\bar{u} = \frac{L}{t_m}$$

$L$ = column length (cm)
$t_m$ = retention time of a non-retained compound (sec)

**Equation 6-2** Average linear velocity ($\bar{u}$).

Inject 1–2 µL of the non-retained compound when using a split injector. Very small injection volumes (0.1–0.2 µL) or dilution of the non-retained compound will be necessary for splitless, direct or on-column injectors. When using a splitless injector, it is easier and slightly more accurate to use the split injection mode when setting the average linear velocity.

In practice, an average linear velocity is selected and the retention time of the non-retained peak is calculated which gives the desired average linear velocity (Equation 6-3). The non-retained compound is injected and a retention time is obtained. The carrier gas head pressure is adjusted until the desired retention time is obtained. The head pressure is raised to decrease the non-retained peak retention time, thus increasing the average linear velocity; the head pressure is lowered to decrease the average linear velocity. A retention time ±0.05 minutes of the calculated value is sufficient in most cases. It is difficult to adjust the head pressure to obtain the exact value. Small errors and changes in the column length used in the calculation render the calculated retention time slightly inaccurate.

$$t_m = \frac{L}{60\,\overline{u}}$$

$t_m$ = retention time of a non-retained compound (min)
$L$ = column length (cm)
$\overline{u}$ = desired average linear velocity (cm/sec)

**Equation 6-3** Non-retained peak retention time to obtain a specific average linear velocity.

Table 6-2 lists the retention times of the non-retained compound peak for a range of average linear velocities and column lengths.

The desired average linear velocity or volumetric flow rate is directly input when using a pressure programmable injector. The accuracy is dependent on the column length and diameter entered into the system. A deviation from the actual values results in an error in the average linear velocity. For this reason, it is still recommended to inject a non-retained compound to determine the actual average linear velocity if an accurate and precise value is needed.

The average linear velocity is dependent on column temperature. For this reason, the same temperature must be used each time the carrier gas velocity is set for a particular method. Using a lower column temperature results in a faster average linear velocity; a higher temperature results in a slower average linear velocity.

**Table 6-2** Non-retained peak retention times (min).

| $\overline{u}$ | Column length (m) | | | | | |
|---|---|---|---|---|---|---|
| | 12 | 15 | 25 | 30 | 50 | 60 |
| 10 | 2.00 | 2.50 | 4.17 | 5.00 | 8.33 | 10.00 |
| 15 | 1.33 | 1.67 | 2.78 | 3.33 | 5.56 | 6.67 |
| 20 | 1.00 | 1.25 | 2.08 | 2.50 | 4.17 | 5.00 |
| 25 | 0.80 | 1.00 | 1.67 | 2.00 | 3.33 | 4.00 |
| 30 | 0.67 | 0.83 | 1.39 | 1.67 | 2.78 | 3.33 |
| 35 | 0.57 | 0.71 | 1.19 | 1.43 | 2.38 | 2.86 |
| 40 | 0.50 | 0.63 | 1.04 | 1.25 | 2.08 | 2 .50 |
| 45 | 0.44 | 0.56 | 0.93 | 1.11 | 1.85 | 2.22 |
| 50 | 0.40 | 0.50 | 0.83 | 1.00 | 1.67 | 2.00 |
| 55 | 0.36 | 0.45 | 0.76 | 0.91 | 1.52 | 1.82 |
| 60 | 0.33 | 0.42 | 0.69 | 0.83 | 1.39 | 1.67 |
| 65 | 0.31 | 0.38 | 0.64 | 0.77 | 1.28 | 1.54 |
| 70 | 0.29 | 0.36 | 0.60 | 0.71 | 1.19 | 1.43 |
| 75 | 0.27 | 0.33 | 0.56 | 0.67 | 1.11 | 1.33 |
| 80 | 0.25 | 0.31 | 0.52 | 0.63 | 1.04 | 1.25 |

Usually the initial temperature of the program is used because it is the most convenient point. In practice any temperature can be used; however, a consistent temperature for measurement is the most important consideration.

### 6.5.2 Column Flow Rate

There are several methods of measuring flow rate and each one gives a slightly different value. The values are different because the flow rates being measured are actually at different points in the column. The most common method of measuring flow rate is to use a flow meter attached to the detector. This provides the flow rate at the exit or back of the column. Flow rates of 1–3 mL/min are usually used for smaller diameter (0.18–0.32 mm i.d.) columns. These low flows are difficult to accurately measure with a flow meter. For Megabore columns (0.53 mm i.d.), a low volume flow meter usually gives a reasonably accurate values since flow rates of 4–10 mL/min are common. Flow meter inaccuracies at low flow rates and difficulties in obtaining a leak free connection with the detector create errors. Also, all detector and makeup gases need to be completely turned off to obtain a correct column flow rate. Note that bubble and digital flow meters may give slightly different values due to the different mechanisms used to measure the flow rates.

Using a flow meter at the detector provides the column flow at the exit of the column. This flow is useful when the carrier gas flow and its impact on detector behavior are important (e.g., when using concentration dependent detectors). If a flow meter is unavailable, the average flow rate of carrier gas can be calculated using Equation 6-4. The retention time of a non-retained compound is the same as the one used in the average linear velocity calculation (Equation 6-2), but it is in minutes not seconds. The flow rate calculated using Equation 6-4 provides an average value and not the flow rate at any specific point in the column. It is not possible to directly measure the column flow into the detector for GC/MS systems; therefore, the average flow rate is often used.

$$\bar{F} = \frac{\pi r^2 L}{t_m}$$

$r$ = column radius (cm)
$L$ = column length (cm)
$t_m$ = retention time of a non-retained compound (min)

**Equation 6-4** Average flow rate ($\bar{F}$).

Like linear velocity, flow rate is dependent on column temperature. For this reason, the same temperature must be used each time the carrier gas flow rate is measured or set for a particular method. Using a lower column temperature results in a faster flow rate; a higher temperature results in a slower flow rate. Usually the initial temperature of the program is used because it is the most convenient point. In practice any temperature can be used; however, a consistent temperature for measurement is the most important consideration.

## 6.6
## Carrier Gas Selection

Hydrogen, helium, nitrogen and argon/methane are used as carrier gases in gas chromatography. Helium and hydrogen are the best ones for capillary columns. Nitrogen is a suitable carrier gas, but it is not recommended except for special cases. Argon/methane should not be used as a carrier gas for capillary columns.

### 6.6.1
### Nitrogen

Nitrogen provides the best efficiency when compared to helium and hydrogen; however, $\bar{u}_{opt}$ is at a very low average linear velocity (Figure 6-2). Longer retention times occur at such low linear velocities. Substantial analysis speed is sacrificed for maximum efficiency. The Van Deemter curve for nitrogen is steep. This means efficiency drops sharply with any deviation away from $\bar{u}_{opt}$, thus the linear velocity range for maximum efficiency is small. Attempts to decrease run times by increasing the carrier gas average linear velocity causes a large loss of efficiency. Small errors in velocity measurements result in significant changes in efficiency.

The Van Deemter curve becomes steeper and the $H_{min}$ value at $\bar{u}_{opt}$ increases as compound retention goes up. The large loss of efficiency as compound retention increases makes nitrogen a poor carrier gas for analyses where the compounds elute over a wide temperature range. The linear velocity decreases as the temperature program progresses. Many of the compounds are chromatographing while the carrier gas is well outside the velocity range of maximum efficiency.

Nitrogen is not recommended as a carrier gas for capillary columns primarily due to the low average linear velocity range and the resulting long run times. Also, the average linear velocity can only be optimized over a small temperature or retention range. For isothermal temperature conditions, short temperature programs and compounds that elute over a small range, nitrogen is a satisfactory carrier gas. For simple analyses (i.e., only a few compound eluting close together), the best efficiency is obtained with nitrogen. Nitrogen is the least expensive carrier gas and readily available.

### 6.6.2
### Helium

The Van Deemter curve for helium shows that $\bar{u}_{opt}$ is at a higher average linear velocity than for nitrogen (Figure 6-2). The $H_{min}$ value at $\bar{u}_{opt}$ is slightly higher than nitrogen, but the difference is small. The higher $\bar{u}_{opt}$ value allows the use of higher linear velocities without sacrificing a large amount of efficiency. In general, helium provides nearly equivalent resolution to nitrogen, but at much shorter retention times (Figure 6-3).

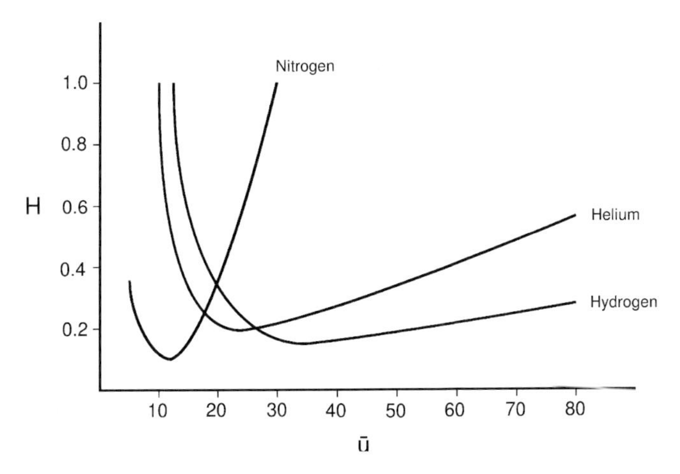


**Figure 6-2** Representative Van Deemter curves for nitrogen, helium and hydrogen.

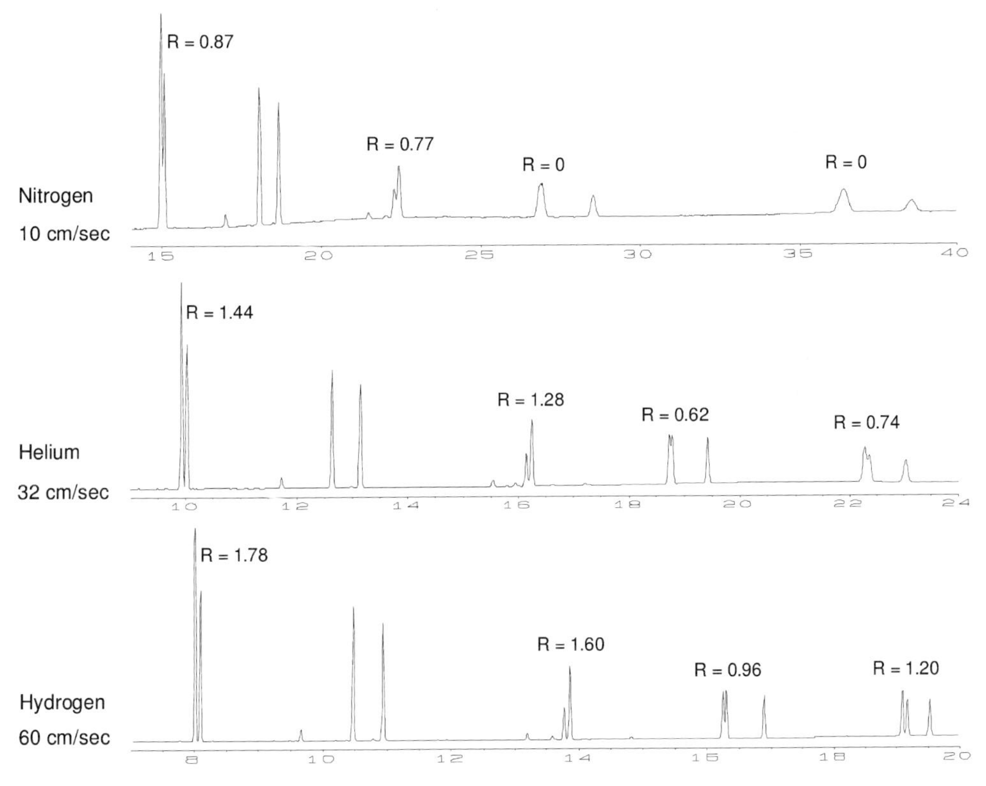


**Figure 6-3** Comparison of resolution and run time for nitrogen, helium and hydrogen.
*Column:* DB-5, 15 m × 0.25 mm, 0.25 μm
*Oven:* 55 °C for 1 min, 55–155 °C at 25°/min, 155–300 °C at 10°/min, 300 °C for 20 min
*Compounds:* Polycyclic aromatic hydrocarbons
The carrier gas average linear velocity was set at the initial oven temperature.

The Van Deemter curve for helium is not as steep as for nitrogen. The flatter curve results in small changes in efficiency as the column temperature changes in a temperature program. The carrier gas linear velocity remains in the range of higher efficiency throughout the temperature program run. There is only a small loss of efficiency as compound retention increases. This makes helium a good carrier gas for compounds that elute over a wide temperature or retention range. Helium is the most expensive carrier gas and may be difficult to obtain at a reasonable price in many parts of the world.

### 6.6.3 Hydrogen

Hydrogen is the best carrier gas for capillary columns. Its Van Deemter curve is very flat and $\bar{u}_{opt}$ is at very high average linear velocity (Figure 6-2). The $H_{min}$ value at $\bar{u}_{opt}$ is slightly higher than nitrogen and equivalent to helium. The very high $\bar{u}_{opt}$ value allows the use of very high linear velocities without sacrificing a large amount of efficiency. Shorter retention times are obtained compared to helium and especially nitrogen while maintaining high efficiency (Figure 6-3). The extreme flatness of the hydrogen Van Deemter curve makes hydrogen vastly superior for compounds eluting over a wide temperature or retention range.

Hydrogen is less costly than helium, but more costly than nitrogen. There are several hydrogen generators commercially available that produce high quality hydrogen suitable for use as a carrier gas. This eliminates the purchasing and changing of high pressure cylinders for high volume users of hydrogen. Most hydrogen generators return their original purchase cost in a few years.

There is a reluctance to use hydrogen as a carrier gas due to its flammability. This a very valid concern, but the actual hazards are not as severe as commonly perceived even though a highly flammable gas and high temperatures may seem like a dangerous combination. There are a variety of reasons that make working with hydrogen less hazardous than it seems; however, operating in the safest manner reasonably possible is still a requirement.

The amount of hydrogen in air must be slightly above 4% before an explosion can occur. Sustained combustion will not occur if the amount of hydrogen in air is much higher or lower than 4%. Hydrogen levels above 4% will explode, but a continuous fire or flame will not occur unless the hydrogen is constantly maintained around the 4–10% level and a source of oxygen is continuously available. Hydrogen is very diffusive in air and the flow rates used in GCs relatively low, thus making it difficult to obtain a high enough buildup of hydrogen to make an explosive mixture with air. Ventilated rooms and the GC fan clear the hydrogen away before it can build up. A completely closed environment and delivery of a very high volume of hydrogen are necessary to build up enough hydrogen for a problem to occur. It usually requires the GC oven fan to be off and a very high flow rate of hydrogen into the GC oven.

Some GC's are equipped with pressure regulators (i.e., forward pressure regulated injectors) to control the total amount of carrier gas entering the GC.

This type of system adjusts the flow of carrier gas to maintain the pressure as set on the head pressure gauge. If the flow path is exposed to the ambient pressure (e.g., a leak or broken column), the pressure controller attempts to pressurize the GC oven to the set pressure. A substantial volume of hydrogen is delivered in a short amount of time. Nearly every incident of a hydrogen explosion has been with a forward pressure regulated injector system and some type of gross operational error. Other GC's are equipped with flow controllers (i.e., back pressure regulated injectors) to control the total amount of carrier gas entering the GC. In the worst case, only 100–200 mL/min of hydrogen can be delivered to the GC. Even with a leak directly into the GC oven, most of the time too little hydrogen is supplied to reach an explosive level. As an additional safety feature, all modern GC's are designed with an explosion absorbing oven door. If an explosion does occur, the door absorbs most of the force of the explosion and it opens instead of becoming a dangerous projectile. Most hydrogen explosions are small and very little to no damage occurs even to the columns installed in the oven. Even with diffusive nature of hydrogen and the safety factors built into modern GCs, the proper precautions still need to be followed to greatly minimize the possibility of a large hydrogen leak.

If hydrogen explosions are still a concern, simple installation of flow controllers in the gas supply lines reduces the possibility of a hydrogen explosion due to leaks into the oven. The flow control can be installed on the hydrogen gas line leading to each GC or at a point further upstream. Each flow controller is set 10–20% above the total volume of hydrogen needed from that hydrogen line. If there is a leak downstream of the flow controller, only the gas volume set with the controller is delivered. This will prevent a buildup of hydrogen at the leak site.

## 6.7 Recommended Average Linear Velocities

Table 6-3 lists the average linear velocity range recommended to obtain maximum efficiency and the OPGV range for the most common carrier gases and column sizes. The nitrogen Van Deemter curve is so sharp that the linear velocity differences between different column dimensions are small; therefore, the same linear velocity range applies to all columns. In most cases, very small efficiency differences are obtained for the average linear velocities within the range.

If maximum efficiency is desired, some experimentation using different linear velocities within the recommended ranges is required. Unless the compounds all elute within several minutes of each other, there is rarely a single average linear velocity that is the best for all of the compounds in a sample. The graph in Figure 6-4 illustrates an example of this type of situation. Maximum resolution for peaks 1 and 2 is obtained at 65 cm/sec, 45 cm/sec for peaks 3 and 4, and 40 cm/sec for peaks 5 and 6, thus maximum resolution for all of the peaks is not possible at a single linear velocity. A linear velocity of 52–53 cm/sec provides the best overall resolution for all of the peaks, but it is not the optimal value for any

**Table 6-3** Recommended average linear velocity and OPGV.

| Diameter (mm) | Length (m) | Film (μm) | $\bar{u}$ | OPGV |
|---|---|---|---|---|
| **Nitrogen** | | | | |
| 0.18–0.53 | 12–60 | 0.10–5.0 | 10–20 | 15–25 |
| **Helium** | | | | |
| 0.20 | 12 | 0.33 | 45–55 | 60–85 |
| 0.20 | 25 | 0.33 | 30–40 | 35–60 |
| 0.20 | 50 | 0.33 | 25–35 | 30–45 |
| 0.25 | 15 | 0.10–0.25 | 40–50 | 50–80 |
| 0.25 | 30 | 0.10–0.25 | 30–40 | 40–70 |
| 0.25 | 60 | 0.10–0.25 | 25–35 | 30–50 |
| 0.25 | 15 | 1.0 | 35–45 | 50–80 |
| 0.25 | 30 | 1.0 | 30–40 | 35–65 |
| 0.25 | 60 | 1.0 | 25–35 | 30.50 |
| 0.25 | 15 | 3.0–5.0 | 25–35 | 40–75 |
| 0.25 | 30 | 3.0–5.0 | 20–30 | 35–60 |
| 0.25 | 60 | 3.0–5.0 | 15–25 | 30–50 |
| 0.32 | 15 | 0.10–0.25 | 40–50 | 50–80 |
| 0.32 | 30 | 0.10–0.25 | 30–40 | 40–70 |
| 0.32 | 60 | 0.10–0.25 | 25–35 | 30–50 |
| 0.32 | 15 | 1.0 | 35–45 | 50–80 |
| 0.32 | 30 | 1.0 | 30–40 | 35–65 |
| 0.32 | 60 | 1.0 | 25–35 | 30–50 |
| 0.32 | 15 | 3.0–5.0 | 25–35 | 40–75 |
| 0.32 | 30 | 3.0–5.0 | 20–30 | 35–60 |
| 0.32 | 60 | 3.0–5.0 | 15–25 | 30–50 |
| 0.53 | 15 | 1.0–1.5 | 30–40 | 50–85 |
| 0.53 | 30 | 1.0–1.5 | 25–35 | 35–70 |
| 0.53 | 60 | 1.0–1.5 | 20–30 | 25–55 |
| 0.53 | 15 | 3.0–5.0 | 20–30 | 50–85 |
| 0.53 | 30 | 3.0–5.0 | 20–30 | 40–75 |
| 0.53 | 60 | 3.0–5.0 | 15–25 | 20–50 |

**Table 6-3** (continued)

| Diameter (mm) | Length (m) | Film (μm) | $\overline{u}$ | OPGV |
|---|---|---|---|---|
| Hydrogen | | | | |
| 0.20 | 12 | 0.33 | 70–80 | 85–150 |
| 0.20 | 25 | 0.33 | 50–60 | 70–105 |
| 0.20 | 50 | 0.33 | 40–50 | 50–75 |
| 0.25 | 15 | 0.10–0.25 | 60–70 | 75–130 |
| 0.25 | 30 | 0.10–0.25 | 50–60 | 70–110 |
| 0.25 | 60 | 0.10–0.25 | 40–50 | 60–80 |
| 0.25 | 15 | 1.0 | 50–60 | 85–130 |
| 0.25 | 30 | 1.0 | 45–55 | 65–110 |
| 0.25 | 60 | 1.0 | 40–50 | 45–80 |
| 0.25 | 15 | 3.0–5.0 | 30–40 | 75–150 |
| 0.25 | 30 | 3.0–5.0 | 25–35 | 60–85 |
| 0.25 | 60 | 3.0–5.0 | 20–30 | 50–65 |
| 0.32 | 15 | 0.10–0.25 | 60–70 | 75–130 |
| 0.32 | 30 | 0.10–0.25 | 50–60 | 70–110 |
| 0.32 | 60 | 0.10–0.25 | 40–50 | 60–80 |
| 0.32 | 15 | 1.0 | 50–60 | 85–130 |
| 0.32 | 30 | 1.0 | 45–55 | 65–110 |
| 0.32 | 60 | 1.0 | 40–50 | 45–80 |
| 0.32 | 15 | 3.0–5.0 | 30–40 | 75–150 |
| 0.32 | 30 | 3.0–5.0 | 25–35 | 60–85 |
| 0.32 | 60 | 3.0–5.0 | 20–30 | 50–65 |
| 0.53 | 15 | 1.0–1.5 | 40–50 | 80–150 |
| 0.53 | 30 | 1.0–1.5 | 40–50 | 70–115 |
| 0.53 | 60 | 1.0–1.5 | 35–45 | 60–100 |
| 0.53 | 15 | 3.0–5.0 | 25–35 | 75–125 |
| 0.53 | 30 | 3.0–5.0 | 25–35 | 65–100 |
| 0.53 | 60 | 3.0–5.0 | 20–30 | 55–100 |

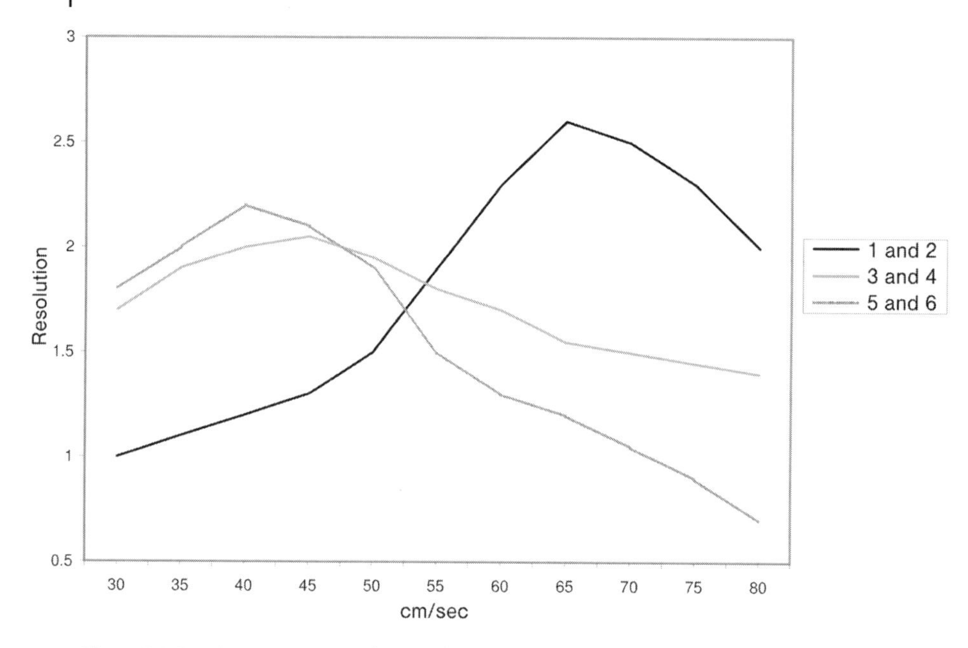


**Figure 6-4** Resolution vs. average linear velocity.
Finding the optimal average linear velocity for three pairs of peaks.

of the peaks. Another approach is to maximize the resolution of one set of peaks at the expense of another set of peaks (e.g., 65 cm/sec in Figure 6-4) to obtain a desirable outcome such as reduced run time.

The goal or purpose of the GC analysis has to be considered when determining which average linear velocity to use. A series of experiments designed to gather the same type of information as shown in Figure 6-4 is recommended to obtain a full understanding of the relationship between linear velocity and resolution for a particular analyses. The greatest range of optimal linear velocities is most common for analyses where the compounds elute over a wide temperature range, thus compromises are often necessary.

## 6.8 Gas Purities

The recommended grades of carrier gas are normally called ultra high purity and oxygen free. These grades typically contain ≤ 2 ppm oxygen, ≤ 10 ppm water and low levels of hydrocarbons (usually reported as THC). The most important gas quality criterion is a very low concentration of oxygen. Oxygen rapidly degrades stationary phases, thus reduces column life. Low hydrocarbon contaminants is desirable and beneficial for most detectors to maintain low noise levels. Several ppm of inert gases rarely have any noticeable effect. Chromatographic, research and similar grade gases are not necessary. They are extremely pure and their very high cost is without any real benefits.

Gas impurity traps are often recommended to further purify carrier gases (see Section 11.10). The lower the grade of the carrier gas, the more important gas impurity traps become. The use of traps has to be balanced against the cost of the traps and the higher cost of better grades of carrier gas. Lower grade, thus lower cost, carrier gases cause the traps to expire more rapidly. The savings in carrier gas cost is usually offset by the higher cost of frequent trap replacement. Better grades of gas do not cause the traps to expire as quickly. The higher cost of the better grade gas is nearly offset by the lower cost of less frequent trap replacement.

The quality of many high purity and oxygen free carrier gases is quite high. Using gas impurity traps may not result in substantial improvement in the cleanliness of the carrier gas. Unless performing trace level analyses or using a detector at its maximum sensitivity level, the use of gas impurity traps with highest quality carrier gases may not be worth the cost and required periodic replacement efforts. The decision should be based on the financial cost (materials and labor) and the benefits received.

## 6.9 Common Carrier Gas Problems

There are only a few problems that are associated with the carrier gas. The most difficult one to isolate and solve is a gas contamination problem. Contaminated gas (i.e., a bad cylinder of gas) is fairly rare. The carrier gas passes through many areas. Volatile materials in any of these areas can be transported by the carrier gas into the column where they are detected as a problem in the chromatogram. The only easy method to determine whether the carrier gas is contaminated is to change the cylinder. If the background problem does not substantially decrease or disappear in several hours, the GC is contaminated and the carrier gas is not the problem. If the problem disappears, the carrier gas may be at fault. It may be a good idea to re-install the original cylinder if that carrier gas seems to be the source of the problem. Sometimes problems disappear just because something was changed. If the problem returns, the gas cylinder is at fault; if the problem does not return, something else was responsible for the problem. Gas cylinders should not be used below 100–200 psi (700–1400 kPa). As the cylinder pressure decreases to 100–200 psi, any contaminants in the cylinder begin to vaporize and contaminate the gas. At higher pressures, the contaminants are liquids and do not have enough vapor pressure to leave the cylinder.

Improper measurement or setting of the linear velocity results in efficiency changes and retention time shifts. If an efficiency changes occurs, but the retention times do not shift, the problem is not with the average linear velocity. If the retention times shift all in the same direction, it is possible the incorrect linear velocity or column temperature is being used. All compounds are equally affected by a change in the linear velocity. If the carrier gas velocity changes, the retention times change by the same amount (absolute not percentage). If a previously run chromatogram is available, calculate the retention factors ($k$) for each peak and compare them to the current chromatogram. If the carrier gas velocity changed, the $k$ values for both chromatograms will be the same. If the $k$ values are not the same, stationary phase behavior or column temperature has changed.

If the carrier gas head pressure does not exceed a particular pressure even though the column head pressure knob is turned up, there may not be enough gas supplied to the GC. Check the pressure of the incoming carrier gas line to make sure it is at the pressure recommended by the GC manufacturer. If the pressure is too low, not enough gas is supplied to the GC to deliver the desired flow. If the head pressure increases when the split line flow is decreased, this is the problem. If the incoming pressure is high enough, check for leaks in the carrier line especially within the GC. Finally, check the flow controller if the GC has a back pressure regulated injector. Many flow controllers have a flow element which limits the range of flows the flow controller can deliver. The flow controller can only deliver the maximum flow as governed by the flow element regardless of the incoming carrier gas pressure. The flow element needs to be changed to a higher flow version to obtain the higher flow rates. Some flow controllers have an adjustable flow element instead of a replaceable one. GCs with programmable injectors do not have adjustable or replaceable flow elements, thus flow limits are built in the systems and can not be altered.

If the linear velocity changes by a large amount with a small turn of the head pressure controller knob, a flow element with a smaller range may be needed. One rotation of the controller knob changes the linear velocity by a smaller amount with a lower range flow element. Finer adjustments to the linear velocity are possible when the range of the flow controller is smaller. This is especially helpful when operating at low carrier gas head pressures.

# 7
# Injectors

## 7.1
## Introduction

The primary goal of sample injection is to introduce the sample into the column; however, there is much more to this operation. It is important to deposit the sample in the column in the narrowest band. The shorter the band at the beginning of the chromatographic process, the shorter it will be as it exits the column. This results in the desirable effect of tall and narrow peaks in the chromatogram.

One of the common assumptions about capillary GC is that the composition of the sample introduced into the column is identical to that of the sample before injection. Except for on-column injection, this assumption is incorrect. The operating conditions, hardware choices and sample composition all influence the similarity between the injected and original sample. Fortunately, the proper use of standards renders the differences unimportant. Several injection techniques have been developed to meet the specific demands of capillary gas chromatography. A universal injector or sample introduction approach has not been developed that can accommodate the wide diversity of capillary columns, samples and automation requirements. Accurate and precise injection results can be obtained if the various benefits and limitations of each injection method are understood and appreciated.

There are two basic types of injectors for capillary columns – vaporization and on-column. Vaporization injectors are more common, and thus more familiar to most analysts. Vaporization injectors include split, splitless and direct. On-column injectors include the cool on-column and direct on-column. Regardless of the type of injector, each has specific advantages and operation requirements.

## 7.2
## The Basics of Vaporization Injectors

All vaporization injectors have similar design features (Figure 7-1). A glass liner resides inside the heated, metal injector body. A carrier gas line supplies carrier gas to the interior of the injector body. Usually the carrier gas enters near the top of the injector. A syringe is used to pierce the septum and introduce the sample into

*The Troubleshooting and Maintenance Guide for Gas Chromatographers, Fourth Edition.* Dean Rood

ISBN: 978-3-527-31373-0

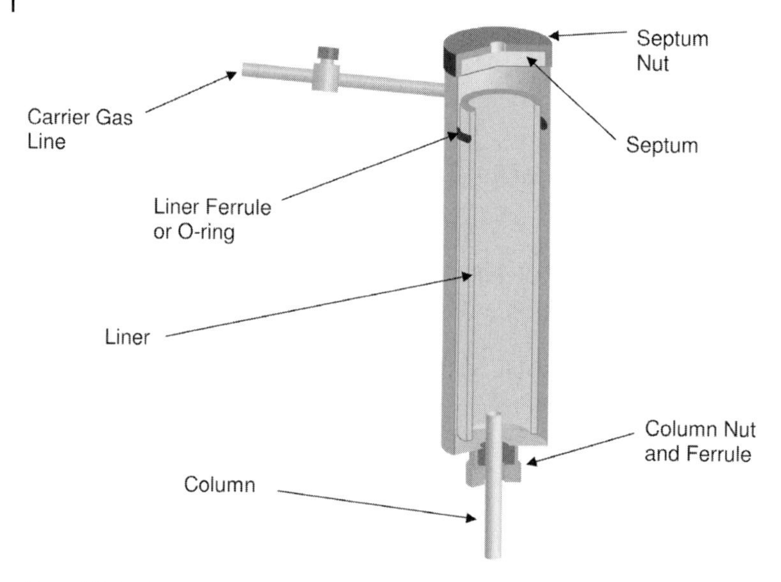


**Figure 7-1** Basic vaporization injector (not to scale).

the injector. The high temperature of the injector causes the volatile components of the sample to rapidly vaporize.

In some cases, a gaseous sample may be directly introduced into the injector. Any vaporized sample mixes with the carrier gas and is carried anywhere the carrier gas travels. Sample is introduced into the column by the movement of the carrier gas into the column. After the sample enters the column, the separation process begins. If there are flow paths other than the column (i.e., gas lines) for the carrier gas, portions of the sample do not enter the column. The presence of these gas lines and the volume of carrier flowing through them is determined by the design and use of the different injectors. The flow of gas through each of the gas lines and column is controlled by flow controllers or pressure regulators.

Most capillary injectors have a septum purge feature. The septum purge is a low flow of carrier gas across the top of the injector immediately below the septum. This gas then leaves the injector via the septum purge line. Purge flows of 0.5–5 mL/min are usually recommended by most GC manufacturers. Septum purge flows of 1–3 mL/min are the best overall settings. Some injectors have a pre-set septum purge flow that can not be easily adjusted.

The septum purge helps to minimizes the condensation of non-volatile or high boiling materials on the exposed portion of the septum. This region of the injector can be more than two times cooler than the set injector temperature. Condensation of sample residues in these areas can result in injector contamination problems. The septum purge also helps to minimize septum bleed problems (Section 7.10.3). The septum purge helps to keep the injector clean and, if used properly, does not have a significant influence on sample vaporization or column introduction processes.

### 7.2.1
### Injector Temperature

The injector temperature should be just hot enough to ensure "instant" vaporization of the entire sample. Excessively high temperatures may cause degradation of thermally unstable compounds. Also, massive expansion of the sample solvent and the more volatile sample compounds is always a concern at higher injector temperatures. This effect is called backflash (Section 7.2.3) and can result in a large solvent front, peak tailing or carryover problems. Excessively low injector temperatures may cause incomplete or slow vaporization of the sample. This may result in broad or tailing peaks especially for compounds with high boiling points. The problem is less noticeable for injectors that transfer the sample to the column very rapidly. Injector contamination is a greater problem at lower injector temperatures. Larger amounts of the high boiling point sample components accumulate in the injector since they are not readily vaporized and carried into the column.

### 7.2.2
### Speed of Sample Transfer

Rapid transfer of the vaporized sample from the injector into the column is desired. This results in the highest efficiency. Rapid sample transfer is obtained when the carrier gas sweeps the volume of the injector liner very quickly. High carrier gas flows and small liner volumes both result in rapid sweep rates. Rapid sample transfer creates an initial sample band that occupies a very short length of column, thus often resulting in peaks that are narrow. If the sample band is spread over a long length of column, the peaks may be excessively wide. The use of excessively high carrier gas flow rates or small liners to obtain high efficiencies can result in other problems. Other factors such as optimal carrier gas linear velocities and the reduction of backflash have to be considered.

Due to their design, some injectors have a low flow of carrier gas in the injector at the time of sample introduction. This results in a slow transfer of the sample into the column. In some cases, the loss of efficiency can be tolerated. In other cases, the wide sample band has to be focused into a shorter band by the use of specific conditions or techniques. Some of the operational requirements for some injectors are necessary to overcome the poor (i.e., slow) sample transfer characteristics of that injector.

### 7.2.3
### Injector Backflash

Upon vaporization in the injector, a sample expands to many times its original volume. The force of this expansion can be quite large and often generates a momentary high pressure pulse greater than the carrier gas pressure in the injector. If the final volume of the vaporized sample exceeds the injector liner volume, some of the vapors flow out of the liner. This event is called backflash.

The more volatile a compound, the more susceptible it is to backflash. The solvent is usually present in substantially greater amounts than the other sample components. Also, it is usually the most volatile sample component. Both of these factors makes the solvent the most likely substance to backflash. As the solvent flashes out of the injector liner, it may carry some of the other volatile sample components along with it.

As the vapors leave the liner, they may travel to a number of places. The vapors that leave the top of the liner flow across the bottom face of the septum and are either swept out through the septum purge line or back into the injector liner. If the pressure pulse is greater than the injector pressure, the vapors may travel into the incoming carrier gas line. The vapors that leave the bottom of the liner are swept out of the split vent line (if present or open).

The various compounds condense on any surface that is below their respective boiling points. These areas include the small portion of the septum exposed to the interior of the injector and the gas lines. Most gas lines are not heated and are at ambient temperature along most of their length. The highest boiling compounds deposit closest to the injector with the most volatile portions traveling the greatest distance. The most volatile compounds may travel over 10 cm before condensing inside the gas lines. Any compounds condensed on the septum eventually re-enter the injector liner. This may occur in two ways. Small amounts of the compounds may slowly volatilize and be carried back into the injector. Depending on the column temperature, these compounds either contribute to the background noise or are evident as peaks. If the column is maintained at higher temperatures, increased background noise is most common; interfering or ghost peaks are evident if the column is maintained at lower temperatures. Unless the contamination is very large, the background noise does not have any significant negative impact. The other route for entering the injector is from the continuation of backflash. Solvent vapors from any backflashing solvent passes over the septum and dislodges some of the condensed compounds. If the vapors are carried back into the injector, these compounds then chromatograph just like an injected sample, resulting in interfering or ghost peaks. The larger the amount of condensed compounds, the greater the problem caused by the background noise or extra peaks.

Any sample that has condensed in the septum purge or split vent lines normally does not cause any significant problems unless the contamination is very severe. Contamination of the incoming carrier gas line poses the same problem as compound condensation on the septum. Small amounts of the contaminants can be carried into the injector by the carrier gas. Also, sample solvent entering the gas line from backflash can carry contaminants back into the injector, resulting in background noise or extra peaks.

In addition to injector contamination problems, backflash can also cause peak tailing or broadening. Since some of the sample leaves the liner, additional time is now required for the carrier gas flow to sweep that portion of the sample back into the injector liner and column. The sample is introduced into the column in a much broader band (i.e., longer period of time). If only a small portion of the

**Table 7-1** Solvent expansion volumes.

| Solvent | Boiling point (°C) | Approximate volume (μL) of vapor | |
|---|---|---|---|
| | | 1 μL | 2 μL |
| *i*-Octane | 98–99 | 110 | 220 |
| *n*-Hexane | 69 | 140 | 280 |
| Toluene | 110–111 | 170 | 340 |
| Ethyl acetate | 76.5–77.5 | 185 | 370 |
| Acetone | 56 | 245 | 490 |
| Dichloromethane | 40 | 285 | 570 |
| Carbon disulfide | 46 | 300 | 600 |
| Acetonitrile | 81–82 | 350 | 700 |
| Methanol | 65 | 450 | 900 |
| Water | 100 | 1010 | 2020 |

Injector temperature: 250 °C; pressure: 20 psig.

sample left the liner, peak tailing is usually evident. If a large portion of the sample left the liner, peak broadening is more prevalent. The peak tailing or broadening usually becomes less severe as peak retention increases. This is due to the greater backflashing of the more volatile sample compounds and focusing at the front of the column of the less volatile compounds.

The area outside of the injector liner is primarily stainless steel. Compounds with hydroxyl or amine groups are particularly sensitive to interaction with heated metal surfaces. Peak tailing or loss of peak size may be evident if a compound interacts with a metal surface. For these cases, it may be especially important to eliminate or minimize any backflash.

The amount of sample expansion primarily depends on the sample solvent, injector temperature and column head pressure. Greater expansion volumes occur at higher injector temperatures and lower head pressures; compound boiling point is not directly correlated to expansion volumes. Table 7-1 lists some common solvents and the approximate amount of sample expansion for a typical injector temperature and head pressure. If backflash is a problem, the use of a sample solvent with a low expansion volume helps minimize the problem.

### 7.2.4
### Injector Discrimination

It is common with vaporization injectors for the entire sample not to be fully vaporized at the time of sample introduction into the column. Vaporization becomes faster and more complete as a compound's volatility increases. Since only the vaporized portion of the sample enters the column in significant amounts, a different percentage of each compound enters the column. A greater percentage of the more volatile compounds enters the column than the less volatile compounds. This effect is called injector discrimination. It is usually evident as progressively decreasing peak size for the later eluting compounds in the sample (Figure 7-2). The relative responses for the peaks in Figure 7-2 are shown in Table 7-2. A relative response factor of 1.00 would be a lack of injector discrimination.

There are a number of factors that influence the amount of injector discrimination. The main factors being the sweep time of the carrier gas, amount of mixing of the vaporized sample prior to column introduction and the volatility range of the compounds. The faster the sample is swept into the column, the less time that is allowed for vaporization. The more volatile compounds have enough time to become substantially vaporized while the less volatile compounds do not; therefore, greater amounts of the more volatile compounds enter the column. This makes discrimination greater at high carrier gas sweep rates. Liners designed to mix the vaporized sample prior to introduction into the column aid in reducing

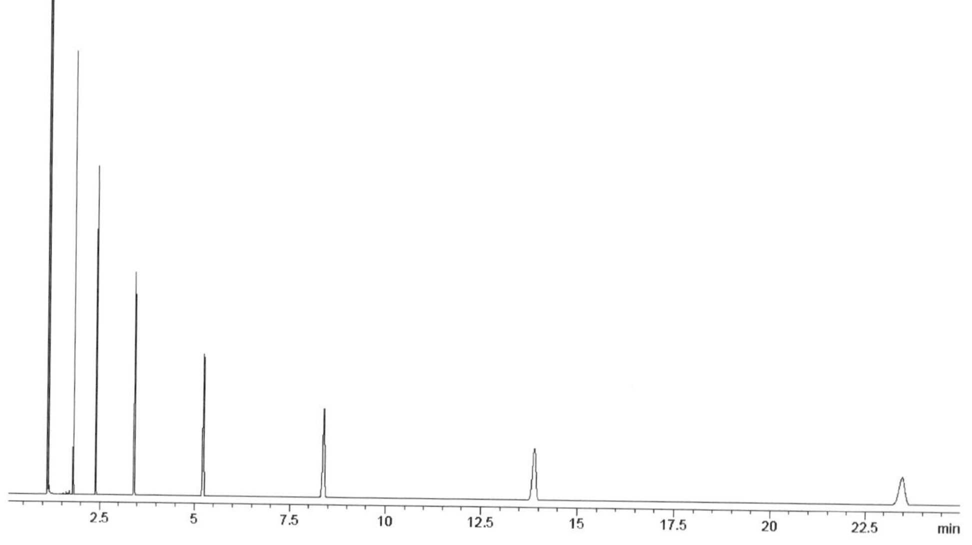


**Figure 7-2** Example of injector discrimination.
*Column:* HP-1, 30 m × 0.25 mm, 0.25 μm
*Injector:* Split 1 : 100
*Oven:* 120 °C
*Carrier:* Hydrogen at 60 cm/sec
*Sample:* *n*-$C_{10}$–$C_{16}$ in hexane at 500 ng/μL each

Table 7-2 Discrimination comparison (for Figure 7-2).

| Compound | Boiling point (°C) | Response relative to *n*-decane |
|---|---|---|
| *n*-Decane | 174 | 1.00 |
| *n*-Undecane | 196 | 0.92 |
| *n*-Dodecane | 215–217 | 0.83 |
| *n*-Tridecane | 234 | 0.77 |
| *n*-Tetradecane | 252–254 | 0.71 |
| *n*-Pentadecane | 270 | 0.67 |
| *n*-Hexadecane | 287 | 0.63 |

Response was measured using peak areas.
Relative responses were corrected for detector response differences.

the amount of discrimination. Greater amounts of sample mixing result in lower discrimination. These types of liners are primarily useful for high sweep rate situations. The larger the volatility differences between the sample compounds, the greater the extent of discrimination. Later eluting peaks can be over 10 times smaller than earlier eluting peaks in a chromatogram when the compounds have boiling points differing by 200–300 °C.

Factors including injector temperature, injection size and technique, and injector design have minor influences on the amount of discrimination. The magnitude of their influence is unique to each sample and can be difficult to predict. Usually the amount of discrimination influenced by these factors is small and rarely considered when selecting operating conditions. Detector response can affect the appearance of discrimination, thus altering the apparent amount of injector discrimination.

## 7.3 Split Injectors

### 7.3.1 Description of a Split Injector

The concentrations of many samples easily exceed the capacities of most conventional capillary columns. Limiting the amount of sample reaching the column by splitting it prior to column introduction prevents column overloading. A split injector splits the sample into two unequal portions with the smaller fraction going to the column and the larger fraction being discarded. Split injectors are used for more concentrated samples since only a very small portion of the sample enters the column. Typical sample concentrations for split injections are usually 0.1–10 μg/μL per sample component.

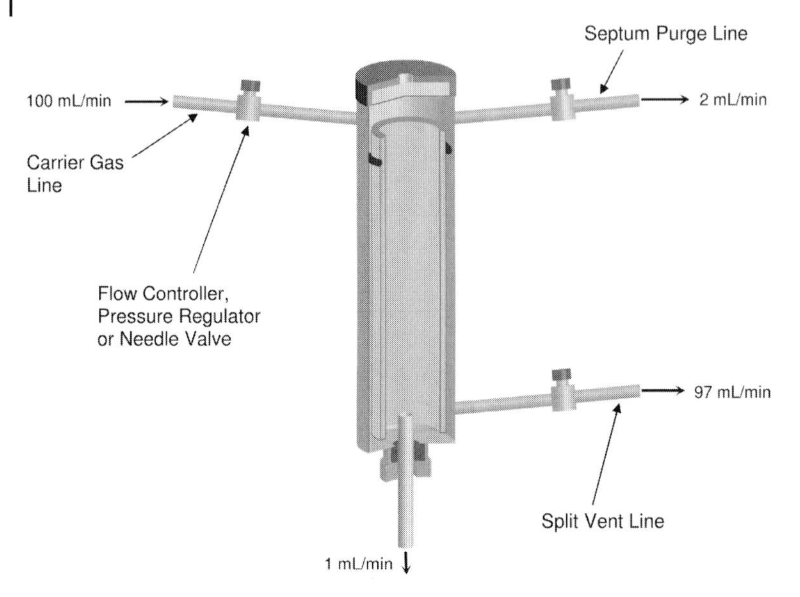


**Figure 7-3** Split injector (Not to scale. Some details are omitted for clarity). The flow values shown are examples.

A split injector consists of the heated injector, liner, incoming carrier gas line, split line and septum purge line (Figure 7-3). After entering the injector, the carrier gas exits by way of the column, split line and septum purge line. Since the septum purge flow is very low and above the region where sample vaporization occurs, very little sample is lost via the septum purge line. Any sample loss from the septum purge line is usually ignored when considering sample splitting.

After injection and during vaporization, the sample vapors are mixed with the carrier gas. The vaporized sample travels wherever the carrier gas goes. The carrier gas can travel down two pathways – the column and the split line. The volume of carrier gas entering the column is small compared to the split line, thus only a small fraction of the sample enters the column. This splitting of the carrier gas flows results in the splitting of the sample also. The amount of splitting is represented by the split ratio.

### 7.3.2
### Split Ratio

The split ratio is the volume of carrier gas entering the column versus the volume leaving via the split line. The ratio of these two gas volumes is called the split ratio. The split ratio is determined by measuring or calculating the gas flows out the split vent and column. Split ratios are reported as the column flow normalized to one. There are two slightly different viewpoints on the measurement of the split ratio, and thus two equations (Equations 7-1a and b). The differences are small and of no consequence.

a) Ratio of sample amount entering the column and the total sample

$$\text{Split ratio} = \frac{F_{\text{column}} + F_{\text{split vent}}}{F_{\text{column}}}$$

b) Ratio of sample amount entering the column and the sample amount leaving the split vent

$$\text{Split ratio} = \frac{F_{\text{split vent}}}{F_{\text{column}}}$$

$F_{\text{split vent}}$ = Gas flow from the split vent (mL/min)
$F_{\text{column}}$ = Gas flow from the column (mL/min)

**Equation 7-1** Split ratio calculations.

Calculating and setting the split ratio in a reproducible manner is more important since the amount of sample splitting does not directly correlate to the split ratio. This topic will be explored in more detail in a later section. Selecting a split ratio is primarily based on sample considerations, column diameter and injector capabilities.

The sum of the column and split vent flow rates should be greater than 20 mL/min. Lower total flow rates may result in broadened peaks especially for early eluting peaks. Large diameter columns use high carrier gas flows, thus low split vent flows (resulting in low split ratios) can be used without dropping the total flows below 20 mL/min. Small diameter columns use low carrier gas flows, and the split line flows therefore need to be high to keep the total flow above 20 mL/min. The lowest possible split ratios for small diameter columns are higher than for a larger diameter column. Lowest recommended split ratios for various column diameters are shown in Table 7-3.

The highest split ratios are dictated by the total amount of carrier gas available to the GC and injector hardware limitations. The maximum total injector flow rate is limited by the design of the injector's flow controllers or pressure regulators. The total varies by model of GC, but the maximum total flows are usually 500–1000 mL/min.

**Table 7-3** Lowest recommended split ratios.

| Column diameter (mm) | Lowest split ratio |
|---|---|
| 0.10 | 1 : 100 – 1 : 150 |
| 0.18–0.20 | 1 : 20 – 1 : 25 |
| 0.25 | 1 : 15 – 1 : 20 |
| 0.32 | 1 : 10 – 1 : 12 |
| 0.53 | 1 : 3 – 1 : 5 |

The split ratio selected is primarily determined by the amount of sample that needs to be introduced in the column. Low split ratios introduce a greater amount of sample into the column than a higher split ratio. Lower split ratios are used when sample concentrations are low and/or detector sensitivity is lacking. Higher split ratios are used for highly concentrated samples that may overload the column or detector. Injector efficiency and discrimination are two other factors that need to be considered when selecting a split ratio.

At higher split ratios, the flow of carrier gas in the injector is high. This sweeps the sample from the injector and into the column within a very short time. Since the sample is introduced into the column so quickly, high injector efficiency is obtained. Also, at high split ratios, the high boiling (less volatile) sample compounds have very little opportunity to completely vaporize during their residence time in the injector. The low boiling (more volatile) compounds are more thoroughly vaporized within the same time period. The difference in the amounts of high and low boiling compounds introduced into the column can be large. This results in high injector discrimination.

At low split ratios, the flow of carrier gas in the injector is lower than for high split ratio situations. The residence time of the sample in the injector is much longer than at a higher split ratio. The higher boiling compounds have a better opportunity to vaporize more thoroughly. The lower boiling compounds vaporize so rapidly that longer times in the injector do not substantially increase the amount vaporized. The difference in the amounts of low and high boiling compounds introduced into the column is much smaller. This results in lower injector discrimination. In general, injector discrimination increases as the split ratio becomes larger.

A proper balance between amount of sample introduced into the column, discrimination and efficiency must be reached. Using a low split ratio places more sample in the column and minimizes discrimination; however, efficiency losses may be evident. Using a high split ratio places less sample in the column and discrimination is more severe; however, greater efficiency is gained. In most cases, the amount of sample introduced into the column is the most important factor. Losses or gains in efficiency with changes of split ratio are usually small and are not noticed unless large split ratio changes are made. Discrimination differences are noticeable when the split ratio is changed, but the effect is counteracted by the proper use of calibration standards.

Split injectors use high volumes of carrier gas especially at high split ratios. It may be desirable to use a slightly lower split ratio when possible to save on gas consumption. For example, dropping to a 1 : 50 split ratio from 1 : 100 reduces gas consumption by half. Approximately two times the sample enters the column (which in many cases is not a problem), but a substantial savings in carrier gas costs results.

A common assumption is that the sample is split in the same proportions as the split ratio indicates. Due to discrimination and errors in flow measurements, the actual amount of sample introduced into the column is not exactly what the split ratio indicates. At a 1 : 50 split ratio, it seems that 1 part of the sample enters the column for every 50 parts discarded via the split vent. The actual amount of each compound entering the column deviates from the expected 1 to 50 ratio depending on a large number of variables. These variable include compound volatility, injector temperature and design, type of liner, column diameter, carrier gas flow, column position in the injector, injection size, and sample solvent. If all of these variables remain constant, very reproducible splits will be obtained. Each compound may be affected differently by change to any of these variables.

In addition to the split ratio not being an accurate measure of the amount of sample splitting, the split ratio is also non-linear. For example, the amount of sample entering the column with a 1 : 50 split ratio is not two times the amount with a 1 : 100 split ratio. Each compound in a sample may deviate from the expected amount by a slightly different amount. Larger deviations are also expected with larger difference between two split ratios.

### 7.3.3 Septum Purge for Split Injectors

The septum purge line can act as a miniature split line if sample reaches the top of the injector during the vaporization and expansion process. The more volatile sample components more readily vaporize and expand into a greater volume. This makes them more likely to travel up into the septum purge gas stream and be carried out of the injector. Fortunately, the rapid sweeping of the sample into the column, even for low split ratio situations, does not permit much time for the sample to remain at the top of the injector. Since the sample spends little time in the region of the septum purge flow, only a small fraction of the sample is lost. For this reason, higher septum purge flows (e.g., 3–4 mL/min) can be used with split injector with minimal loss of the more volatile sample components. In any case, a greater loss of sample occurs for large injection volumes, high injector temperatures, small volume liners, and low carrier gas flows.

### 7.3.4 Split Injector Liners

There are many liners available for split injectors. The primary function of any liner is to provide an inert environment for sample vaporization. It is important to use a liner that is deactivated or active compounds may absorb onto the liner surface. Peak tailing or loss of peak size may be evident if absorption is occurring. Most liners are supplied already deactivated, thus new liners are suitable for immediate use. Liner deactivation decreases with use especially after injections with strongly acidic or alkaline samples. Liner activity is much greater problem for compounds containing hydroxyl or amine functional groups. Compounds lacking these groups are rarely affected. Liner activity is not a major problems for split injectors since the vaporized sample is swept into the column (and out of the injector) so quickly. There is little opportunity for adsorption to occur in such a short time period.

Split liners are available in a variety of designs ranging from simple, straight tubes to ones with complex inner surfaces; some may be packed with glass wool. Split liners with complex inner surfaces are designed to thoroughly mix the vaporized sample before it is split. The carrier gas mixed with vaporized sample does not flow directly to the bottom of the liner but is mixed in a turbulent fashion. This extensive amount of mixing helps to minimize discrimination by making the sample more homogeneous. Liners that reduce discrimination can be especially useful for high discrimination situations such as high split ratios or samples with a broad range of compound volatilities. Greater discrimination is seen with liners without some type of mixing feature. The inverted cup, "cyclo", and fritted liners are examples of liners that exhibit lower discrimination.

Some liners are packed with glass wool. The glass wool must be silylated or severe activity problems (i.e., peak tailing or size reduction) occur. Since it is very difficult to silylate glass wool, it is strongly recommended to purchase pre-silylated glass wool. A small amount of glass wool placed in the proper position reduces discrimination especially for the straight tube liners. The carrier gas has to travel through the bundle of glass wool which causes a thorough mixing of the vaporized sample and carrier gas. This mixing, just like for the complex surface liners, reduces the amount of discrimination. The glass wool also acts as a filter to minimize the amount of non-volatile materials reaching the column, thus reducing the possibility of contamination. The use of glass wool comes with several precautions. The plug of glass wool should not be too tightly packed. The carrier gas has difficulty flowing through the dense plug of glass wool. This causes the sample to reside in the injector too long with a subsequent broadening of the peaks. The pressure of the carrier gas may push a dense glass wool plug towards the bottom of the liner. A plug of glass wool at the bottom of a liner usually causes irreproducible peak size, blockage of the column or split line, or distorted peak shapes (i.e., tailing, rounded, broad or split). When forming the plug of glass wool, breakage of the individual fibers should be minimized. While the surface of the glass wool is deactivated, the interior is not. Any ends created upon fiber breakage are potential active sites. The glass wool plug should be formed by gently rolling the fibers into a loosely packed ball about the size of the liner's inner diameter. Excessive mashing or crunching of the glass wool plug into a ball creates numerous ends. Conversely, a glass wool plug that is excessively loose does not mix the carrier gas enough to reduce discrimination or trap non-volatile materials. The plug of glass wool should be gently and carefully pushed into the proper position in the liner. For maximum injection reproducibility, the same person should install the glass wool or liners with pre-installed glass wool should be purchased. Surprisingly, even small differences in the amount of glass wool and how tightly it is packed affect the degree of discrimination.

There are differing opinions and studies concerning the best position for glass wool in a split liner. The most common conclusion is to have the glass wool below the end of the syringe needle. Other studies have shown better results if the needle passes through the glass wool. It is suspected that the glass wool wipes residual sample from the needle and provides better injection reproducibility. In any case, the most important factor is consistent placement of the glass wool.

Probably, the most important aspect of selecting a split injector liner is consistency. Each type of liner provides different results especially when discrimination is considered. If maximum reproducibility from analysis-to-analysis or lab-to-lab is desired, the same type of liner needs to be used each time. There are many claims stating that a specific liner is especially good for a certain type of sample. While this may be true, there are so many variables that affect discrimination, linearity, sensitivity, efficiency and peak shape. Liners are often only one of many variables that influence the final results. In general, liners with the complex inner surfaces are more costly. They will often cost 2–4 times more than a straight tube or very simple liner. This may be an important consideration if the liners need frequent replacement. If liners are re-used, the complex liners are more difficult to clean and properly re-silylate. In either case, the more costly or difficult to clean liners may not be worth the reduced discrimination they provide. Essentially, a good liner is one that provides satisfactory results.

### 7.3.5 Column Position in Split Injectors

Each model of GC has a distance to which the column is inserted into the injector. This value is usually found in the column installation instructions for the GC. The distance recommended by the GC manufacturer is probably the best one for the widest variety of samples; however, it may not be the best one for a specific sample. Errors in the insertion distance, especially if the distance is too short, may result in severe problems. The column insertion distance primarily influences efficiency and discrimination. Sometimes, even several millimeters difference has a pronounced impact on these parameters. The best distance varies depending on the nature of the sample and the specific liner.

The GC manufacturer's usually have more than one liner available for each GC. Also, there may be a number of different liners available from other sources. The manufacturer's column insertion distance may not be suitable or optimal for some of these other liners. This is due to the position of any flow disruption devices or an unique design. In general, the end of the column should be near the bottom of the liner. It is quite important that the column end is not completely below the liner and in the ferrule or fitting area of the injector. If the column end resides in this area, poor peak shape, reduced peak size or severe discrimination may occur.

Regardless of the type of liner, the position of the column in the liner may have a significant effect on efficiency and discrimination. Experimenting with column position, within the previously discussed guidelines, may result in the best efficiency or minimal discrimination for a particular sample. If optimizing these two parameters is not extremely important, using the GC manufacturer's recommended insertion distance is satisfactory. Most importantly, consistent column position is critical if precise reproducibility is required.

### 7.3.6 Common Problems with Split Injectors

Split injectors are relatively simple, thus there are only a few problems that ever arise. If a problem with a previously working method occurs, the following areas should be checked for deviations or errors.

1. *Column, split and septum purge gas flow rates:* Most problems in these areas result in changes in peak sizes (i.e., change in the split ratio or sample loss through the purge vent). Tailing peaks or solvent fronts may occur if the split flow is extremely low.
2. *Inlet purge on/off setting:* The purge function should be on at the time of injection. If the purge is off, a very large solvent front is obtained.
3. *Injector temperature:* Wide or tailing peaks may occur at low temperatures (below ~150 °C). Changes in absolute or relative peak sizes (i.e., change in injector discrimination) may be a sign of a large change (> 50 °C) in injector temperature.
4. *Liner:* Check the liner for contaminants, debris, breakage and proper installation. Liners become more active with use and may need replaced. The type of liner and glass wool position and amount are also important factors. Poor peak shapes or loss of peak size are the most common symptoms of a liner problem.
5. *Column installation:* The position of the column in the injector may affect peak shape and size.
6. *Leaks:* A leak in the injector may result in unstable retention times, changes in peak sizes and rapid column deterioration. Old septa (i.e., needs to be changed) and damaged or loose fittings are the usual leak sources.
7. *Contamination:* Residues from injected samples can contaminate nearly every part of the injector including liners, seals, fittings, injector body and gas lines.

## 7.4 Splitless Injectors

### 7.4.1 Description of a Splitless Injector

A splitless injector does not split the sample, thus most of the vaporized sample is introduced into the column. This makes splitless injections suitable for trace level analyses, but not for high concentration analyses. The lowest concentration level depends on the sensitivity of the detector being used. The highest concentration level is primarily limited by the capacity of the column and sometimes the linear or mass range of the detector.

Most injectors are capable of operating in the split and splitless modes. Split only injectors exist, but they are usually only found in very low cost or specialty instruments. There is a heated injector, liner, incoming carrier gas line, split line and septum purge line. The split line has a solenoid which opens and closes this line to gas flow. After entering the injector, the carrier gas exits by way of the column, split line and septum purge line. The septum purge flow is very low and above the region where most of the sample vaporization occurs, thus very little sample is lost via the septum purge line.

Splitless injectors have two operational modes which occur in sequence during an analysis. At injection and for a short time afterwards, the solenoid closes the split line and there is no gas flow through the line (Figure 7-4a). This mode is usually called purge off. Upon sample injection and vaporization, the sample mixes with the carrier gas and is transported towards the column. Since the split line is closed at this time, the only place the sample can travel is into the column. Sample transfer into the column is very slow since the carrier gas flow in the injector is the same as the column flow (1–3 mL/min) during purge off. The resulting poor sample transfer is compensated by use of the solvent effect or cold trapping (Sections 7.4.3 and 7.4.4). After a preset time period (called the purge activation time or purge on time; Section 7.4.2), the split line is opened by the solenoid. The carrier gas flow that previously by-passed the injector during purge off flows through the injector and out through the split line (Figure 7-4b). This mode is usually called purge on. The splitless injector now behaves just like a split injector. Any residual sample still in the injector is swept out of the split line (except for an inconsequential amount that enters the column). The inlet purge flow (not the septum purge flow) is usually set between 10 and 25 mL/min.

The inlet purge function is necessary to keep the solvent front from being excessively wide. The slow carrier gas flow rate in the injector during purge off requires an excessively long time to flush all of the sample from the injector and into the column. Since almost all of the sample is solvent, most of the residual sample in the injector is solvent. Failure to purge the injector of this solvent results in a very broad and tailing solvent front. Too short purge activation times results in very small amounts of sample entering the column; too long purge activation times places more sample in the column, but creates a very large solvent front.

One injector design (back pressure regulated) diverts the split line gas flow through the septum purge line during the purge off time. This results in a very high gas flow (septum purge + split line) across the top of the injector during the time of maximum sample flashback. Potentially, large amounts of the most volatile sample components can be lost via the septum purge line. For this reason, the split line flow with back pressure regulated injectors used in the splitless mode should be kept fairly low (10–25 mL/min). Another injector design (forward pressure regulated) maintains the same septum purge flow throughout the analysis and turns off the split line flow during the purge off time. Sample loss via the septum purge line is much less for these types of injectors and often lower discrimination is evident also. Higher split line flows can be used to sweep forward pressure regulated injectors without extensive sample loss.

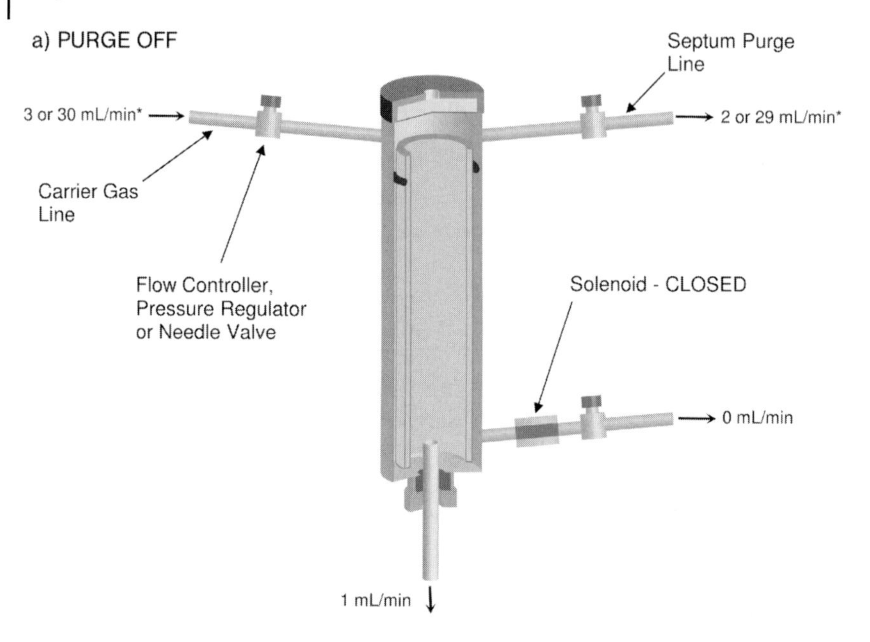


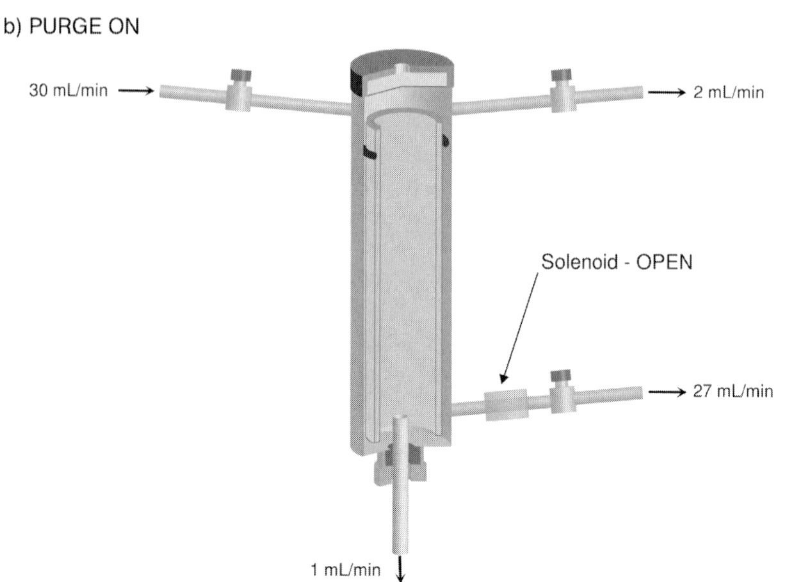


**Figure 7-4** Splitless injector (Not to scale. Some details are omitted for clarity). Flow values shown are examples.
* Flow depends on injector design.

Some programmable splitless injectors use a hybrid format. When in the purge off mode (i.e., at and slightly after sample injection), the injector is configured in the forward pressure regulated mode and switches to the back pressure regulated mode during the purge on mode. This design utilizes the strengths of both injector designs for optimal performance.

### 7.4.2
### Selecting Purge Activation Times

The best purge activation time depends upon the carrier gas flow rate and the volatility of the sample compounds. In general, higher boiling point (lower volatility) compounds require longer purge activation times than lower boiling point compounds. More time is needed to thoroughly vaporize higher boiling point compounds and introduce them into the column in sufficient amounts. Lower boiling point compounds do not require as much time to thoroughly vaporize, thus short purge activation times are satisfactory.

As a guideline, purge on should occur when about 1.5 volumes of carrier gas has swept the injector. This is calculated by dividing the liner volume by the flow rate of the carrier gas. Most liners have a volume of 0.5–1 mL; most carrier gas flows are 1–3 mL/min. This results in typical purge activation times of 15–90 seconds. Experimenting with purge activation times above and below the calculated time is recommended. The goal is to maximize analyte peak size and minimize solvent front size. Excessively long purge activation times increase solvent front size without an appreciable increase in peak size; too short purge activation times result in smaller than desired analyte peaks. For most applications, it has been found that purge activation times of 40–90 seconds provides the best balance of analyte peak and solvent front sizes.

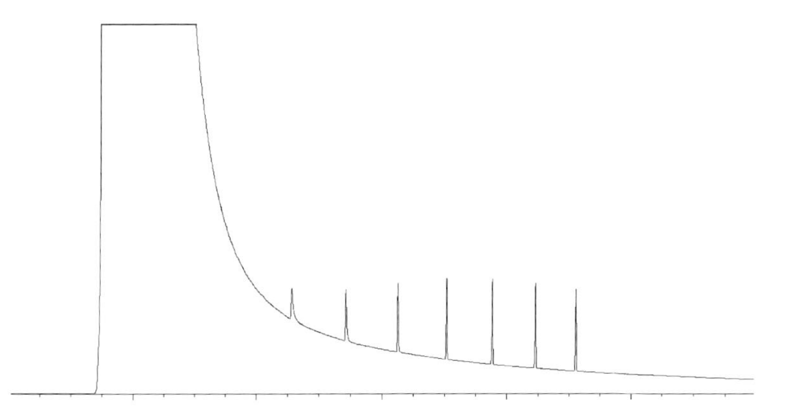

**Figure 7-5** Lack of a purge function example.
*Column:* HP-1, 30 m × 0.25 mm, 0.25 μm
*Oven:* 50 °C for 1 min, 50–150 °C at 10°/min
*Injector:* Splitless; no purge on time
*Sample:* $n$-$C_{10}$ to $n$-$C_{16}$ hydrocarbons in hexane

If the purge on function does not activate or occurs after a very long time, a very broad and tailing solvent front is obtained (Figure 7-5). If the purge is on during the injection, the solvent front and analyte peaks are much smaller than expected. Essentially, a split injection is made.

### 7.4.3
### Solvent Effect for Splitless Injectors

Sample transfer into the column occurs during purge off (before the purge activation time). Sample transfer occurring over such a long time results in the vaporized sample being spread over a long length of column with a subsequent loss of efficiency. This problem is overcome by focusing the sample in a short band at the front of the column. The solvent effect is used to focus the sample in splitless injections.

The solvent effect depends on the condensation of the sample solvent at the front of the column (Figure 7-6). This is accomplished by maintaining the initial temperature of the column at least 10 °C below the boiling point of the sample

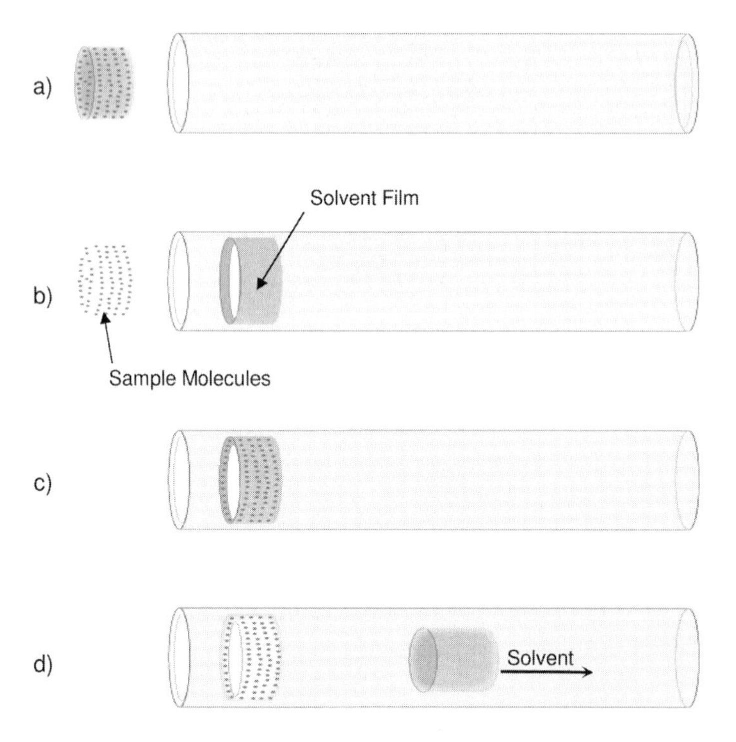


**Figure 7-6** Solvent effect:
(a) Injected sample.
(b) Solvent film forms upon condensation of the sample solvent.
(c) Compounds trapped in the solvent film.
(d) Solvent elutes from the column leaving the compounds behind in a narrow band.

solvent. Since the column is below the boiling point of the solvent, it condenses in a thin film at the front of the column (Figure 7-6b). The solvent comprises the overwhelming majority of the sample and is usually the most volatile component present. As the remaining sample components enter the column after the solvent, they are trapped in the solvent film (Figure 7-6c). This focuses the injected sample over a short length of column and overcomes the poor sample introduction problem. As the column is heated during the temperature program, the solvent volatilizes first and the remaining sample components are left behind in a short band (Figure 7-6d). As the column temperature continues to increase, the sample components begin to chromatograph as desired.

The solvent effect does not occur if the sample solvent does not condense in the front of the column (i.e., the initial oven temperature is too high). A solvent effect violation is fairly easy to detect since the peaks are broad, rounded or distorted with the earlier eluting ones being the worst (Figure 7-7). Even though the initial oven temperature is too high and a solvent effect violation has occurred, some late eluting peaks in the chromatogram may be well formed (Figure 7-7). This result is caused by cold trapping (Section 7.4.4).

In some cases, better peak shapes are obtained when the initial column temperature is 20–30 °C below the solvent boiling point instead of the normally recommended 10 °C. This is often improves the shapes of peaks eluting immediately after the solvent front. Also, the solvent should to be the most volatile sample

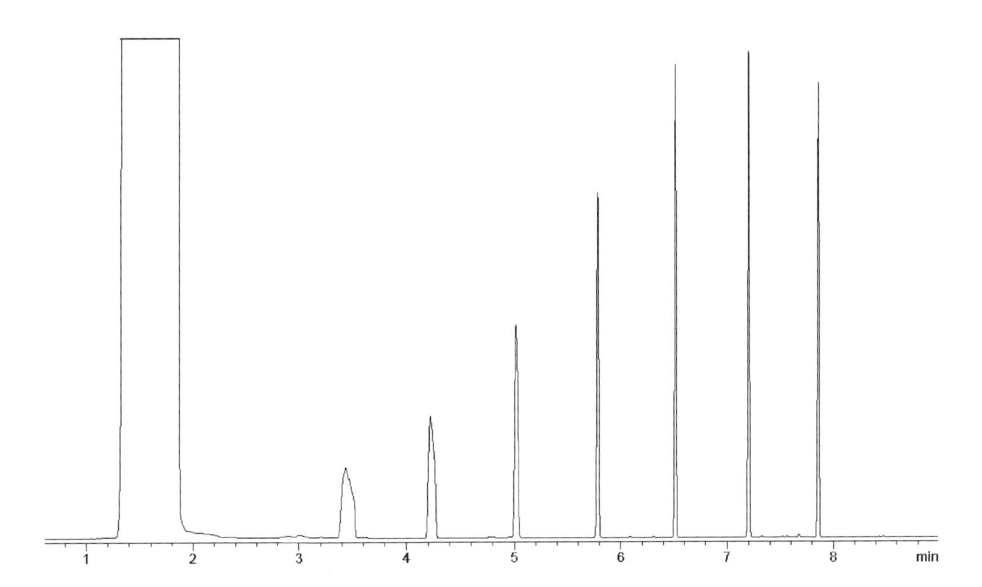


**Figure 7-7** Solvent effect violation and cold trapping example.
*Column:* HP-1, 30 m × 0.25 mm, 0.25 μm
*Oven:* 90 °C for 1 min, 90–225 °C at 15°/min
*Injector:* Splitless, 30 sec purge on time
*Sample:* $n$-$C_{10}$ to $n$-$C_{16}$ hydrocarbons in $n$-hexane (boiling point = 69 °C)

component. Any peaks eluting before the solvent peak will be broad and distorted. If these peak are not of interest, their poor shapes are not important.

If possible, use a sample solvent with a boiling point at least 25 °C less than the most volatile sample component. This will ensure the solvent being the earliest eluting sample component and improve the shapes of any analyte peaks eluting immediately after the solvent front. Solvents with extremely low boiling points require the use of low initial oven temperatures which may require extremely long times for the GC to achieve. Oven temperatures below 35 °C may not be possible without the use of liquid nitrogen or carbon dioxide to cool the GC oven. Rapid evaporation of highly volatile solvents from the sample vials is another potential problem. Using the lowest boiling point solvent (e.g., pentane, dichloromethane) may not always be the best choice, and sometime compromises have to be made to obtain the best peak shapes and tolerable analysis conditions.

Injection size has a direct effect on the length of the solvent film deposited as part of the solvent effect. The length of the solvent film increases with increased injection volumes. Small losses in efficiency may be observed as injection sizes start to exceed 2 μL. The solvent film length is shorter for larger diameter columns since the surface area per unit length of column is greater. In some cases, the reduced solvent film length may offset the decreased efficiency of a larger diameter columns. This effect primarily occurs for the peaks that elute closest to the solvent front. As peak retention increases, the benefits of the larger diameter column diminishes until no gains are obtained.

### 7.4.4 Cold Trapping for Splitless Injectors

Some compounds focus at the front of the column regardless whether a solvent film is present. Compounds with boiling points at least 150 °C above the initial column temperature undergo cold trapping and focus at the front of the column. The initial column temperature is low enough so that the sample compounds condense in a tight band at the front of the column – a narrow solvent film is no longer needed to focus the sample. Peaks with very narrow widths are obtained by cold trapping and they are often better than those obtained by the solvent effect (Figure 7-7). It is common for some of the later eluting compounds to successfully cold trap while some of the earlier eluting compounds to require the solvent effect. In these cases, the solvent effect rules have to be followed to obtain satisfactory peak shapes for the earlier eluting peaks.

### 7.4.5 Septum Purge for Splitless Injectors

The septum purge line can act as a miniature split line if the sample reaches the top of the injector during the vaporization and expansion process. The more volatile sample components are more likely to flash towards the top of the injector. Due to the lower sweep rate of splitless injectors, there is greater opportunity for the more volatile sample components to reach the top of the injector. The amount of sample loss varies depending on the design of the injector. The sample loss occurs during the purge off time (before the purge activation time) with most of the loss occurring within the first several seconds after injection. A septum purge flow of 0.5–3 mL/min is recommended for splitless injections.

### 7.4.6 Splitless Injection Liners

Splitless injection liners are usually straight tubes with some having a restriction at one or both ends. The restrictions act as backflash barriers. They help to keep the sample inside of the liner during the expansion portion of sample vaporization. This helps to keep the injector lines and septum face much cleaner, which helps to minimize carryover or ghost peak problems. A bottom restriction acts as a guide to keep the column centered in the liner and away from the walls. Erratic peak shapes and sizes may be encountered if the column end is touching the inner wall of the liner. An upper restriction acts as a needle guide and helps to direct the syringe needle to same location for each injection. An upper restriction will require better alignment of autosamplers and precise manual injections to prevent damaging the tips of the needles.

The lowest cost liners are the straight tube liners. The cost increases if there are any restrictions. The double restriction liners are more difficult to clean since the openings are quite small. The most common splitless liner inner diameters are 2 and 4 mm. The 2 mm i.d. liners are more susceptible to backflash problems, but they provide slightly better efficiency. In most cases, the small gain in efficiency is not worth the increase in backflash (injector contamination) problems. Also, injection volumes are limited to 1 μL or significant backflash problems occur.

Using glass wool in a splitless liner is not recommended. The glass wool causes excessive mixing and dilution of the vaporized sample with carrier gas. This may cause broader or smaller peaks. In cases where the injected sample contains a large amount of high boiling point or non-volatile compounds, the use of glass wool may be warranted. The slight losses in efficiency or peak size may be worth the benefit of the filtering effect of the glass wool. Thoroughly deactivated liners are more important with splitless injections. The sample spends a large amount of time in the liner, thus active compounds have greater opportunities to suffer from adsorption (peak tailing or size loss).

### 7.4.7 Column Position in Splitless Injectors

Each model of GC has a distance to which the column is inserted into the injector. This value is usually found in the column installation instructions for the GC. The distance recommended by the GC manufacturer is probably the best one for the widest variety of samples; however, it may not be the best one for a specific sample. Errors in the insertion distance, especially if the distance is too short, may result in severe problems. The column insertion distance primarily influences efficiency and discrimination. Sometimes, even several millimeters difference has a pronounced impact on these parameters. The best distance varies depending primarily on the nature of the sample compounds.

For most splitless applications, the manufacturer's recommended column position is usually near the bottom of the liner. If a bottom restriction is present, it is suggested that the column end be placed slightly above the opening of the restriction. It has been reported that there should be 1–1.5 cm between the end of the syringe needle and the column end. This requires the column to be inserted farther into the injector than normally recommended by the GC manufacturer. This column position seems to work best for samples containing very high boiling compounds (e.g., PAH's, steroids). Regardless of the type of liner, the position of the column in the liner may have a significant effect on efficiency and discrimination. Experimenting with column position, within the previously discussed guidelines, may result in the best efficiency or minimal discrimination for a particular sample. If optimizing these two parameters is not extremely important, using the GC manufacturer's recommended insertion distance will be satisfactory. Most importantly, consistent column position is critical if a high degree of reproducibility is required.

### 7.4.8
### Other Aspects of Splitless Injectors

The use of a retention gap (Section 11.8) will often improve the peak shapes for compounds that rely on the solvent effect for focusing. The peaks closest to the solvent front are often slightly rounded or broad. A retention gap sometimes helps to sharpen these peaks with a potential increase in resolution.

If a peak shape problem occurs that can not be attributed to a solvent effect problem, a polarity mismatch may be the cause. This usually occurs when the solutes and solvent are similar in polarity while the stationary phase is different. Instead of condensing into an even, thin film at the front of the column, the solvent beads up or smears. The solutes are dispersed into the uneven film, thus they are not focused into a narrow band and the peak shapes are poor. Only compounds that are very soluble in the solvent are affected by the polarity mismatch. For example, only the polar alcohols in a mixture of solvents in water exhibit peak shape problems when using a relatively nonpolar column; the immediate and nonpolar solvent peaks are well formed. The polar alcohols are very soluble in the very polar water while the other solvents are much less soluble. Changing the solvent to one that is closer in polarity to the stationary phase improves the peak shapes. If the sample solvent can not be changed, the use of a retention gap sometimes improves the peak shapes.

Since splitless injections deposit a large amount of sample in the column, a retention gap also acts as a guard column. All of the sample is focused in the retention gap instead of the analytical column. Non-volatile residues and compounds that can damage the stationary phase are not concentrated on the stationary phase.

A peak shape problem with a low concentration compound eluting close to a high concentration compound is another problem encountered in splitless injections. Instead of partitioning only into the stationary phase and carrier gas, the low concentration compound also partitions into the band of high concentration compound in its vicinity. As the band of high concentration compound moves through the column, it may pull or push a portion of the low concentration compound along with it. This often causes a tail or shoulder on the side of the peak closest to the high concentration compound. The amount of peak distortion is dependent on the concentration of the high level compound. A shift in the low level compound's retention time may also be observed as the concentration of the high level compound changes. High and low concentration sample compounds with similar polarities exhibit the greatest peak shape problems; compounds with large polarity differences have slight or no peak shape problems.

### 7.4.9 Common Problems with Splitless Injectors

Splitless injectors are more complex than split injectors, thus more problems are usually encountered. If a problem with a previously working method occurs, the following areas should be checked for deviations or errors:

1. *Inlet purge on/off settings:* The absence or too long of a purge on time results in an extremely large solvent front. Very small peaks and a small solvent front is an indication of the purge being on at the time of injection.
2. *Initial oven temperature:* If the oven temperature is too high, broad or distorted peak shapes may occur especially for early eluting peaks. A solvent effect violation has probably occurred.
3. *Column, purge and septum purge flows:* Purge (split line) flows below 10 mL/min may cause an abnormally large solvent front. Extremely high septum purge flows may cause a loss of the more volatile or earlier eluting analyte peaks.
4. *Injector temperature:* Wide or tailing peaks may occur at low temperatures (below ~150 °C). Changes in absolute or relative peak sizes (i.e., change in injector discrimination) may be a sign of a large change (> 50 °C) in injector temperature.
5. *Liner:* Check the liner for contaminants, debris, breakage and proper installation. Liners become more active with use and may need replaced. Poor peak shapes or loss of peak size are the most common symptoms of a liner problem.
6. *Column installation:* The position of the column in the injector may affect peak shape and size.
7. *Sample solvent:* Poor peak shapes for select compounds may be caused by a polarity mismatch between the column, solvent and analytes.

8. *Leaks:* A leak in the injector may result in unstable retention times, changes in peak sizes and rapid column deterioration. Old septum (i.e., needs to be changed) and damaged or loose fittings are the usual leak sources.

9. *Contamination:* Residues from injected samples can contaminate nearly every part of the injector including liners, seals, fittings, injector body and gas lines.

## 7.5 Direct Injectors

### 7.5.1 Description of a Direct Injector

Depending on the GC manufacturer, different names are used for injectors specifically configured to work with 0.53 mm i.d. columns. The term direct injector will be used to refer to capillary injectors specifically intended for use with 0.45–0.53 mm i.d. capillary columns. While 0.45–0.53 mm i.d. columns can be used with split and splitless injectors, smaller diameter columns (≤ 0.32 mm i.d.) can rarely be used successfully used with a direct injector.

Direct injectors are quite easy to use and suffer from relatively few problems. A direct injector introduces nearly all of the vaporized sample into the column. Due to this injector design and higher capacity of larger diameter columns, samples over a wide concentration range are suitable for 0.53 mm i.d. columns. Trace to high concentration sample analyses are possible with direct injectors. Like all injectors, the lowest concentration level depends on the sensitivity of the detector being used; the highest concentration level is primarily limited by the capacity of the column and, sometimes, the linear or mass range of the detector.

Direct injectors consist of a heated injector, liner, incoming carrier gas line and, sometimes, a septum purge line (Figure 7-8). After entering the injector, the carrier gas exits by way of the column or the septum purge line (if present). The septum purge flow is very low and above the region where most of the sample vaporization occurs; thus very little sample is lost via the septum purge line.

After injection and vaporization, the sample vapors are mixed with the carrier gas. The carrier gas travels into the column, thus transporting the sample into the column. Since there are no other outlets for the carrier gas, nearly all of the vaporized sample enters the column. The sample is swept into the column at a moderately high rate.

Typical carrier gas (helium or hydrogen) flow rates with Megabore columns are 4–10 mL/min. This results in sweep rates of 10–100 seconds for most direct injectors. Smaller diameter columns have flow rates of 1–3 mL/min. These flow rates result in sweep rates of 30–400 seconds. Such long sweep times result in very broad peaks especially the solvent front. This is the reason columns with diameters of 0.32 mm or less are usually incompatible with direct injectors.

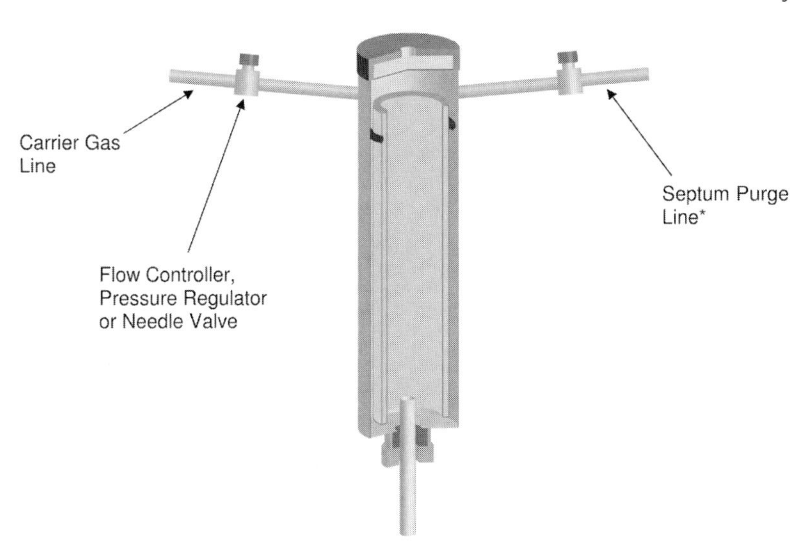


**Figure 7-8** Typical direct injector (Not to scale. Some details are omitted for clarity).
* May not be present on some injectors.

### 7.5.2
### Direct Injection Liners

The main difference between different direct injectors is the liner. The type of liner has a major impact on the quality of the chromatography obtained with a 0.53 mm i.d. column. The primary liner styles are the straight tube and the direct flash. The type of the liner may be limited by the design of the direct injector. Some require specific liners as designated by the manufacturer of the GC.

The most common and lowest cost liner is the straight tube. As the vaporized sample is carried towards the column, some of it can travel past the column. Also, any sample that backflashes can travel out of the top of a straight tube liner. These vapors now have to be swept by the carrier gas from these areas and back towards the column. This requires extra time and results in considerable tailing of the solvent front. The severity of the solvent front tail is more pronounced for large injection volumes, volatile sample solvents, low carrier gas flows and very hot injectors. Some straight tube liners are very small diameter sleeves that fit inside of a metal insert. The very low volume of this type of direct injector amplifies the amount of solvent front tailing.

The other type of direct injector liner is the direct flash vaporization. It is distinguished by the restriction at the top and another restriction several centimeters further down. The exact position of the bottom restriction varies depending on the source of the liner. The 0.53 mm i.d. column is pushed up into the liner until it can not move any further. The columns fit into the tapered region of the bottom restriction and forms a leak-free seal. The polyimide column coating is compressible and forms a tight seal between the column and the liner.

This prevents any vaporized sample from flowing around or past the column. The restriction acts to funnel all of the sample into the column. The outer diameters of smaller diameter (≤ 0.32 mm i.d.) columns are too small and these columns pass through the lower restriction. The column and liner do not form a tight seal, thus defeating the purpose of this type of liner. The upper restriction acts both as a needle guide, and more importantly, as a backflash barrier. If any sample flashes to the top of the liner, it is prevented from leaving by the upper restriction. Containing the vaporized sample within the liner chamber results in minimal backflash and a fast sweeping of the sample into the column. This results in better peak efficiencies, and more dramatically, reduces the width and amount of tailing of the solvent front. This improvement is most noticeable when compared to a straight tube liner. A direct flash vaporization liner is recommended for direct injectors, especially for injection volumes above 2 μL or low boiling sample solvents. There are two variations on the direct flash vaporization liner. One has the upper restriction is removed so that the liner can be easily cleaned or packed with silylated glass wool. The benefits of the backflash barrier created by the upper restriction is no longer present, but the ease of cleaning or use of glass wool may outweigh the loss.

The other type of direct injection liner is called a hot on-column liner. The restriction where the column forms the seal with the liner is at the top of the liner. A standard 26 gauge syringe needle fits inside a 0.53 mm i.d. column. The injection is made directly into the column, thus by-passing the liner. In some ways, this is not a vaporization technique. The injector is usually maintained at an elevated temperature, but at one lower than normally used for vaporization techniques. Injector temperatures of 100–200 °C are recommended with the lower range more commonly used with low boiling point solvents. Excessively high temperatures cause the sample solvent to significantly expand within the column. This results in severe peak and solvent front broadening or backflash problems. Since it is easy to flood the column with sample, the maximum sample volume for hot on-column liners is 0.5 μL. Injecting larger sample volumes results in distorted peak shapes and large solvent fronts. Even with small injection volumes, broad solvent front and peaks are common with hot on-column injection liners. The advantage of using a hot on-column liner is minimal discrimination and decreased temperature induced sample decomposition. Improved peak shapes are often obtained if a retention gap is used with a hot on-column injector liner. The added benefit of longer column lifetime is usually realized. The sample is directly deposited in the column, thus exposing it to a large amount of non-volatile materials. These can interfere with proper chromatography and reduce column life. The retention gap helps to minimize the problems associated with the buildup of non-volatile materials at the front of a capillary column.

### 7.5.3
### Septum Purge for Direct Injectors

Some direct injectors do not have a septum purge function. This is most common for packed column injectors converted to a direct injector. The septum purge line can act as a miniature split line if the sample reaches the top of the injector during the vaporization and expansion process. The more volatile sample components are more likely to flash towards the top of the injector. This will be a greater problem for the liners without some type of upper restriction. Losses of volatile sample compounds are minimized with a liner having an upper restriction. Septum purge flows of 0.5–5 mL/min are still recommended for direct injectors. Lower flow rates are best for the open top liners.

### 7.5.4
### Column Position in Direct Injectors

For the straight tube type of direct injection liner, the column should be located near the bottom of the liner. If the column opening is above the bottom portion of the liner, the carrier gas has to sweep the liner volume many times before all of the sample vapors are transferred into the column. The solvent peak (and perhaps the solute analyte peaks) become very wide and have an extreme tail if the column is positioned too high in the liner.

Improper installation of the column in a direct flash vaporization liner results in a broad and tailing solvent front. The shape of the peaks may also suffer. The correct sealing of the column in the liner restriction is critical. Any leakage of carrier gas around the seal will result in a broad and tailing solvent front. The worse the leak, the worse the solvent front. A column that has not been inserted far enough into the liner does not seal, but leaks. A column that is pushed into the restriction with excessive force may chip or break; it will not seal properly, but will leak instead. Leaks also occur if the column end is poorly cut. The column end needs to be even, without any burrs or chips, and at a 90° angle to the column wall. Also important is the cleanliness of the sealing surface of the liner. Any debris in the restriction prevents the proper sealing of the column. Small pieces of septa may fall into the restriction and become lodged when a column is removed. On occasion, a small portion of column or polyimide coating adheres to the walls of the liner restriction upon column removal. This prevents proper sealing. Repeated column sealing difficulties warrant the removal of the liner for inspection since its interior is not visible during installation.

### 7.5.5
### Other Aspects of Direct Injectors

The low efficiency of 0.53 mm i.d. columns and direct injectors make closely eluting peaks more difficult to resolve. One method of potentially improving injector efficiency is to use a retention gap (Section 11.8). The retention gap helps to focus the sample in the front of the column. Any focusing of the sample helps to decrease peak widths, and thus improves resolution. The retention gap also acts as a guard column. This is beneficial since direct injections place most of the sample into the column. Non-volatile residues and harmful compounds are kept away from the analytical column. Improved column life and performance and less column maintenance may result.

Using an initial column temperature at least 10 °C below the boiling point of the sample solvent may help to focus the injected sample. The solvent effect (Section 7.4.3) occurs and this often improves peak widths. The greatest improvement occurs for peaks that elute closest to the solvent front.

### 7.5.6
### Common Problems with Direct Injectors

Direct injectors are relatively simple, thus problems are usually easy to diagnosis and repair. If a problem with a previously working method occurs, the following areas should be checked for deviations or errors.

1. *Column and septum purge flows:* Extremely low carrier gas flows can result in a broad and tailing solvent front. Extremely high septum purge flows may cause a loss of the more volatile or earlier eluting analyte peaks.

2. *Injector temperature:* Wide or tailing peaks may occur at low temperatures (below ~150 °C). Changes in absolute or relative peak sizes (i.e., change in injector discrimination) may be a sign of a large change (> 50 °C) in injector temperature.

3. *Liner:* Check the liner for contaminants, debris, breakage and proper installation. Liners become more active with use and may need replaced. Poor peak shapes or loss of peak size are the most common symptoms of a liner problem.

4. *Column installation:* The position of the column in the injector may affect peak shape and size. A very large solvent front may be normal for a straight tube liner, but an indication of improper column installation with for a direct flash liner.

5. *Leaks:* A leak in the injector may result in unstable retention times, changes in peak sizes and rapid column deterioration. Old septum (i.e., needs to be changed) and damaged or loose fittings are the usual leak sources.

6. *Contamination:* Residues from injected samples can contaminate nearly every part of the injector including liners, seals, fittings, injector body and gas lines.

## 7.6 Cool On-Column Injectors

### 7.6.1 Description of an On-Column Injector

On-column injection eliminates the vaporization processes as the mechanism for sample introduction into the column. On-column injectors deposit a liquid sample directly into the column without going through a vaporization step. On-column injection eliminates many of the problems associated with vaporization injectors. Eliminating sample vaporization eliminates backflash problems. Since the sample is not vaporized and subject to expansion, it does not flow out of the injector. On-column injection does not subject the sample to high temperatures which makes it ideal for heat sensitive compounds. On-column injection places all of the sample in the column. Discrimination problems are avoided since the amount of each compound entering the column is dependent on its concentration and not its volatility. On-column injection makes possible the analysis of some high boiling compounds by capillary GC that are not possible using a vaporization injector. Many high boiling point compounds do not vaporize fast enough in vaporization injectors to permit their rapid introduction into the column. This is especially true for the fast injection modes such as split. Since the entire sample is placed directly in the column, the high boiling point compounds have the opportunity to chromatograph and separate. The characteristics of cool on-column injections make it ideal for several types of compounds or samples – very high boiling point compounds, heat sensitive compounds and samples that require minimal discrimination effects.

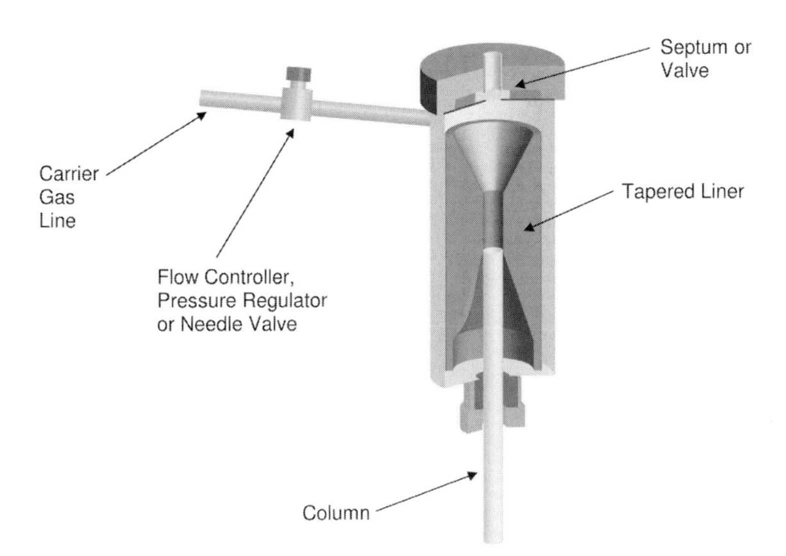


**Figure 7-9** Basic cool on-column injector (Not to scale. Some details are omitted for clarity).

A dedicated on-column injector is needed to make on-column injections (Figure 7-9). Depending on the injector, columns with inner diameters as small as 0.25 mm can be used. Most on-column injectors require a syringe specifically designed for the injector or column diameter. Usually there is some type of insert that helps to guide the syringe needle into the column and to keep the carrier gas from escaping around the column. The syringe needle is inserted through some type of valve or specialized septum which is used to keep the carrier gas from leaking from the needle entrance and air from entering the column. Since all of the sample is directly deposited into the column, split or septum purge lines are not needed, thus absent.

A short section at the front of the column is usually maintained at a lower temperature to obtain better peak shapes for on-column injections. An injector heater is used to keep the injector temperature below the oven temperature. Some on-column injectors use a fan to blow ambient air through the injector to cool the part of the column inside of the injector. By cooling only a small section of column outside of the GC oven, the entire oven does not need to be maintained at a low temperature which reduces analysis run times.

### 7.6.2
### Solvent Effect and Cold Trapping for Cool On-Column Injectors

A 1–2 μL injection of a liquid can flood a large region of the column. This is a greater problem with smaller diameter columns. The large flooded zone results in a broad sample band at the front of the column which usually causes broad peaks. The sample needs to be focused into a short band to prevent poor peak shapes. The solvent effect (Section 7.4.3) or cold trapping (Section 7.4.4) is used to focus the sample in cool on-column injections.

Utilizing the solvent effect requires the column to be at least 10 °C below the boiling point of the sample solvent. The solvent condenses at the front of the column in a short band since the column temperature is below the boiling point of the solvent. This film of solvent traps the rest of the sample and focuses it in a short band. Any sample component more volatile than the solvent enters the column prior or at the same time as the solvent. The peaks representing those compounds may be broad and distorted to varying degrees. If those peaks are not of interest, this may not a problem.

In some cases, cold trapping can be used to focus the sample. Cold trapping occurs when the column temperature is at least 150 °C below the boiling point of a compound. Any sample compounds that fulfill this condition focus in a short band at the front of the column. These compounds have good peak shapes without the aid of the solvent effect. To obtain good results with a cool on-column injector, either the solvent effect or cold trapping needs to occur. After the sample has entered and focused in the column, the column temperature is raised to facilitate separation. A temperature program is almost always used to start the column at a lower temperature and to raise the temperature higher enough to elute all of the compounds within a short time period.

### 7.6.3
### Secondary Cooling

Since the solvent effect or cold trapping is necessary to obtain good peak shapes, the initial value of the temperature program is usually relatively low. This temperature is often much lower than that needed to adequately separate the sample compounds. Most on-column injectors have some type of a secondary cooling function that eliminates the necessity of cooling the entire column down to a low temperature.

Secondary cooling is may be accomplished by setting the on-column injector temperature at least 10 °C below the sample solvent boiling point. The secondary cooling temperature should be controlled, thus relying on ambient temperature to consistently cool the column is not recommended. If the on-column injector is temperature programmable, the on-column injector is often heated starting a few minutes after injection. Heating the on-column injector helps to remove any low volatility sample components, thus reducing the amount of any sample carry-over and injector contamination. Some on-column injectors are not heated and use a small fan to force room temperature air around a small section at the front of the column. After injection of the sample, the secondary cooling is discontinued and the column heats up to the oven temperature.

The oven temperature should not be increased during sample injection or while secondary cooling is occurring. To potential minimize peak broadening, using a temperature program with an 0.5–1 minute initial hold time is recommended. While on-column injections can be made without secondary cooling by using a low enough initial oven temperature, the main advantage to secondary cooling is the reduction of time between analyses. Shorter analysis times are possible since the temperature program can be started at a higher initial temperature. Also, less time is needed to cool down the oven between runs. One disadvantage of secondary cooling is the probable need for a retention gap.

### 7.6.4
### Retention Gaps and Cool On-Column Injectors

Whenever secondary cooling is used, a retention gap is usually needed to maintain good peak shapes (Section 11.8). Any time a sample band moves from a region of high retention to a region of low retention, band broadening occurs. Band broadening creates broad peaks. This occurs for cool on-column injections with secondary cooling when a retention gap is not used. The sample band moves from a region of higher retention (e.g., the portion of the column maintained at a lower temperature) to one of lower retention (e.g., the portion of the column in the GC oven). The use of a retention gap corrects this problem.

A 1–5 meter piece of deactivated fused silica tubing is an adequate retention gap. Tubing of equal or slightly larger diameter than the column is often best for the retention gap. Since there is no stationary phase in the retention gap, there is no compound retention in this section. When a retention gap is installed, the portion of column in the on-column injector has no stationary phase. The sample now travels from a region of no retention (i.e., the retention gap) into one of much higher retention (i.e., column containing stationary phase). This occurs even though the portion of the column containing stationary phase is at a higher temperature than the front section of the retention gap. Failure to use a retention gap for cool on-column injections with secondary cooling usually results in very broad peaks. If secondary cooling is not used, a retention gap is not necessary, but still recommended. It will usually improve the peak shapes and help to protect the column from non-volatile sample components.

### 7.6.5 Other Aspects of Cool On-Column Injectors

The entire sample is deposited in the column with on-column injection. The biggest problem is with the non-volatile components in the sample. These non-volatile materials rapidly accumulate in the column and create chromatographic problems. In general, column contamination is a greater problem with on-column than vaporization injectors. The use of a retention gap, in addition to providing better peak shapes, helps to keep the non-volatile materials away from the column. This reduces the frequency of problems related to non-volatile residue deposition and may result in increased column lifetime.

Cool on-column injection volumes are limited to 0.5–2 µL. Substantial peak broadening or distortion may occur upon injections greater than 2 µL. The large volume of solvent spreads over a large length of column. Even though the film may be dispersed uniformly, the excessive solvent film length causes the broad peaks.

If a retention gap is not used, stationary phase-solvent polarity mismatch may be a problem. If the solvent is of significantly different polarity than the stationary phase, distorted peak shapes may occur. The problem is compounded if the solutes are similar in polarity to the solvent. The use of a retention gap may improve peak shapes when a polarity mismatch occurs.

### 7.6.6 Common Problems With On-Column Injectors

On-column injectors are often more susceptible to problems than most vaporization injectors. If a problem with a previously working method occurs, the following areas should be checked for deviations or errors.

1. *Contamination:* The most common problem with on-column injections is the rapid fouling of the column. Problems with peak shape and size, separation and baseline disturbances are the most common symptoms of a contaminated column. Peak tailing is usually more severe for active compounds. A retention gap helps to minimize the onset of column contamination problems. The retention gap will have to be trimmed or changed as the non-volatile materials accumulate.

2. *Secondary cooling function or temperature:* The absence of second-ary cooling or using too high of an injector temperature may cause wide or tailing peaks especially for earlier eluting peaks.
3. *Oven temperature:* Using too high of an initial oven temperature reduces the effectiveness of the solvent effect or cold trapping. Even though secondary cooling is used, lowering the initial oven temperature may improve peak shapes. The lack of an initial hold temperature may also cause peak shape problems.
4. *Leaks:* Leaks in the valve or septum may results in erratic peak sizes. Introduction of air into the column may also cause rapid column damage usually evident as high column bleed or peak tailing of active compounds.
5. *Column installation:* An incorrectly installed column may allow sample to escape around the column. Erratic peak sizes and retention times, sample carry-over or ghost peaks, and unstable baselines are common symptoms of a poorly installed column. Also, difficulties with inserting the syringe needle into the column may occur.
6. *Retention gap:* The use of a retention gap is usually needed to obtain good peak shapes. A poorly installed retention gap union may cause peak tailing with the earlier eluting peaks exhibiting the most severe tailing.
7. *Sample solvent:* Using a mixed sample solvent may cause broad or split peaks. This usually occurs when the sample compounds are significantly more soluble in one of the solvents. The problem becomes worse as the difference in the boiling points of the two solvents increases. Little can be done to correct this problem besides eliminating or removing one of the solvents.

## 7.7 Pressure and Flow Programmable Injectors

### 7.7.1 Description of Programmable Injectors

The first commercially available GC's with electronic control of gas flows were introduced in the mid 1990's. Up to this time GC systems used mechanical pressure regulators or flow controllers to manipulate carrier gas flow rates or pressures. With these mechanical systems, the desired column head pressure, split and septum purge flow rates are obtained by manually adjusting the appropriate control knobs. Flow rates are measured using a volumetric flow meter and the column head pressure is directly read from an analog gauge. The carrier gas average linear velocity is calculated from the retention time of a non-retained compound. Electronic control systems enable the direct setting of gas pressures, flows and velocity rates without using manual measurements such as flow meters and calculations. This is done via the GC keyboard or software. The desired pressure, flow rate or linear velocity is input and the GC system adjusts the appropriate regulator or controller to obtained the set value. Electronic pressure control not only makes it easy to set the gas flows and pressures, the various gas parameters can be stored in a method for easy retrieval and duplication. Most system also have programming capabilities.

The carrier gas average linear velocity or flow rate is dependent on the column length and diameter, carrier gas identity and column temperature. These parameters are required by the electronic pressure control system to set the carrier gas velocity or flow rate. The desired carrier gas pressure, velocity or flow rate is directly input and manual measurements are not needed. An algorithm is used to calculate the column head pressure required to obtain the set carrier gas flow rate or linear velocity. Conversely, the average linear velocity or flow rate is calculated from the head pressure. The flow rate or velocity is not directly measured, but calculated using the inputted column, carrier gas and temperature characteristics. Split, septum purge and detector gas flow rates are also directly entered and set without the use of a flow meter.

### 7.7.2 Constant Pressure Mode

Mechanical systems are limited to one mode of operation which is the constant pressure mode. The column head pressure is maintained at the same pressure during the entire GC run. Hydrogen, helium and nitrogen are compressible, and their viscosity and diffusion coefficients increase with temperature. These properties have implications when operating in the constant pressure mode.

Carrier gas viscosity increases as the column temperature increases. This causes the average linear velocity to decrease as the column temperature increases during a temperature program run. Depending on the carrier gas, column dimensions and amount of temperature change, the average linear velocity can differ by 3–4 times between the start and end of the temperature program.

Most GC analyses are run using the constant pressure mode due to the simplicity and ease of transferring methods to any comparable GC system. Operating in the constant pressure often requires efficiency compromises since the average linear velocity shifts during the temperature program. Regardless of the average linear velocity used, it will not be optimal for some of the sample compounds.

### 7.7.3 Constant Flow or Velocity Mode

Pressure programmable injectors have a built in program to automatically increase the carrier gas pressure to maintain a constant flow rate or velocity throughout a temperature program run. There is a decrease in the average linear velocity when operating in the constant pressure mode. The most obvious impact of operating in the constant flow mode is the decrease in retention times especially for the later eluting compounds. Compared to constant pressure, run time reductions of 10% are typical; however, the actual amount depends on the column and the GC parameters.

Depending on the column, carrier gas and its average linear velocity, a resolution change may occur in the constant flow mode compared to the constant pressure mode. It is somewhat difficult to predict the amount of resolution change and some trial and error is usually involved. If resolution loss occurs it is usually limited to 2–3%, thus the operating in the constant flow mode is often beneficial from a time perspective.

Another advantage of operating in the constant flow mode is related to the behavior of some detectors. Some detectors are concentration dependent and their response is proportional to the volume of carrier gas flowing through them. As the carrier gas flow changes during a temperature program run, the chromatogram baseline drifts proportionally to the change in the carrier gas flow. The drift may be upward or downward depending on the detector and the amount of column bleed. If the carrier gas flow is kept constant, flow induced baseline drift will not occur. Carrier gas related baseline drift problems are the most noticeable with NPD, TCD and GC/MS systems. Mass spectrometer (MS) reproducibility is better at a constant ion source pressure which is primarily dictated by the carrier gas flow rate. Also, the ion source pressure affects the MS tune, thus constant carrier gas volume into the source maintains the same tune performance for all compounds in the sample.

### 7.7.4 Pressure Program Mode

Pressure programmable injectors can be programmed to change the carrier gas pressure, velocity or flow to set values at specific times. The format of carrier gas pressure programs are very similar to column temperature programs. There is an initial pressure, hold time, ramp rate, final pressure and final hold time. Programs with multiple ramps and holds are also possible.

One of the uses of a pressure program is to adjust the carrier gas average linear velocity to its maximum efficiency value during the course of a GC run. Optimal average linear velocity ($\bar{u}_{opt}$) varies with compound retention ($k$). Also, the carrier gas velocity decreases during a temperature program run in the constant pressure mode. Regardless of the carrier gas velocity selected, it is not optimal for all of the sample compounds and some efficiency sacrifices usually occur. Temperature programs with a wide range between the initial and final temperatures requires the greatest carrier gas velocity compromises. Pressure programming makes it possible to change the carrier gas velocity during the GC run to obtain optimal efficiency for all compounds.

Finding the optimum pressure values often involves some trial and error, thus requiring a time investment. Pressure programming may increase, decrease or not alter resolution. It is dependent on the temperature program and column dimensions. Accurate predictions are difficult and uncertain, thus some experimentation is necessary.

In general, $\bar{u}_{opt}$ increases as compound $k$ increases, thus pressure programs ramp from lower to higher pressures. While the optimal velocity can be experimentally determined for each compound, it is a substantial amount of work for a potentially small gain. Trying several different final pressures and then several ramp rates is a quick and fairly efficient approach to finding a suitable pressure program. While the best pressure program may not be found, but one very close to the optimal is often quickly discovered. The constant flow mode is one very specific type of pressure program.

In addition to maintaining optimal efficiency during the GC run, pressure programming can be used to further reduce the analysis time in some cases. For analysis where some compounds are highly retained, but well resolved, increasing the carrier gas velocity during the later portion of the temperature program will decrease their retention. Even though some measurable resolution loss occurs, it is inconsequential since there originally was excessive resolution of the peaks. Reducing the analysis time with pressure programming may also prolong column life by decreasing the time a column is exposed to elevated temperatures.

The later part of some temperature programs are intended to elute highly retained sample matrix components (i.e., no analytes) from the column. Increasing the carrier gas velocity immediately after the last eluting analyte will decrease the time needed to elute all of the highly retained matrix components, thus shortening the total run time.

### 7.7.5 Pulsed Pressure Mode

Pressure pulsing is used as a means to improve injector efficiency. In general, high carrier gas flow rates in the injector at the time of injection results in smaller peak widths. Pressure pulsing involves using a high flow rate (pressure) during the time of injection followed by a quick reduction to a lower flow rate. This allows for better injector efficiency during injection and better column efficiency while the sample is migrating through the column. Both injector and column efficiency can be optimized without sacrificing one for the other.

Splitless injections tend to benefit the most from pressure pulsing; however, split injections can benefit from pressure programming also. Very low split ratios can not normally be used due to the excessively low carrier gas flow rates in the injector. If the carrier gas flow is increased for only a few seconds at the start of the run, very low split ratios can be used since the total flow in the injector is high enough (Section 7.3.2). The momentary high flow rate in the injector increases the injection efficiency.

Pressure pulsing decreases the residence time and expansion volume of the sample in the injector. Any compound degradation induced by the high temperature of the injector is reduced since the injected sample is rapidly transferred into the column which is usually much cooler. Loss of volatile compounds due to backflash is decreased since the sample expansion volume is minimize due to the higher pressure and the lower residence time of the sample in the injector.

For large volume injections, better peak size reproducibility sometimes occurs with pressure pulsing.

Peak broadening is a possible outcome of pressure pulsing especially for early eluting compounds. Using a retention gap (Section 11.8) helps to minimize peak broadening when using a pressure pulse. Limiting the initial (pulsed) pressure to 2–3 times the final pressure used for the column minimizes the amount of peak broadening. Pulse times above 1 minutes may also cause significant peak broadening. Pulse times around 0.25 minutes are often work the best and pulse times above 0.5 minutes often result in excessive peak broadening.

### 7.7.6
### Gas Saver Mode

Many GCs with programmable injectors have a gas flow reducing or gas saver feature. The split flow rate is reduced to a lower value after a period of non-use or after the start of the GC run. Since the split flow is most critical at the time of injection, reducing the spilt flow several minutes after sample injection has no impact on the resulting chromatogram. A reduced split flow of 20–25 mL/min at 2–3 minutes after start is a common setting. Using zero or a very low split flow may cause the programmable injector to malfunction since it may not be able to set such a low split flow and still maintain the desired head pressure. More leakage may occur at lower pressures for some injectors, thus avoiding extremely low head pressures also helps to increase column life by minimizing the amount of air entering the column at high temperatures.

### 7.7.7
### Other Aspects of Programmable Injectors

Electronic pressure control system require periodic calibration. Usually there is a calibration procedure outlined in the GC manual. Part of regular GC maintenance should include electronic pressure control system calibration. Out of calibration system results in inaccurate pressure, flow or velocity values. Calibrating electronic pressure control system may also be called “zeroing”.

If a flow meter is used to measure the flow from the split line, a slightly different value may be obtained than the set gas flow value. This is not unusual and differences up to 10% can be obtained. Flow meter accuracies, differences in flow measurement points (e.g., flow in the injector versus flow at the split line exit point), and numerous other variables are responsible for the apparent discrepancies.

The column diameter and length is input into the electronic pressure system so the proper carrier gas flow and velocity can be calculated. Any differences between the actual and set values often results in inaccuracies in the reported flow, velocity or pressure values. Calculating the average linear velocity using a non-retained compound often produces a slightly different value than the one reported by the GC. This is normally caused by the difference between the actual and input column diameter and length. Columns with smaller diameters and shorter lengths are the most sensitive to these differences. Diameter differences of only 0.02 mm can result in a surprising large discrepancies between the calculated and reported average linear velocities.

The valves used in electronic pressure controlled systems are more prone to plugging and sticking than their mechanical counterparts. Gases free of particulate matter is more important with the electronic valves. The electronic valves are more susceptible to contamination by high molecular weight sample contaminants, thus minimizing backflash and better sample preparation may be advisable.

Electronic pressure control systems are more sensitive to leaks than mechanical systems. If a pressure set point can not be reached due to a large leak or any other gas related fault, most systems will sound an alarm and/or shut off. If there is insufficient gas supplied to the GC (e.g., too low of a delivery pressure or an empty gas cylinder), the system assumes a leak is present and acts accordingly. Failure to turn off injector or detector gas flows when changing a column may result in the triggering of GC alarms or shut down. A small leak in the injector (e.g., septum, fittings) may become evident as retention time or pressure fluctuations.

The inability of the GC to reach or hold a pressure may be evidence of a leak. However, very low or high pressures may not be obtained due to the flow limits of the system. For example, very low split flows may not be possible if the head pressure is very high, and the high split flows may be difficult to obtain at low head pressures.

## 7.8 Injection Techniques

### 7.8.1 Syringe Filling Techniques

It may seem trivial, but the method used to fill the syringe with sample can have a surprising impact on the results. Erratic peak sizes are the most common problem with poor syringe technique. Peak shape problems can also occur, with spilt or fronting peaks being the most common.

The best results are usually obtained when the syringe needle does not contain any sample especially for slower injection speeds. When the needle is inserted into the hot injector, the needle is heated very rapidly. Volatile compounds in the needle, including the solvent, may begin to volatilize and leave the needle before the syringe plunger is depressed. This introduces a small amount of sample into the column before the rest. This usually creates a small peak immediately in front of each peak; in some instances, a tail on the front of the peak results. The separation between the two peaks or the severity of the front side tail increases as peak retention becomes longer.

The markings on the syringe barrel rarely account for the volume of the needle. When measuring sample volume, this factor needs to be considered. Most needles on 10 µL syringes have a volume of 0.8–1 µL. If 1 µL of liquid is measured in the syringe barrel, the actual volume is about 2 µL. A 2 µL volume as measured in the syringe barrel is actually about 3 µL which is not double the measured 1 µL volume.

Accurate and precise injections requires more than just pulling a volume of sample into the syringe. There are two techniques that aid in making accurate volume injections. The air plug method involves pulling 1–2 µL of air into the syringe before the sample is pulled into the syringe. Additional air is pulled into the syringe following the sample to ensure that no sample is in the needle (Figure 7-10a). The first plug of air helps to push all of the sample out the syringe needle and into the injector. Care needs to be taken to make sure that any residue of previous samples or solvents are not in the syringe. Removing traces of solvent from the syringe can be difficult and time consuming. Another syringe technique that eliminates this problem is the solvent plug method. This method first involves pulling about 1 µL of clean solvent into the syringe. Ideally, the solvent should be the same as the sample solvent. This clean solvent is followed by a 0.5–1 µL plug of air then the sample. Additional air is pulled into the syringe following the sample to ensure that no sample is in the needle (Figure 7-10b). The solvent and air helps to push all of the sample out the syringe needle and into the injector. The solvent also helps to clean the syringe needle. The extra solvent does increase the size of the solvent front, but this is usually not a problem.

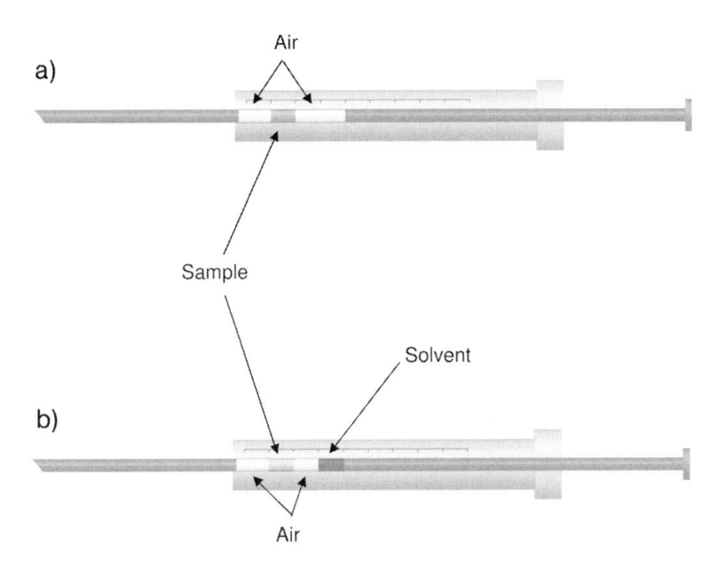


**Figure 7-10** Syringe filling techniques: (a) air plug; (b) solvent and air plug.

### 7.8.2
### Injection Speed

The different injection modes require different syringe techniques when making manual injections. The difference is primarily the speed of injection. Injection speed can be thought of as the speed of plunger depression and insertion/ withdrawal of the needle from the injector. In general, relative injection speeds follow the order of split > on-column > direct > splitless.

The high carrier gas flows in the split injector rapidly sweeps the vaporized sample into the column. For this reason, the entire sample needs to be introduced into the injector very quickly. A fast injection is needed to obtain the best efficiency. The entire process of needle insertion, plunger depression and needle withdrawal should take 1–3 seconds. Slow plunger depression may cause slightly broadened peaks or, in some severe instances, split peaks. The plunger depression should also be uniform in speed and the same from injection to injection. As soon as the plunger is fully depressed, the syringe should be immediately withdrawn from the injector. Slightly broader or tailing peaks can result if the syringe remains in the injector too long.

The slower carrier gas flow rates in direct injectors allow a slightly slower injection speeds than with split injectors. Even though slower rates are satisfactory, it is recommended to make injections in the 1–3 second range. Slower injection speeds, unless very slow (e.g., > 5 seconds), does not have a noticeable effect on the peak shapes.

Unlike split injections, the sample is swept into the column very slowly for splitless injections. Rapid sample introduction into the injector is not nearly as critical since it can take several seconds to transfer most of the sample into the column. It has been reported that syringe plunger depression at about 1 second per μL of sample is best for splitless injections. Faster injection rates do not seem to have significant detrimental effects.

The injection rate for on-column injections needs to be fast. With slow injections, a portion of the sample may be lost because it adheres to the inner walls or tip of the syringe needle. Fast injections help to spray the sample into the column away from the needle, thus avoiding partial sample introduction. Using a solvent plug helps to displace the entire sample out of the needle; however, the added solvent may slightly increase peak broadening.

A change in injection speed may be warranted when analyzing samples with high boiling point compounds. With any injection without a solvent plug, there is a small portion of sample that remains behind in the syringe needle after plunger depression. With vaporization injectors, the syringe needle rapidly heats and volatilizes some of the more volatile compounds remaining in the needle. These can now enter the injector, and thus also the column. Some of the higher boiling compounds may still remain behind on the walls of the needle. This portion of the sample is not introduced into the injector and contributes to the loss of high boiling point compounds. This type of compound loss is called syringe discrimination. It is more severe with fast injection speeds such as those used for split injections. The use of the hot needle technique may help to decrease syringe discrimination. This technique requires that there is no sample, only air, in the syringe needle. The needle is inserted into the injector and plunger depression occurs 1–2 seconds afterwards. This allows thorough heating of the needle by the hot injector. This gives the higher boiling point (i.e., slower vaporizing) compounds a better opportunity to vaporize and be introduced into the injector. It is important to have no sample in the needle or a small portion is prematurely introduced into the column. This normally results in split peaks with the first one of the pair being very small. More frequent septum changes may be required when using the hot needle injection technique. A better seal around the needle is needed when it is left in the septum for longer times.

## 7.9 Autosamplers

Autosamplers are primarily used to allow unattended sample injections. Another benefit is the greater reproducibility of autosampler injections over manual injections. Injection speed and sample amounts are controlled very precisely, resulting in peak size reproducibility superior to manual injections. With these benefits comes the need to consider other factors usually not present with manual injections.

Depending on the model of autosampler, the numerous aspects of sample injection can be controlled. Most autosampler default settings are for fast injections, but the injection speed is one of the controllable variables. The sample, injector and liner dictate which injection speed is the best. Experimenting with several of the possible speed settings may be worth the effort. Improved peak shape or size, better reproducibility or less discrimination may be obtained at a particular injection speed.

There are usually only a few autosampler options for syringe loading methods. The ones available depend on the model of the autosampler. Whether a solvent or air plug can be used, whether there is sample in the needle and available sample volumes are controlled by the flexibility of the particular autosampler.

If syringe solvent wash steps are available with an autosampler, they should be used. This prevents residual amounts of the previous sample from contaminating the next sample. This is a particular problem when a low concentration sample follows a high level one. Even small traces of a high concentration sample are significant when compared with a lower level sample. One of the wash solvents should be the sample solvent. This solvent is usually the most effective in removing previous sample materials. If more than one solvent is used for syringe washing, make sure that each successive solvent is miscible. For example, do not use methanol and hexane in succession; they are immiscible and this hinders the cleaning of the syringe. The solvent wash containers should be frequently cleaned. Traces of the numerous previous samples can accumulate in these vials upon evaporation of the wash solvent. If the vials are only refilled with clean solvent, the trace amounts of contaminants slowly build up to higher levels. After prolonged use of the vial, these compounds can build up to levels where they can contaminate the syringe instead of cleaning it.

When using volatile sample solvents or analyzing volatile compounds extra care needs to be taken to completely seal the autosampler vials. As the samples sit in the autosampler tray, the more volatile sample components can evaporate from a poorly sealed vial. Repetitive injections from a single autosampler vial can also lead to sample evaporation since the sample loss occurs from the holes created in the septum. If the sample solvent evaporates, the sample compounds becomes more concentrated. The peaks are larger than they should be, thus higher concentrations than the actual values are reported. If some of the more volatile sample compounds escape from the vial, a reduction in the peak size for these compounds occur. The more volatile the compound, the greater the loss from the vial. Also, there is a time factor that has to be considered. The longer the vial remains in the autosampler tray before injection, the greater the loss of solvent or compound. The error for the first sample injection is negligible while the error for the sample injected after 10 hours in the autosampler tray is more severe. The room temperature also has an effect on the amount of sample evaporation. Greater loss occurs when the room temperature is higher. Variations in room temperature throughout a day can alter the influence of room temperature on sample loss from the vial. The combination of sample solvent loss, volatile compound loss, time in the vial, and room temperature variations can make peak size irreproducibility problems extremely variable. By using good quality vials and caps and properly sealing the vials, these types of problems can be avoided. Slow sample concentration increases or volatile compound losses also occur with repetitive injections from the same autosampler vial.

Larger diameters (i.e., 22 gauge) syringe needles are often used with autosamplers. This helps to prevent bending of the needles by the autosampler which is more common with the smaller diameter needles (i.e., 26 gauge). Also, the larger needles are less prone to become plugged with particulate matter such as pieces of septa. Since autosamplers are often left unattended, a plugged needle is not detected until the chromatograms are examined. This may result in many wasted hours and the need to re-inject the samples. The larger needle pierce large holes in the septum. This usually necessitates more frequent septum changes. Even though the sample tray holds 100 vials, only 50 can be run during one session due to the need to change the septum after 50 injections. The use of a septumless injector may solve this problem providing this style of injector is compatible with the samples.

Some autosamplers use a tapered needle which is thinner at the tip than at the base. This needle style has the strength benefits of a larger needle while making smaller holes in the septum. Pre-cored septa also available which enables the use of syringes with rounded tips instead of beveled or pointed. Over 200 injections can be consistently made with the combination of a well maintained round tip needle syringe and properly installed pre-cored septa.

## 7.10
## Injector Septa

### 7.10.1
### Introduction

The septum's primary function is to seal the injector, thus preventing vaporized sample and carrier gas from leaving and air from entering the injector at the point of syringe introduction. Since a syringe needle repeatedly punctures the septum, it has to be periodically changed to prevent leaks. A septum leak may result in retention times and peak size variations which makes sample analysis more difficult and inaccurate. More serious is the leakage of air into the injector which exposes the column to high concentrations of oxygen. Oxygen accelerates the degradation of the stationary phase resulting in high column bleed and activity, and finally, short column life.

Lifetime, ease of piercing, sliver formation, bleed and cost can all be important considerations when selecting septa. Most of these characteristics are determined by the material used to made the septa and the method of septa usage. While there seems to be a pattern concerning septa color, there is no color coding used to indicate septa characteristics.

### 7.10.2
### Septa Hardness

There is a range of septa hardness. The softer septa are easier to pierce with the syringe needle. This makes injecting samples slightly easier and reduces the frequency of bent syringe needles. Softer septa makes it easier to obtain a leak free seal, but it is more difficult to determine when the septum nut is on too tight. Over tightening a septum nut reduces septum life and increases the severity of any leaks.

Softer septa have the tendency to produce tiny slivers of septa upon injection. These tiny pieces of septa fall into the injector. Depending on the design of the liner and injector, these particles can interfere with the flow of carrier gas out of a split line or through a liner. Tailing or broad peaks, irregularities with the split vent flow, or baseline disturbances may occur when septa particulates accumulate in the injector. The worst problem occurs if one of the particles falls into the column. The probability of this problem occurring increases dramatically with larger diameter columns. Severe tailing of nearly all peaks occur if solid particles become lodged in the column.

### 7.10.3
### Septa Bleed

Septa contain a small amount of volatile materials originating from the manufacturing process. These compounds are released from the septum when it is heated to high temperatures in the injector. Many of them accumulate in the column during the times when it is held at cooler temperatures. A column maintained at a lower temperature will accumulate more septum bleed materials. When the column is heated such as during a temperature program, the septum bleed materials chromatograph much like an injected sample. The resulting chromatogram can take on a variety of appearances. Usually there will be some type of wide hump or blob often associated with sharper peaks. The size and appearance of the bleed depends on the brand of septum, the injector temperature, the column, the oven temperature, time left at the lower temperature, and the rate and amount of temperature elevation. The amount of septum bleed usually decreases slowly with time – the septum becomes conditioned upon heating. Most of the bleed reduction occurs 15–60 minutes after installation.

The type of injector influences the severity of a septum bleed problem. Injectors that use a high flow rate of carrier gas (e.g., split and splitless) reduces the amount of septum bleed materials reaching the column. Most of the bleed materials are swept out of the injector via the split line during all or most of the analysis. Injectors such as direct without any split flow direct all of the bleed products into the column. The presence of a septum purge flow helps to reduce septum bleed problems. The proper use of the septum purge function helps to keep the bleed materials from reaching the column. Higher flow rates sweep away more bleed materials, but a loss of the more volatile sample compounds via the septum purge line may occur.

GC systems operated at more sensitive levels respond to the septum bleed materials to a greater extent. Any baseline disturbances appear to be quite severe and can overwhelm small peaks. Some selective detectors do not respond to septum bleed compounds, thus no septum bleed may be seen even though the bleed compounds are present. GC/MS poses an additional problem with septum bleed. The mass spectrum of many of the septum bleed compounds appear to be very similar, if not identical, to many non-polar stationary phase bleed mass spectra. It is easy to misidentify the septum bleed compounds as stationary phase bleed compounds. The presence of peaks, humps or blobs rules out the possibility of column bleed being responsible for the baseline disturbances.

In general, higher quality (i.e., higher cost) septa have substantially lower bleed than low cost ones. At injector temperatures of about 200 °C or less, the differences in bleed between the various septa are negligible. As the injector temperature becomes higher, especially above 300 °C, the differences in bleed become quite apparent. The higher cost for low bleed septa are well worth the price when operating at high injector temperatures. Septum bleed materials are only a few of the many substances that can accumulate in a column at low temperatures. The elution of the other materials have the same appearance as septum bleed, thus it is difficult to determine whether the baseline disturbances arise from septum bleed problems. There is a method to determine whether the septum is the source of the contaminants. Before proceeding, make sure the injector and liner are clean. Wrap one layer of aluminum foil around a new septum. Install the septum by placing the side with the continuous piece of foil towards the injector. Be careful as to not expose any of the septum by tearing the foil. Setup the injector and column with the previous conditions. Leave the septum in place for at least 3–4 hours, but do not make any injections. Run a temperature program as previously used and collect the blank chromatogram. Compare with the previous baseline problems. If the severity of the baseline disturbances are noticeably better, the septum is probably the source of the contaminants. If the problem is about the same, the septum is not at fault and the source of the contaminants is elsewhere.

There are several ways to keep septum bleed interferences to a minimum. Lower injector temperatures reduce the amount of septum bleed; however, the injector still needs to be hot enough to thoroughly and rapidly vaporize the sample. Maintaining the column at higher temperatures reduces the amount of materials that accumulate in the column. This decreases the severity of the baseline disturbances, but at the expense of reduced column life. A finned septum nut keeps the septum at a lower temperature. This reduces the bleed levels, but this type of septum nut is not available for many models of GC's. Pre-treating the septum with solvent or heat helps to remove some of the contaminants, but this can reduce the life of the septum.

### 7.10.4
### Handling Septa

There are several method to clean and condition septa prior to use. Septa can be soaked in solvents to extract some of the contaminants from the surface. This removes contaminants and sometimes reduce septum bleed. Submerging the septa for 15–30 minutes in a beaker of solvent is usually sufficient. Halogenated solvents such as chloroform or methylene chloride should be avoided since many septum swell in these types of solvents. Thoroughly dry the septa before use by heating for 10–15 minutes at 50–100 °C.

Septa can be preconditioned by baking for several hours at or slightly above the injector temperature. Do not exceed the upper temperature limit for the septa. Placing the septa in a clean beaker and heating them in the GC oven is the easiest method. Sometimes the septa are left in a beaker in the GC oven while it is being used for analyses. A septum is removed as needed for replacement. If septa are heated or solvent extracted before use, reduced septum life is usually experienced.

Septa do not require special handling during installation or storage. The use of clean gloves or tweezers is often recommended, but not required providing clean hands are used. Residues from materials such as lotions, food and cosmetics can contaminate a septum so care must be taken to ensure there are none present. Provided clean hands are used, any materials transferred to the septum upon handling quickly vaporize and leave the injector before the next analysis is run. In addition, only a small area of the septum is exposed to the interior of the injector, thus very little of the contaminants will be transferred. Tweezers are recommended when performing trace level analyses at the detection limits of the GC system. Even tiny traces of residues or contaminants are visible as interferences in the chromatograms or mass spectra.

## 7.11
## Injector Maintenance

### 7.11.1
### Cleaning Injectors

Injectors require periodic maintenance and cleaning. They become fouled with non-volatile materials and backflashed materials condensed in cool areas. Seals and gaskets need periodic replacement.

Regardless of the types of samples injected, injectors eventually need cleaning. Samples containing large amounts of non-volatile materials foul injectors much quicker, thus more frequent cleaning is needed. Frequent occurrences of backflash require frequent injector cleaning, also. Baseline disturbances, ghost peaks, tailing peaks or reduced size peaks are the most common symptoms of a dirty injector.

Cleaning an injector involves more than just replacing the liner. Cool the injector to room temperature. Remove the liner, dismantle the various fittings, and remove any of the seals, washers or gaskets. Place a beaker inside the GC oven directly below the injector opening. Rinse the interior of the injector with 20–30 mL of several different solvents. Methanol, acetone and hexane remove nearly all contaminants from the injector; however, other organic solvents will not harm the injector. If biological samples such as blood, urine or plant extracts have been injected, water should be used as the first solvent. After each solvent has passed through the injector, scrub the injector with a small test tube brush. Rinse the injector with an additional 10 mL of the solvent after each scrubbing. Rinse any of the reusable injector parts with the same solvents. Ultrasonication of the parts may help remove contaminants from small crevices and ridges. Thoroughly dry all of the parts before reassembling the injector. New seals or gaskets should be used each time the injector is reassembled. Heat the injector and establish gas flows before installing a column.

In situations where backflash has occurred, the gas lines may need cleaning. Take apart the injector and remove all of the fittings, seals, washers or gaskets. Place a beaker inside the GC oven directly below the injector opening. Label each gas line and disconnect it at the fitting closest to the injector. Some of the lines may be up to 0.5 meters long. Use a large syringe (an adaptor to connect the syringe hub to the line fitting will be necessary) and flush each gas line with 5–10 mL of several different solvents. Methanol, acetone and hexane removes nearly all contaminants from the lines; however, other organic solvents can be used. Do not get any solvents in the flow controllers, pressure regulators or solenoids since this usually damages them. After the last solvent, flush each line with carrier gas for 5–10 minutes. Heat the lines with a hair dryer or heat gun to remove the last traces of solvent. Reattach the lines and reassemble the injector. Heat the injector and establish gas flows before installing a column. In cases of severe contamination, several cleaning may be necessary.

Sometimes it is easier to remove the entire injector from the GC. Consult the GC instruction manual on the proper method to remove the injector. Flushing the lines may be easier with the injector out of the GC. If the injector is unattached, a HPLC or similar type of pump can be used to flush solvents through the lines. This is easier than using a syringe and can be left unattended. To further facilitate cleaning, the injector can be baked out at a high temperature by placing it in an oven. If possible, have a flow of gas through the lines while heating. This improves and speeds up the cleaning process.

It can be difficult to remove the entire injector in some GCs due to the location of the gas lines, flow controllers and pressure regulators. To facilitate injector removal, carefully cut the gas lines about 5–10 cm from the injector body. Leave enough tubing to install a stainless union and still have enough slack to move the tubing around. The injector can be easily removed by disconnecting the unions. Forcing the gas lines through the tangle of other tubing, fittings and flow controllers is avoided. Color coding the unions or using an unique union on each line prevents connection of the wrong gas lines.

### 7.11.2 Injector Traps

Most GCs have a trap located on the split line to catch any of the compounds leaving the split vent. Sometimes a trap is located in the carrier gas line leading to the injector. These traps are often inside of the GC and not readily visible. Split line traps can become clogged or plugged with use. This creates back pressure in the split line. This may change split ratios and alter column carrier gas flows in some systems. Inspection and replacement of these traps is important. It is difficult to accurately predict when a split line trap will become plugged, thus regular replacement is recommended. Many of these traps are about the size of a pencil while others are contained in a larger diameter cylinder.

Small traps are available for the exit port of the split vent. These are designed to trap any of the compounds that leave the split vent. Sample solvent comprises most of the material leaving the split vent. Most sample components condense somewhere along the split line which is usually at room temperature. These split line traps primarily help to keep volatile compounds such as solvents from contaminating the lab air. Depending on their structures, some of these compounds can be detected by sensitive GC systems as part of the background. For example, methylene chloride contamination of the lab air is a common problem for some laboratories performing low level halocarbon analyses. Installing an external split vent trap may interfere with electronic pressure controlled injectors especially if the traps becomes plugged even partially.

### 7.11.3 Cleaning Injector Liners

Depending on the types of samples injected, injector liners need periodic cleaning or replacement. The analysis of dirty samples or active compounds (e.g., -NH and -OH containing compounds), especially at low levels, requires more frequent liner maintenance. It is primarily an economic factor that determines whether liners are cleaned and reused or replaced with a new one. Eventually, all liners have to be replaced or resilylated.

A liner without visible deposits or discoloration is cleaned by rinsing with solvents. A squirt bottle or pipet can be used to rinse the inside of the liner with 10–20 mL of methanol, acetone and hexane. If there are deposits or discolorations, the liner can be ultrasonicated to remove the materials. If desired, a soft bristled brush can be used to scrub the inner surface. It is important not to scratch the surface of the liner. This will expose untreated glass which will adsorb active compounds. If the liner is scratched, it has to be discarded or resilylated. Liners with complex inner surfaces are much harder to clean than simple straight tube types of liners.

A liner becomes more active with use. The surface is no longer deactivated and the liner is not suitable for use with most samples. Samples containing active compounds are much more susceptible to liner activity problems. Low concentration samples magnifies the problem. Inactive compounds such as hydrocarbons are fairly immune to liner activity problems. Liners exhibiting unacceptable activity for the analysis of active compounds may have adequate performance for inactive compounds.

When a liner becomes active, it needs to be replaced or resilylated. Resilylation requires an investment in time and reagents, but liners can be recycled. Replacement is easier and does not require handling of solvents and reagents. The cost of new liners, labor expenses, frequency of liner replacement and complexity all have to be considered when determining whether cleaning, resilylation or replacement makes more sense.

### 7.11.4
### Silylating Injector Liners

Untreated glass contains silanol groups that interact with most compounds. These silanols contribution to peak tailing or adsorption problems for active compounds. Silylation converts the silanols to groups that do not interact with active compounds. Before the liner can be silylated it needs to be properly cleaned and treated.

Sample residues, trace metals in the glass matrix and the action of previous silylations need to be removed before a liner can be silylated. This is done with an acid cleaning or leach step. Place the liner in a clean test tube. Cover the liner with a 1N HCl or $HNO_3$ solution and soak for at least 8 hours (overnight is the most convenient). If the acid solution is highly discolored after use, replace it with clean acid solution and continue to soak until color change is noted. Do not leave the liner is the acid solution for more than 24 hours. Carefully decant the acid solution and thoroughly rinse the liner with deionized water followed by methanol. Dry the liner at 100–150 °C. Do not exceed 150 °C.

The acid solution removes prior silylation treatments and converts the surface to all silanols groups. These free silanol groups readily react with silylation reagents to form an inert surface suitable for sample injection. The liner is not suitable for use after acid treatment; it must first be silylated. Before silylating liners (or any glass), there are several precautions that must be taken. Silylating reagents (silanes) are very reactive towards small molecular weight alcohols (e.g., methanol, ethanol, *i*-propanol), primary and secondary amines (e.g., dimethyl- and diethylamine) and water. At no point should the silylation reagents come in contact with these types of compounds. The reagents preferentially react with alcohols, amines or water and the liner deactivation will not be successful. The reagents need to be kept dry and exposure to air kept to a minimum. It is best to flush the container with nitrogen and stored it in a freezer. A Teflon lined cap is essential; rubber or paper decomposes in the presence of most silylating reagents. Silylating reagents are very volatile, flammable and irritants, thus they must be handled in a hood by qualified individuals using proper safety equipment and precautions.

There are two silylation methods – a gas phase and a solution method. The gas phase method provides a slightly better deactivated liner; however, it is a more difficult and labor intensive procedure. The solution method is much easier; however, the liner is not as thoroughly deactivated. Except for the most demanding conditions (very active compound at low levels), the solution method will be satisfactory.

*Gas Phase Silylation Procedure*

1. Place the liner in a glass test tube that can be easily flame sealed.
2. Heat the neck of the test tube until a 2–3 mm opening remains.
3. Add 2–3 drops of diphenyltetramethyldisilazane to the test tube.
4. Immediately flush the test tube with a stream of dry nitrogen or argon.
5. Immediately flame seal the test tube (remember the silane is very flammable).
6. Heat the test tube for 3 hours at 300 °C.
7. Allow to cool to room temperature.
8. Carefully open the test tube and thoroughly rinse the liner with pentane or hexane.
9. Dry the liner at 75–100 °C.

*Note:* Liners can be stored in the sealed test tube until needed. Several liners can be placed in one test tube; however, rotate the test tube several times during heating to ensure all surfaces are exposed to the silane gas.

*Solution Silylation Procedure*

1. Place the liner in a screw cap test tube.
2. Cover the liner with 10% trimethylchlorosilane or dimethyl-chlorosilane in toluene.
3. Tightly seal the test tube with a Teflon lined cap.
4. Let stand for at least 8 hours.
5. Remove the liner from the solution and thoroughly rinse with toluene then methanol.
6. Dry the liner at 75–100 °C.

*Note:* Liners can remain in the silane solution for longer than 8 hours. Several liners can be placed in one test tube; however, rotate the test tube several times to ensure all of the liner surfaces are exposed to the silane solution.

It is important to rinse excess silylating reagent or solution from the liner before use. Residual reagent may cause an excessively high baseline that often requires hours to days to stabilize. This baseline often mimics column bleed especially if examined by GC/MS. The silylating reagent will not harm columns except for PEG stationary phases.

# 8
# Detectors

## 8.1
## Introduction

As compounds elute from the column, they interact with the detector. This creates an electrical signal whose size is related to the amount of the corresponding compound. The detector electronics (electrometer) processes the signal and sends it to a recording device. The plot of the signal size versus time results in the chromatogram.

There is no single detector that is best for all GC analyses, this being the reason for the variety of choice. Each GC manufacturer offers the most common ones, but their design and performance may differ. Since the principle of operation for each detector is the same, there is a general set of rules and characteristics that apply to all.

## 8.2
## Detector Characteristics

### 8.2.1
### Detector Dead Volume

In an ideal situation, the compounds are detected the moment they emerge (elute) from the column. Due to the design limitations of some detectors, there is space between the end of the column and the point of detection. This space is called dead volume. It requires long times for the carrier gas to sweep the compounds from the end of the column into the detector. This delay in reaching the detector can lead to broad, tailing or reduced size peaks. These problems occur for several reasons.

The surfaces between the end of the column and the detection zone are usually metal. Some compounds react with these hot, metal surfaces. Large dead volumes provide a large amount of surface area for the compounds to contact before entering the detector. Also, the long times spent in this area provides ample opportunity for these compounds to interact with the metal surfaces. This type of interaction can result in tailing or reduced size peaks. If the interaction is very

*The Troubleshooting and Maintenance Guide for Gas Chromatographers, Fourth Edition.* Dean Rood

ISBN: 978-3-527-31373-0

short lived, some of the compound molecules are slightly delayed before reaching the detector. The delay of some of the molecules results in a tailing peak. If the interaction is long lived or permanent, some of the solute molecules never reach the detector. The lower number of compound molecules reaching the detector results in a smaller peak.

For compounds that do not interact with the metal surfaces, another problem may occur. It takes a long time for the carrier gas to sweep all of the compound molecules from the dead volume into the detector. From a detector standpoint, peak width is related to the difference in time when the first and last compound molecules enters the detector. Longer sweep times translate into broader peaks. Compounds that elute from the column in tight, compact bands (i.e., narrow peaks) are converted into wide, spread out bands (i.e., broad peaks) by the presence of excessive dead volume. Dead volumes are reduced or eliminated by the proper installation of the column in the detector. The proper use of makeup gas helps to rapidly sweep the compounds through any dead volume and often improves the peak shapes.

### 8.2.2
### Detector Makeup or Auxiliary Gas

Most GC detectors require 20–40 mL/min of gas flow to obtain the best sensitivity and peak shapes. The carrier gas flow rates for capillary columns are well below these desired values. A gas is usually supplied to the detector to make up the difference between the actual carrier gas flow and the desired detector flow. This gas is called the makeup gas or sometimes the auxiliary gas. The makeup gas does not flow through the column, but is added at the exit of the column.

Makeup gas helps to rapidly sweep the compounds from the column through the detection zone. This helps to minimizes the effect of any dead volume. No or too little makeup gas can result in tailing or broad peaks since the compounds are not rapidly swept to the detector. Excessively high makeup gas flows usually do not negatively affect peak shape, but may reduce sensitivity. The makeup gas flow becomes more critical as the amount of dead volume increases. Some detectors have more dead volume than others, thus the magnitude of the peak shape problems are different when the makeup gas flow is incorrect.

Some detectors require a specific gas to function. This gas is often used as the makeup gas, thus serving two purposes. Loss of sensitivity and broad peaks may occur when there is not enough makeup gas. The detector does not function properly without enough gas. Excessively high makeup gas flows usually reduces sensitivity, but improves peak shapes. Smaller peaks occur since the compounds are swept through the detector too quickly for the detector to completely respond to their presence.

The best makeup gas depends on the specific detector. Some detectors can properly function with several different makeup gases. One may provide better performance than another, but often the differences are small. Using the same type of gas for carrier and makeup is convenient since the same gas cylinder and regulator can be used. The best makeup gas flow rate depends on the specific detector and often differs between GC manufacturers. The GC instruction manual provides recommended flows for the makeup gas. A range of flows is often provided. If the best possible sensitivity or peak shapes are desired, experimenting with the makeup gas flows within the specified range is recommended. Using a test sample (the more like the actual samples, the better), comparing peak shape and sensitivity at several different makeup gas flows helps to determine the best makeup gas flow rate. If long term, consistent detector response is important, the makeup gas flow needs to be set at the same value. Different flows alters detector sensitivity and causes inconsistent detector response problems.

Makeup gas flows are measured with a flow meter at the exit of the detector. Make sure that all of the other detector gases are turned off. Measuring the total flow of two gases simultaneously then subtracting the flow of one from the total flow can lead to errors in the measurements. Since two flow measurements have to be made anyway, it best to measure each gas separately and avoid potential errors. To protect the column, the carrier gas flow is often left on during detector gas flow measurements. This flow needs to be subtracted from the measured makeup gas flow. The carrier gas flow is usually small and does not contribute any significant error to the makeup gas flow value.

### 8.2.3 Detector Temperature

A detector's temperature can have a pronounced influence on its performance. Some detectors are very sensitive to changes in their temperature while others perform satisfactorily and consistently as long as the temperature is high enough. The detector has to be hot enough to prevent the condensation of any compounds eluting from the column on the surfaces of the detector. Rapid contamination of the detector occurs at low temperatures. Increased noise, loss of sensitivity and shifts in selectivity or linear range are some of the more common symptoms of detector contamination.

The detector does not have to be maintained at a temperature that is higher than the column temperature. While this is desirable, it is not necessary. It is not required to maintain the detector temperature below the column's upper temperature limit since no significant damage to the column occurs. A darkening of the polyimide coat of the column may be observed at detector temperatures above 325 °C. This section of column inside the detector may be slightly weakened if detector temperatures above 350 °C are used. Occasional trimming of the section is recommended to prevent breakage of the column in this area during use.

Temperature sensitive detectors should be kept away from heating and cooling vents or other sources of temperature change. When used at their more sensitive settings, these detectors may respond to the sudden change in the ambient temperature. Usually, a sudden shift in the baseline, drift of the baseline or change in the sensitivity is noticed if a sudden temperature change occurs.

### 8.2.4 Detector Sensitivity

The most basic definition of sensitivity is the smallest amount of a compound that can be detected. The actual amount will vary depending on the compound, detector, peak width, and gas flow. A value called the signal to noise ratio (S/N) is often used to definite sensitivity. The signal is the height of the peak and the noise is the height of the baseline noise (Figure 8-1). There are many different definitions of signal to noise; however, this one is the most practical for chromatographers. Be aware of any scale changes when measuring peak and noise heights. Sometimes, to measure the noise, the vertical scale has to be expanded. The peak is then too large to be measured and the scale has to be reduced to view the entire peak. The scale differences (which may be not linear) have to be accounted for in order to obtain an accurate measurement.

The S/N ratio is often used as a measure of a detector's response. For example, 10 pg of a specific compound gives a S/N of 10. Often minimum detection limits (MDL) are quoted as S/N ratios. For example, a peak has to have a S/N greater than 4; a peak with a lower S/N value is too small to be confidently assigned to a compound. Many regulatory agencies set MDL's at a S/N of 2–4. This range is probably more applicable to analytical standards in clean solvents. For the average gas chromatographic detector and normal operating conditions, a S/N value of 4–8 is more realistic for extracted samples. Peak assignments become suspect at very low S/N values. As the complexity (i.e., dirtiness) of a sample increases, the S/N values should be higher to compensate for possible interferences (co-eluting compounds) and higher background.

The response of some detectors change with even small flow rate changes of the carrier or makeup gases. The response of other detectors are fairly independent of the gas flows. Detectors that respond independently of the carrier or makeup gas flow rates are called mass flow rate detectors. The detector response is proportional to the mass of compound passing through the detector per unit time. In theory,

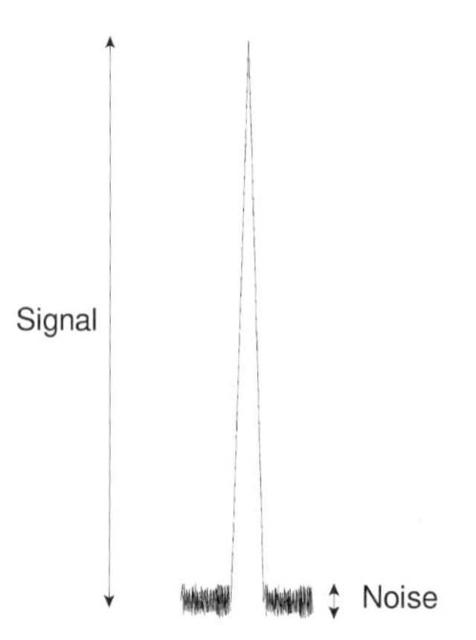


**Figure 8-1** Signal to noise measurement.

these detectors are not affected by small changes in the carrier and makeup gas flow rates. In practice, changes in gas flows greater than 25% have a significant effect on detector behavior. Detectors that are affected by the volume of gas flowing through the detector are called concentration dependent detectors. The detector response is proportional to the mass of solute per unit volume of gas. In other words, the concentration of the compound in the gas. The same mass of compound eluting from the column generates a peak of different size depending on the flow of the carrier and makeup gases. At higher flow rates, the compound becomes diluted by the carrier or makeup gas. This reduces the concentration of the compound, and the peak size thus decreases. Lower flow rates increase the peak size, thus sensitivity improves. Excessively low flow rates create dead volume problems; therefore, there are lower flow limits for each detector.

Drifting baselines related to the temperature program are common for concentration dependent detectors. As the column temperature increases during a temperature program, the carrier gas flow rate drops. Since the volume of carrier gas is decreasing, the background signal also changes. Because the baseline is a function of the background signal, this gives rise to a drifting baseline that follows the changes in the temperature program. Detectors equipped with electronic flow controllers can be programmed to increase the makeup gas flow rate to compensate for the decrease in carrier gas flow rate. This helps to minimize the amount of baseline drift. Operating in the constant carrier gas flow rate mode is another option when using a concentration dependent detector. Mass flow rate detectors do not exhibit the baseline drift behavior since they are not very sensitive to changes in the gas flows.

### 8.2.5 Detector Selectivity

Only a few detectors respond to every compound eluting from the column. In most cases, only particular types of compounds elicit a response from a detector. For some detectors, this is a very broad range of compounds, and for others, only a narrow range of compounds. A detector is highly selective if it responds to only a narrow range of compounds. This means that peaks are obtained only for compounds whose structures contain the characteristic necessary for interaction with the detector. The other compounds pass through the detector; however, very small or no peaks are obtained for these compounds. A less complex (i.e., fewer peaks) chromatogram is achieved. It is now much easier to resolve the peaks of interest from the matrix or background peaks since there are fewer interfering peaks present.

It is usually an atom, functional group or structural feature that elicits a selective detector's response. For example, a detector responds only to compounds containing halogens; all other compounds have little to no response. Detector selectivity can be viewed as either enhanced response for a specific class of compounds or a lack of response for another class of compounds. Most selective detectors have an enhanced response for one class of compounds and a poor response towards others. Most detectors have substantially better sensitivity for their selective compounds than the general purpose (i.e., not very selective) detectors. A few detectors only filter out the nonselective compounds. Their sensitivity is not much better than some of the general purpose detectors, but the filtering out of the unwanted peaks makes it easier to discern and measure the peaks of interest.

Detector response or behavior may be affected by the presence of non-responsive compounds. Alterations in the peak size or linear behavior (Section 8.2.6) are the most common problems related to the presence of non-responsive compounds. Problems usually occur when a high concentration of a non-responsive compound elutes very near or co-elutes with a responding compound. Even though the detector does not produce a peak response, the detector is influenced in a manner that affects its behavior. In most cases, the peak of interest becomes reduced in size. This effect is often called quenching.

Selectivity is usually measured by comparing the response of the selective compound to a reference compound (Equation 8-1). Hydrocarbons are the usual reference compounds for the highly selective detectors.

$$\text{Selectivity} = \frac{\text{Peak Size of Selective Compound}}{\text{Peak Size of Reference Compound}}$$

**Equation 8-1** Detector selectivity.

### 8.2.6
### Detector Linear Range

Any change in the amount of a compound in an injected sample should result in a corresponding change in the compound's peak size. For example, if the amount of compound is doubled, the size of the resulting peak should also double. This behavior occurs only over a certain range of compound amounts – this range is called the linear range of the detector. If the amount of compound is outside this range, the response of the detector does not directly reflect the amount of compound actually passing through the detector.

The linear range of a detector is determined with a simple experiment. A series of standards of the compounds of interest are made over a wide concentration range. The lowest should be at the minimum detection limit (Section 8.2.4), and the highest at least $10^5$ times higher. The minimum number of standards should be five with a greater number providing more accurate information. Inject each standard at least 3 times and plot the average of the peak size versus the amount of the compound. The linear range is the region where a straight line can be drawn through the points (Figure 8-2a). Another method is to divide the peak areas by the corresponding compound amounts. These values are plotted and the linear range is the region of the straight line (Figure 8-2b). Usually a deviation of ±5% is allowed before the linear range is considered to be exceeded.

Depending on the column and detector, the column's capacity may be exceeded before the maximum value of the detector's linear range is reached. For some

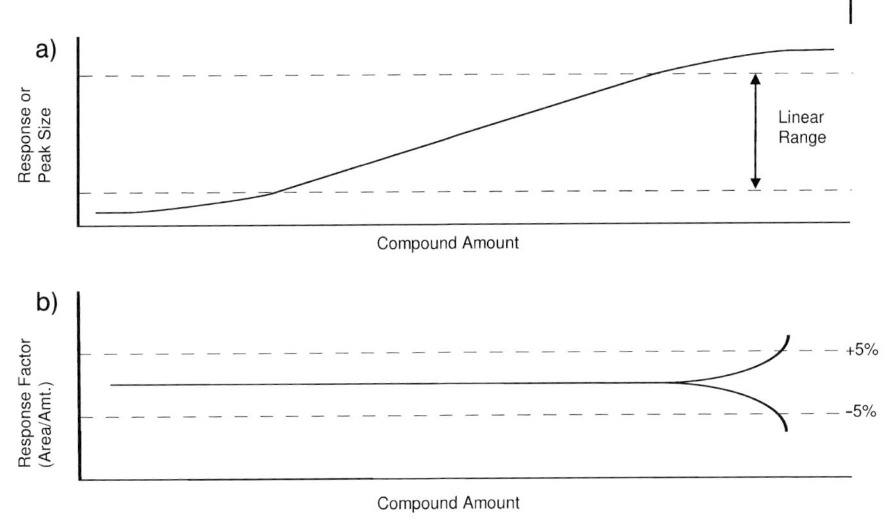


**Figure 8-2** Linear range curves.

detectors, the linear range varies by compound and operating conditions, thus the linear range for each compound and operating condition may need to be individually determined. All injector conditions need to be kept constant to ensure a true measure of a detector's linear range. Variations can affect the amount of compound entering the column, and reaching the detector.

## 8.3 Flame Ionization Detector (FID)

### 8.3.1 FID Principle of Operation

A simplified drawing of a FID is shown in Figure 8-3. A voltage is applied to the flame jet and the collector. A hydrogen and oxygen (usually supplied as an air mixture) flame is maintained at the tip of the jet. Carrier gas exiting from the column is delivered into the flame. Organic compounds in the carrier gas are burned in the flame. This combustion process creates ionic species which are attracted to the charged collector. The movement of the ions from the flame to the collector produces a small current. This current is measured as the signal from the detector. A background or baseline signal is always present and it is a result of trace levels of impurities such as gas and system contaminants and normal column bleed. The introduction of an organic compound into the flame results in the compound's ionization and an increase in the detector current above the background level. This detector signal is processed by the recorder into a corresponding peak in the chromatogram. Greater amounts of the compound generate a larger signal, thus giving a larger peak.

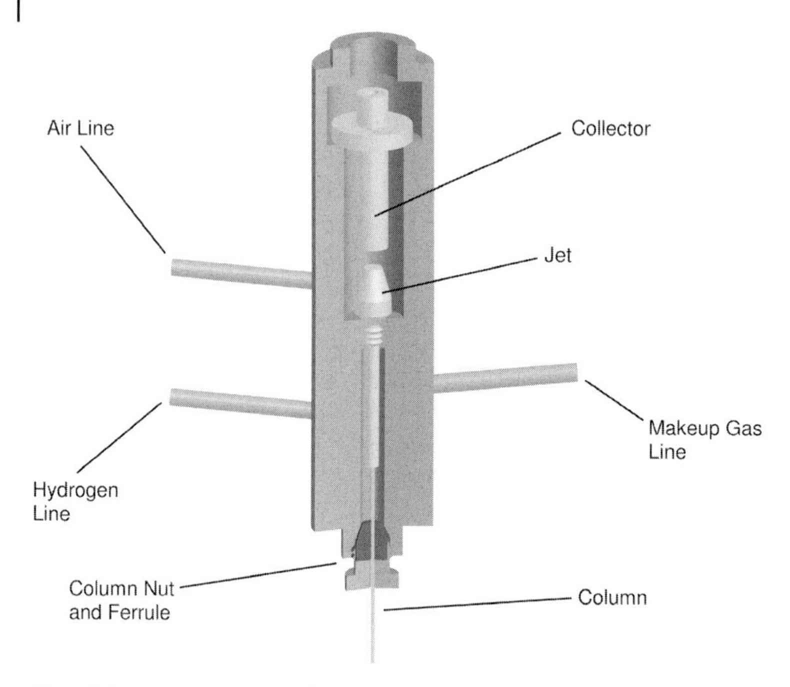


**Figure 8-3** Major components of a FID.

### 8.3.2 FID Gases

Hydrogen and air are used as the combustion gases. Usually, the air flow is about 10 times the flow of hydrogen (30–40 mL/min hydrogen and 300–400 mL/min for air). Each model of detector is unique, thus the exact flow values are found in the instruction manual. The flows of the hydrogen and air combustion gases affects FID sensitivity. The same flow values are needed to maintain consistent performance. If the gas flows are too far away from the recommended values, difficulties in maintaining or lighting the flame occur.

Nitrogen is a better makeup gas than helium. Using nitrogen as the makeup gas increases detector sensitivity by 10–20% over helium. Recommended makeup gas flow rates are found in the GC's instruction manual. Makeup gas flow rate affects sensitivity, thus a correct and consistent flow is important for best performance. Experimenting with adjustments (5–10%) to the makeup gas flow may improve sensitivity, but often the gains are quite small and not worth the effort. Hydrogen is not recommended as a makeup gas. Since hydrogen is one of the combustion gases, its presence alters the flame behavior. Sensitivity problems and flame lighting difficulties are the most prevalent problems.

### 8.3.3
### Column Position in a FID

The proper column position in the detector is essential for optimal performance. The GC instruction manual is a good source of information on the positioning the column in the detector. In most cases, the column end needs to be $\leq 1$ mm below the tip of the flame jet. If the end of the column is too far below jet tip, dead volume problems occur. If the column is positioned too far into the jet, difficulty in lighting the flame, peak shape problems or excessive background noise may occur. If pushed too far and hard into the jet, A small piece of fused silica tubing can be lodged in the tip. Difficulties in lighting the FID or peak shape problems may occur.

### 8.3.4
### FID Temperature

The temperature of a FID is not critical to its performance. The detector needs to be hot enough to prevent the condensation of high boiling compounds. A temperature of 225–325 °C is typical. The higher temperature range is recommended to keep the detector from rapidly becoming dirty. Most FIDs will not light if the temperature is below 150 °C.

### 8.3.5
### FID Selectivity

The FID responds to compounds containing a carbon-hydrogen bond. The exceptions are formaldehyde and formic acid which have very poor responses. Fully halogenated hydrocarbons give weak responses even though there are no carbon-hydrogen bonds present. The FID does not respond to permanent gases ($O_2$, $H_2$, $N_2$, He, Ar, Xe, etc.), oxides of nitrogen (NO, $NO_2$, $N_2O$, $N_2O_2$), sulfur compounds ($H_2S$, $SO_2$, $CS_2$, COS, etc.), $CO_2$, CO, and water. Large solvent peaks are nearly eliminated for samples in $CS_2$ or water. Baseline disruptions in the area where these compounds elute are common. In practice, $CS_2$ gives a larger response than expected and can be of appreciable size at sensitive settings.

### 8.3.6
### FID Sensitivity and Linear Range

The practical minimum detectable quantity is 0.05–0.5 ng of compound. There is an increase in response with respect to the number of carbons – the more carbons atoms, the better the response. Response decreases as the amount of substitution with halogens, sulfur, oxygen, nitrogen and phosphorus atoms increases. For example, chloropropane has a poorer response than propane. The response loss due to substitution is variable and is dependent on the exact groups and their position on the molecule. Carbonyl carbons do not contribute to the response.

A FID is a mass flow rate detector. Changes in the carrier and makeup gas flow rates do not significantly affect the sensitivity. Sensitivity changes do occur; however, large changes in the gas flow rates are necessary.

The linear range of $10^5$–$10^6$ for the FID is among the largest of all GC detectors. This makes the FID suitable for use with a very wide range of sample concentrations. Three point calibration curves are often acceptable for FID due to their large linear range.

### 8.3.7 Verifying Flame Ignition of a FID

The signal output as displayed on the data signal or GC display can be used to verify flame ignition. Check the signal prior to lighting the FID. The signal is higher when the flame is burning. By comparing the two values, it is easy to determine if the flame is lit. Different GC's and data systems report detector output in different units or values. This is not a problem since only a comparison between the lit and unlit detectors are needed. The signal difference between the unlit and lit condition can be entered in some data systems to determine whether the FID is lit.

To physically verify that the flame is lit, place a cool, shiny object (e.g., wrench, knife blade, etc.) over the vent of the detector. After several seconds, condensation should be visible on the shiny object after it is removed from the vent. If no condensation is visible, the detector flame may not be lit.

If there are doubts about whether the flame is lit, inject about 1 µL of a detectable solvent. A large peak should be obtained after a few minutes providing the column temperature is high enough to elute the solvent. The solvent will not harm the column even if it has not been conditioned.

### 8.3.8 FID Maintenance

FID's require little maintenance to keep them performing at satisfactory levels. The hydrogen, air and makeup gas flows should be occasionally measured. They can drift over time or be unintentionally changed without knowledge of it occurring. Each gas flow should be measured independently to obtain the most accurate values.

The FID requires periodic cleaning. In most cases, this only involves the collector and the jet. Assembly and cleaning instructions can be found in the GC manual. FID cleaning kits are available containing brushes and wires that simplify the cleaning of all of the detector parts. The brushes are used to dislodge particulates clinging the metal surfaces. Be careful cleaning in the vicinity of the ignitor since it is fragile and can be easily damaged. A fine wire is used to clean the jet opening of particles. Do not force too large a wire or probe into the jet opening or the opening will become distorted. A loss of sensitivity, poor peak shape and/or lighting difficulties may result if the opening is deformed. The various parts can be ultrasonicated after cleaning with a brush. Eventually the jet needs replacing, but this should be a fairly infrequent necessity.

## 8.3.9
## Common Problems with a FID

### 8.3.9.1 Change in FID Sensitivity

A loss of sensitivity is the result of a loss in peak size and/or an increase in the amount of noise. Losses of peak size or increases in noise are usually from improper detector gas flows or a dirty detector. Sensitivity is dependent on the flow rates of hydrogen, air and the makeup gas. Incorrect flow rates alter FID sensitivity. In some cases, only the peak size or the noise is affected. Some compounds may suffer a larger response shift than others. Combustion and makeup gases that are improperly set can cause a loss of peak size. An increase in the noise may occur when the combustion gas flows are incorrect, but this usually occurs only when the gas flows are significantly away from the proper flow rates. A dirty detector may affect sensitivity also. Deposits form on the jet and collector which may partially inhibit the formation or detection of the ions. A reduction in the number of ions generated or detected results in a loss of response. Contaminants in the detector can contribute to the background noise. They slowly reach the collector and elevate the background signal.

For collectors that can be disassembled, an inadequately tightened collector may cause noise. Some jets have an opening that is larger than the outer diameter of the column. If the column is pushed too far into the detector, the column can protrude into the flame. The polyimide coating decomposes and causes high background signals and noise problems. This is evident from the melted end of the column, the lack of polyimide coating (clear glass is visible) or black discoloration on the end of the column upon removal from the detector. Usually peak shape problems occur also. Columns forced into the tip of the jet can leave debris behind which causes peak shape problems or difficulties in lighting the FID.

Contaminated detector gases or gas lines can cause excessive noise problems. These problems are hard to distinguish from other sources of contamination, thus carefully examination of the GC and related devices is required. The use of a jumper tube test may help to isolate the source of the contamination (Section 12.4.1).

### 8.3.9.2 Difficulty in Lighting the FID Flame

Difficulties in lighting the FID flame can usually be attributed to improper gas flows or a dirty jet. The incorrect mixture of hydrogen and air will not sustain a flame. A popping noise may be heard, but this does not mean the flame is lit. The proper ratio is needed to maintain a stable flame. Sometimes, if the makeup gas flow is off, the flame will not light. A dirty or clogged jet may prevent the proper flow of hydrogen from reaching the combustion area. Particulate matter, broken pieces of column or a column completely blocking the jet opening may prevent hydrogen from flowing at the proper rate.

Make sure that the FID igniter is working. It usually is located and visible near the top of the FID. It should have an intense orange glow to it when on. Lighting the flame with a match is not recommended since foreign matter may fall into the detector giving rise to noise. A lighter is a better choice of alternative ignition.

Hydrogen carrier gas is used at high flow rates. For Megabore columns, the flow of hydrogen can be quite high. This hydrogen is additive with the detector hydrogen and alters the total amount of hydrogen reaching the combustion area. This may be enough of a change to make it difficult to light the flame. Reducing the flow of detector hydrogen by the same amount as the column flow usually solves this problem.

Some FID's become difficult to light with use. Sometimes it is a specific jet that causes the problem. If no problems are evident, it may be just a quirk with that particular FID. A difficult to light FID can sometimes be coaxed into lighting. Lower the air flow to 50–100 mL/min. Activate the igniter then slowly raise the flow of air. Usually a popping sound will be heard, but continue to raise the air flow to its recommended value.

#### 8.3.9.3 Peak Shape Problems Attributed to the FID

Improper column installation in the detector or too low makeup gas flows are the most common source of broad or tailing peaks caused by the detector. If a column is forced too far into the detector, a broken column end may result. Upon tightening the column nut, the column is forced upward and its end may be crushed in the tip of the jet. The broken column end and resulting debris disrupts the flow at the end of the column. This turbulent flow may cause broad or tailing peaks. When a column is too far away from the tip of the jet, there is excessive dead volume. Tailing or broad peaks may result with the problem being more severe for very volatile compounds. Too low or nonexistent makeup gas flows often causes tailing or broad peaks.

It is possible to push the column through the opening of some FID jets. In addition to excessive noise, peak shape problems may also occur. Depending on the design of the FID and jet, it may be possible to miss the jet. The column slides around the outside of the jet instead of into it. This creates a tremendous amount of dead volume and the usual occurrence of tailing peaks.

#### 8.3.9.4 Miscellaneous Problems with a FID

A drifting baseline is sometimes obtained when using hydrogen as the carrier gas. This occurs with temperature programs and is more common at very high carrier gas flow rates. The carrier gas flow decreases as the column temperature increases. This changes the amount of total hydrogen (column and combustion) in the detector. This change causes a slight shift in detector baseline or background signal. The baseline shift is more severe when there is a large difference between the initial and final temperatures in a temperature programmed run. Also, the shift becomes more visible when using more sensitive settings.

Spikes in the chromatogram can be caused by particulate matter falling into the detector. This generates a sudden surge in the signal, and thus a spike in the chromatogram. Particulates usually accumulate on the collector or upper regions of the FID. Some PLOT columns are known to emit particles, thus spiking is common with these columns.

Large volumes of nonflammable solvents may extinguish the FID flame. This is a particular problem with water as the sample solvent. If this problem occurs, changing the sample solvent is the easiest solution. In cases where a change is not possible, the options are limited. Sometimes changing the hydrogen and air flows may help to keep the flame lit. Usually increasing the amount of hydrogen is the most successful. Adding a small amount (5–10%) of methanol or acetone to the sample may help. These are flammable solvents and may help to sustain flame combustion in the presence of water. Sample concentrations have to be adjusted to account for the increase volume of the sample. Also, a larger solvent front is obtained due to the much better response of the added solvent.

## 8.4 Nitrogen-Phosphorus Detector (NPD)

### 8.4.1 NPD Principle of Operation

A simplified drawing of a NPD is shown in Figure 8-4. A voltage is applied to the jet and the collector. A bead or cylinder (usually rubidium silicate) resides above the jet. The bead is electronically heated to 600–800 °C by application of a current to the bead. A plasma is sustained in the region of the bead by the addition of hydrogen and air. The flow of hydrogen is too low to support a flame. Nitrogen or phosphorus containing compounds react in the gaseous boundary layer around the surface of the bead to produce specific ions. A small current is produced by the movement of these ions from the plasma to the charged collector. This current is measured as the signal from the detector. A background or baseline signal is always present, and it is a result of trace levels of impurities such as gas and system contaminants and normal column bleed. The introduction of a nitrogen or phosphorus containing compound results in its ionization and an increase in the detector current above the background level. This signal is processed by the recorder into a corresponding peak in the chromatogram. Greater amounts of the compound generate a larger signal, thus giving a larger peak. Since a flame is not sustained in a NPD, hydrocarbon ionization does not occur as with a FID. This results in the high selectivity to nitrogen and phosphorus containing compounds and the lack of response for other compounds. The ionization efficiency of a NPD is about 10,000 times better than a FID. This means that the sensitivity of a NPD for nitrogen and phosphorus compounds is about 10,000 times better than a FID.

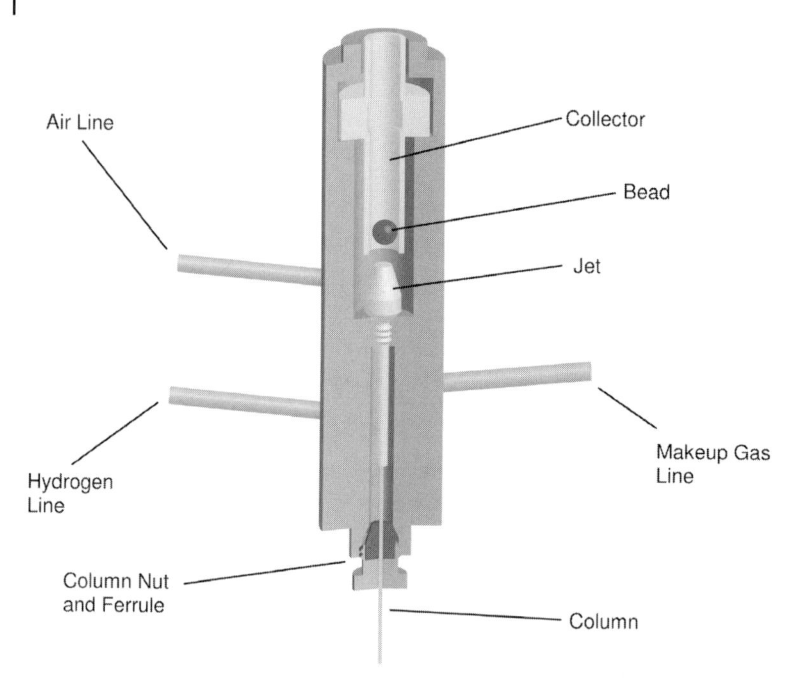


**Figure 8-4** Major components of a NPD.

## 8.4.2
## NPD Gases

Hydrogen and air are used as the gases to support the plasma. Flow rates of 2–6 mL/min for hydrogen and 60–200 mL/min for air are typical. NPD selectivity and sensitivity is very dependent on the hydrogen flow rate; the flow of air is less critical. Older NPD's use a different hydrogen flow for the nitrogen and phosphorus modes (1–5 mL/min for nitrogen and 20–30 mL/min for phosphorus) and 70–100 mL/min of air.

Helium is the usual makeup gas for NPD's since it provides the best sensitivity. The flow rate is usually 30–40 mL/min. The GC's instruction manual recommends the best flow rates for a specific model of NPD. Sensitivity, especially for nitrogen compounds, is affected by the makeup gas flow rate. Too low or no makeup gas results in very poor peak shapes and sensitivities. Nitrogen can also be used as a makeup gas. Actually, it provides more stability to the NPD.

## 8.4.3
## Column Position in a NPD

The proper column position in the detector is essential for optimal performance. The column end needs to be $\leq 1$ mm below the tip of the jet. If the end of the column is too far below jet tip, dead volume problems occur. If the column is positioned too far into the jet, peak shape problems or excessive background noise may occur. The GC instruction manual is a good source of information on the positioning the column in the detector.

### 8.4.4 NPD Temperature

Selectivity and especially sensitivity are affected by the detector temperature. A stable temperature environment is critical to maintain consistent NPD behavior. The detector temperature for a NPD should be as high as reasonably possible. Temperatures of 275–325 °C are routinely used. High temperatures help to stabilize the NPD. Large concentrations of a compound, such as a sample solvent, often create large baseline disturbances. A high detector temperature helps to minimize the severity and length of the baseline disturbance. A high detector temperature is also recommended when high carrier or makeup gas flows are used. The higher flow of gas has a tendency to cool the interior of the detector. This may reduce detector sensitivity; a higher temperature can thus be used to compensate for the loss of response. Higher detector temperatures allow the use of lower bead currents, thus increasing the life of the bead.

### 8.4.5 NPD Selectivity

The selectivity of a NPD for nitrogen over carbon is $10^3$–$10^5$ and for phosphorus over carbon is $10^4$–$10^{5.5}$. Phosphorus specificity is easily obtained over a relatively wide range of gas flow rates and bead currents. Higher hydrogen flows increase the specificity of phosphorus over nitrogen. Nitrogen specificity is significantly more dependent on gas flow rates and bead current. For this reason, the NPD is usually optimized for nitrogen specificity.

### 8.4.6 NPD Sensitivity and Linear Range

The practical minimum detectable quantity is 0.1–0.5 pg for phosphorus compounds and 0.5–1 pg for nitrogen compounds. The sensitivity is primarily dependent on compound structure, bead current and age, gas flow rates, and detector temperature. There are known instances where sensitivity is very dependent on compound structure. The response for compounds with the nitrogen atom bound to a carbon with a hydrogen is better than those without (e.g., amides, nitro compounds). Operating the bead at high currents produces the best sensitivity, but at the expense of increased noise and reduced bead life. As the beads become older, a higher current is usually necessary to maintain previous sensitivity levels. Eventually, the noise become too high and renders the detector unusable. Consistent gas flows, especially hydrogen, are necessary for constant response. Small flow changes can result in large sensitivity changes primarily for nitrogen compounds. Increasing the hydrogen flow into the higher recommended flow range tends to increase the phosphorus compound sensitivity, but usually at the expense of decreased bead life. A higher detector temperature usually provides slightly better sensitivity especially at high carrier and makeup gas flow rates.

A NPD is a mass flow rate detector. Small changes in the carrier and makeup gas flow rates do not significantly affect the sensitivity. Nitrogen compounds are much more susceptible to flow related response changes than phosphorus compounds. Larger changes in the gas flows affect the temperature in the detector. It is this temperature change than mainly affects sensitivity.

The linear range for a NPD is typically $10^4$–$10^5$. This range is dependent on the compound, gas flows, bead current and age, and detector temperature. Five point calibration curves are recommended with NPD's (due to their temperamental nature) even though their linear range is good.

### 8.4.7
### NPD Maintenance

NPD's are temperamental and require frequent maintenance. Small changes in any of a number of parameters can significantly change the performance characteristics of a NPD. The bead requires the most maintenance. It needs changing frequently, thus a spare is a necessity. The beads have to be kept dry which limits their storage life to about 6 months. When a new bead is installed, slowly raise the detector temperature and bead current. Rapid heating can crack or break the bead especially if it has been stored under humid conditions. It has been observed that higher hydrogen flows and bead currents decrease bead life. If the NPD is not in use, the hydrogen flow and bead current should be reduced or turned off to increase bead life. Make sure there is some type of gas flow in a heated detector or when there is current to the bead.

The hydrogen, air and makeup gas flows should be frequently measured. They can drift over time or be unintentionally changed without knowledge of it occurring. Each gas flow should be measured independently to obtain the most accurate values. NPD's are very sensitive to changes in the gas flows and consistent flows are necessary to maintain performance levels.

The NPD requires periodic cleaning. In most cases, this only involves the collector and the jet. Assembly and cleaning instructions can be found in the GC manual. Cleaning kits are available containing brushes and wires that simplify the cleaning of all of the detector parts. The brushes are used to dislodge particulates clinging the metal surfaces. A fine wire is used to clean the jet opening of particles. Do not force too large a wire or probe into the jet opening or the opening will become distorted. A loss of sensitivity or poor peak shape may result if the opening is deformed. The various parts can be ultrasonicated after cleaning with a brush. Eventually the jet will need replacing, but this should be a fairly infrequent necessity.

## 8.4.8 Common Problems with a NPD

### 8.4.8.1 Change in NPD Sensitivity

NPD's are noted for their instability with the sensitivity being the most variable feature. Practically every detector parameter affects the sensitivity. Due to the erratic nature of the bead, the sensitivity can frequently and rapidly change without warning or alteration of any of the operating parameters. Most cases of sensitivity changes are due to some type of change in the bead.

When the detector is first turned on, the sensitivity and the signal level changes slowly over several hours. For applications that require very stable operation (e.g. trace level analysis), leave the detector on overnight for maximum stabilization. If the signal continues to drift, the bead may need replacing.

Sensitivity shifts occur with flow rate changes in the detector gases especially hydrogen. Consistent flows are necessary to maintain constant or stable sensitivity. Depending on the compound and the direction of the flow rate change, the sensitivity may increase or decrease. Large changes in the gas flows change the temperature of the detector. This shift in detector temperature causes a change in the sensitivity. Obviously, a change in sensitivity is seen when the detector body is heated to a different temperature.

The primary cause of sensitivity changes is the bead. The gradual but normal depletion of the bead results in a drop in the sensitivity. Usually the bead current is increased to compensate for the loss. Increasing the current decreases bead life. Eventually the bead is depleted to a point where the sensitivity is inadequate or the noise is too high.

Another bead related sensitivity problem is due to contamination of the bead. Detector response is strongly influenced by the amount of surface contamination of the bead. Common sources of detector contamination are high boiling sample residues, fingerprints and leak detection fluids. Excessive column bleed products can coat the bead with contaminants and reduce the sensitivity. Excess silylation (derivatization) reagents in samples should be avoided for the same reason. Scratched or damaged beads adversely affect sensitivity and increase noise levels.

Avoid exposing NPD samples to glassware that have been washed in phosphorus based detergents. The residue phosphorus compounds can be detected by the highly sensitivity NPD. High baseline signals or noise can occur with this background of phosphorus compounds.

### 8.4.8.2 Peak Shape Problems Attributed to the NPD

Improper column installation in the detector can cause broad or tailing peaks. If a column is forced too far into the detector, a broken column end may result. Upon tightening the column nut, the column is forced upward and its end may be crushed in the tip of the jet. The broken column end and resulting debris disrupts the flow at the end of the column. This turbulent flow may cause broad or tailing peaks. When a column is too far away from the tip of the jet, there is excessive dead volume. Tailing or broad peaks may result with the problem being more severe for very volatile compounds.

Too low or nonexistent makeup gas flows may cause tailing or broad peaks. The error in the makeup gas flow also affects sensitivity, thus a loss of peak size accompanies a peak shape problem due to makeup gas flow problems.

Peak tailing can sometimes be caused by the bead, especially if dirty samples are frequently injected. Older beads are also more susceptible to this problem. Phosphorus compounds appear to be more sensitive to bead induced peak tailing. Increasing the bead current or the hydrogen flow often reduces the amount of tailing; however, a shift in sensitivity usually occurs.

#### 8.4.8.3 NPD Baseline Problems

A drifting baseline is obtained when using a temperature program. The problem is more severe with hydrogen as the carrier gas. The carrier gas flow decreases as the column temperature increases. This changes the amount of total gas in the detector. If helium is the carrier gas, a temperature change occurs in the detector. If hydrogen is the carrier gas, a change in the hydrogen flow and the temperature of the detector changes. This change causes a shift in detector baseline or background signal. The baseline shift is more severe when there is a large difference between the initial and final temperatures in a temperature programmed run. Also, the shift becomes more visible when using more sensitive settings. There is very little that can be done to directly compensate for this problem. Using a pressure programmable injector in the constant flow mode to maintain a constant flow throughout the temperature program may help (Section 7.7). Detectors equipped with electronic flow controllers can be programmed to increase the makeup gas flow rate to compensate for a decrease in carrier gas flow rate. This may also helps to minimize the amount of baseline drift.

Some stationary phases contain nitrogen. The degradation products from these stationary phases usually contain nitrogen containing species. Normal column bleed is amplified due to the extreme sensitivity of the NPD to nitrogen. The baseline rise may appear to be severe and excessive when it is normal. As a general rule, nitrogen containing stationary phases should be avoided with NPD's unless a large baseline rise can be tolerated.

A large amount of a compound passing through the NPD may cause a baseline disturbance. This is most common for the solvent front. Negative baseline drops and erratic baselines are common immediately after the solvent front. This even occurs with solvents that are not detected by the NPD. Temporary changes in the detector temperature or bead behavior are the cause of the baseline disturbance.

If the bead is severely contaminated, it is possible to get negative peaks for hydrocarbons. If a negative peak occurs in the same location, especially for multiple injections of the same sample, it may be related to bead contamination. If the negative peak occurs near the solvent front, it may be only the effect of the high concentration of sample solvent.

## 8.5 Electron Capture Detector (ECD)

### 8.5.1 ECD Principle of Operation

A simplified drawing of an ECD is shown in Figure 8-5. A source of electrons from a $^{63}$Ni foil bombards the column effluent passing through the detector. An auxiliary or moderating gas is added to the detector and in many cases functions as the makeup gas as well. Nitrogen or a 95 : 5 ratio of argon/methane (sometimes referred to as P5) is used as the auxiliary gas. Both gases can be ionized by the free electrons from the source. The electron bombardment of the auxiliary gas creates a plasma containing, among other species, thermal electrons. A potential difference is applied to the detector cell which allows the capture of the negatively charged, thermal electrons by the anode. The migration and capture of the thermal electrons creates a background current which is the baseline (background) signal. When a molecule with the ability to capture an electron elutes from the column and enters the detector cell, it captures a thermal electron. The loss of thermal electrons decreases the current in the cell. The amount of the current loss is indirectly measured and is converted into the detector signal. The signal is processed by the recorder into a corresponding peak in the chromatogram. Greater amounts of the compound generates a larger signal, thus giving a larger peak.

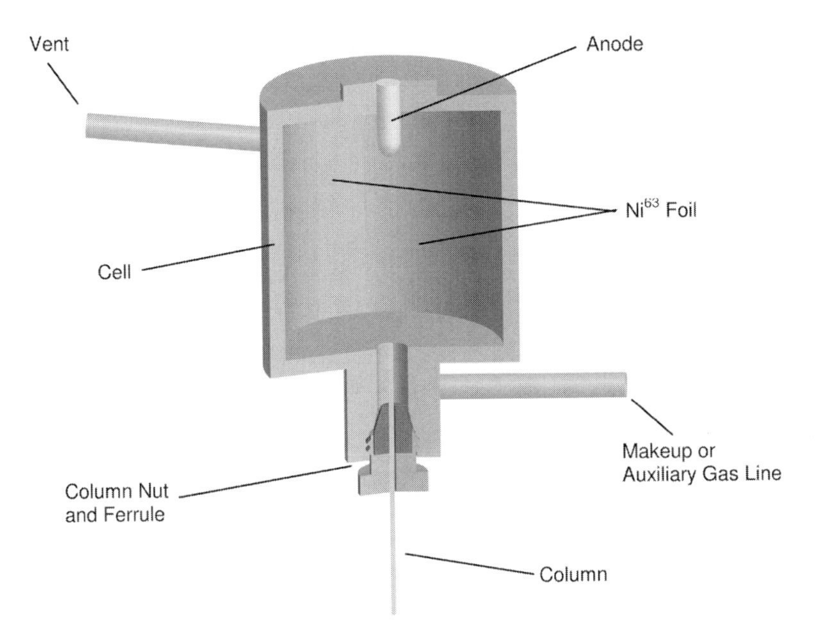


**Figure 8-5** Major components of an ECD.

### 8.5.2
### ECD Gases

Nitrogen or argon/methane (95 : 5) is necessary for an ECD. This gas is called the auxiliary or moderating gas and is often supplied as the makeup gas, thus serving two purposes. In general, most ECD's require a total gas flow rates of 20–60 mL/min. If they are supplied separately, the flows of the auxiliary and makeup gases need to be considered as a total flow. A change in either gas flow significantly affects the ECD's behavior. The exact gas flow rates are found in the GC's instruction manual. ECD's are very sensitive to changes in the flow rate of auxiliary or makeup gas. Changes as little as 1–2 mL/min can be very noticeable. Constant flows are necessary to maintain consistent performance from an ECD.

Better sensitivity is obtained if nitrogen is used as the auxiliary gas. The sensitivity is approximately two times greater than possible with argon/methane. Argon/methane provides slightly better linear range at the sacrifice of sensitivity. The choice of the auxiliary moderating gas primarily depends upon the requirements of the analyses being performed. Samples with possible wide concentration ranges may benefit from the better linear range available with argon/methane. High sensitivity requirements makes nitrogen a better choice. Another consideration is that nitrogen is less expensive per cylinder than argon/methane.

To achieve the best overall results for sensitivity and peak shape, a small amount of experimentation is recommended. Start with the auxiliary gas flow rate about 10% below the recommended flow values. Inject the sample and determine the peak size (area or height) and width. Increase the flow rate by 1–2 mL/min and re-inject the sample. Repeat the tests until a flow rate about 10% above the recommended value is reached. The best results are obtained when 5 or more flow rates are tested. A change in the sensitivity is usually observed. The peak widths narrow as the flow rates increase. Choose the value where the best sensitivity is obtained without a tremendous sacrifice of peak width. For some ECD's, a change of only 1–2 mL/min in the auxiliary gas flow can significantly affect the sensitivity or peak width without any noticeable changes in the other.

Helium is the preferred carrier gas for ECD's. Hydrogen is not recommended by some GC manufacturers since it may reduce the usable lifetime of the detector. If nitrogen is used as the carrier gas, sensitivity changes and baseline drifts may become evident when using a temperature program. The flow rate of nitrogen carrier gas decreases as the column temperature increases. The amount of nitrogen reaching the detector changes during the run, thus the change in the sensitivity and baseline.

### 8.5.3
### Column Position in an ECD

The position of the column relative to the detector cell is important for optimal sensitivity and peak shape. The end of the column should be at the entrance to the cell. If the column is positioned too low, dead volume problems occurs. It is easy to insert the column into the cell for many ECD's. Insertion of the column into the cell itself normally results in the loss of sensitivity, and in extreme cases, some distortion of peak shape. The instrument manual is the best source of information on positioning the column in the detector.

### 8.5.4 ECD Temperature

Most $^{63}Ni$ detectors have a maximum temperature limit of 350–400 °C. They are usually operated in the higher temperature range. ECD temperature affects sensitivity. Sensitivity usually increases with an increase in the ECD temperature; however, there are exceptions with carbonyl compounds being the most common examples. Using higher ECD temperatures has the added benefit of reducing detector contamination caused by the condensation of high boiling sample components.

### 8.5.5 ECD Selectivity

ECD's are selective for halogenated compounds, nitrates, conjugated carbonyls and certain organometallic compounds. Poor response is obtained for hydrocarbons, alcohols, amines, aliphatic substituted aldehydes, thioethers and aromatics. The best selectivity is for halogenated compounds. Table 8-1 lists

**Table 8-1** Relative response of an ECD.

| Compound | Relative response compared to benzene |
|---|---|
| Benzene | 1 |
| Acetone | 8 |
| Di-*n*-butylether | 10 |
| Methyl butyrate | 15 |
| 1-Butanol | 17 |
| | |
| 1-Chlorobutane | 17 |
| 1,4-Dichlorobutane | 250 |
| 1,1-Dichlorobutane | 1250 |
| 1-Bromobutane | $5 \cdot 10^3$ |
| 1-Iodobutane | $2 \cdot 10^6$ |
| | |
| Chlorobenzene | 1250 |
| Bromobenzene | 7500 |
| | |
| Chloroform | $1 \cdot 10^6$ |
| Carbon tetrachloride | $7 \cdot 10^6$ |
| Chlorotrifluoromethane | 100 |
| Dichlorotrodifluoromethane | $9 \cdot 10^5$ |
| Trichlorofluoromethane | $4 \cdot 10^7$ |

some relative responses for a variety of compounds. ECD's are primarily used for the analysis of halogenated compounds due to their large selectivity over most other compounds.

### 8.5.6 ECD Sensitivity and Linear Range

The practical minimum detectable quantity is 0.1–10 pg for multiply substituted halogenated compounds, 10–100 pg for nitrates and mono halogenated compounds and 50–250 pg for carbonyl compounds. Halogen response increases in the order of I > Br > Cl > F. ECD response increases with multiple halogen substitution especially on the same carbon atom. The sensitivity trends concerning the type, number and position of the halogens are readily evident in Table 8-1.

The ECD is a concentration dependent detector meaning sensitivity decreases at higher flow rates of carrier, makeup and auxiliary gases. It appears that low gas flows are preferred to maximize sensitivity. To obtain desired sensitivities, ECD's require a large cell volume and a high total gas flow to effectively sweep the cell. Optimal sensitivity and minimal peak broadening can not be achieved at the same flow rates; therefore, a compromise must be reached. High flow rates that sacrifice some sensitivity are used to maintain narrow peak widths.

The linear range of $^{63}$Ni ECD's is poor and barely exceeds $10^3$ for some compounds. Argon/methane as the auxiliary gas results in a slightly better linear range than with nitrogen. This linear range gain is made at the sacrifice of sensitivity. Nitrogen provides about two times the sensitivity obtained with argon/methane. It is very easy to exceed the linear range for an ECD. Five point calibration curves are recommended since samples outside the linear region of the ECD's response curve are fairly common.

### 8.5.7 ECD Maintenance

ECD's require little maintenance. Due to the slightly radioactive source, the cell is sealed and is not directly accessible. ECD's have a lifetime of 2–5 years. After that period, the detector has to be returned to the GC manufacturer for rebuilding or replacement. Baseline or sensitivity problems are the most common symptoms of an old ECD. The types of samples injected and the amount of oxygen exposure affects ECD life. Lower oxygen exposure and cleaner samples will prolong ECD life. Sending the ECD back is expensive and requires days of GC downtime, thus it should only be done when necessary. Cleaning the ECD should be attempted before returning it for reconditioning. Even if the detector is several years old, a thorough cleaning may extend its usable lifetime by 3–12 months. A hydrogen bakeout is the most common method and is suitable for all $^{63}$Ni ECD's. Hydrogen is supplied to the detector cell either as the carrier gas (at a high flow rate) or as a substitute for the auxiliary gas. The column does not have to be removed prior to a hydrogen cell bakeout. Heat the detector to 350 °C and leave the detector at this temperature for about 24 hours. Do not leave the detector in this configuration for more than 24 hours. After the hydrogen bakeout, return the GC to its normal operating state. After several hours, the ECD can be evaluated to determine the success of the bakeout. A more drastic cleaning method may be required if ECD performance has not returned.

The steam cleaning method appears to contradict some of the general guidelines about ECD's. There is disagreement between individuals whether this method should be used. There are numerous cases of very successful implementation of steam cleaning ECD cells. If there is any doubt, the GC manufacturer should be contacted before proceeding. A hydrogen bakeout should be attempted before a steam cleaning is undertaken. The procedure is simple, but it requires 1–2 days before the ECD is suitable for use. Remove the capillary column from the GC and install a short length of deactivated fused silica tubing or old column. Establish a low rate of carrier gas flow through the tubing. Adjust the ECD temperature to 350 °C while maintaining the usual detector gas flows. Set the injector temperature above 200 °C and the oven temperature above 150 °C. Make 50–100 injections of 10–50 μL of deionized water into the GC. An autosampler is ideal for this procedure. The vaporized water steam cleans the ECD cell. No radioactive materials are removed from the cell. The ECD requires about one day before its background signal stabilizes. Steam cleaning followed by the hydrogen bakeout results in the cleanest detector. Several cleanings are usually possible before the detector has to be returned for reconditioning. The time interval between cleanings decreases as the detector approaches complete failure. Eventually detector performance no longer improves upon cleaning and reconditioning is necessary.

Most ECD's have a swab kit available from the manufacturer to check for leakage of source materials. Areas around the detector are wiped with a moist swab, sealed in a bag and sent to the GC manufacturer. The swabs are checked for radioactivity to make sure the ECD cell is not leaking source materials. This is a rare occurrence, but should be checked. If excessively high levels are found, the GC manufacturer will advise on the proper methods to handle the situation.

## 8.5.8 Common Problems with an ECD

### 8.5.8.1 Change in ECD Sensitivity

ECD response changes with temperature. The effect of a temperature change on the response depends on the compound, the original detector temperature and the direction of the temperature change. Most ECD's are very highly affected by the total flow of gases through the cell. Even a small change in the auxiliary or makeup gas flows causes a noticeable shift in the sensitivity. Changes in the carrier gas flow have a smaller impact. The affect of a flow change on response depends on the compound, the original flow rate, which gas has changed and the direction of the flow change.

A change in the amount of oxygen or water in the gases (carrier, auxiliary or makeup) shifts ECD response. In general, increases in water and oxygen decreases the ECD response; however, under specific conditions oxygen enhances the response of some compounds. Water and oxygen present at levels as low as 10 ppm affects detector sensitivity. Water and oxygen traps are usually installed on the detector gas lines.

As an ECD cell becomes dirty, response (i.e., peak size) may increase. This increase in response is always accompanied by an equal or greater increase in noise. The signal to noise ratio has not increased; therefore, the actual sensitivity of the detector has not improved and may have actually decreased. The response for each compound may not be affected equally by the condition of the cell.

#### 8.5.8.2 Peak Shape Problems Attributed to the ECD

Improper column installation or too low makeup gas flow are the most common source of broad or tailing peaks caused by the ECD. If the column is positioned too low (i.e., below the cell entrance), there is excessive dead volume. If a column is forced too far into the detector, a broken column end may result. It is possible to insert the column into the cell for some ECD's. Having the column a small distance into the cell usually does not cause any peak shape problems; however, a sensitivity problem may occur. For all of these column position problems, tailing or broad peaks may result, with the problem being more severe for very volatile compounds.

Too low or the lack of makeup or auxiliary gas flows may cause tailing or broad peaks. ECD cells have large volumes and require high makeup or auxiliary gas flows to minimize dead volume problems. Usually a large change in the sensitivity occurs with excessively low gas flows. Depending on the original gas flow, no change or a gain in the sensitivity is observed with the loss of peak shape.

#### 8.5.8.3 ECD Baseline Problems

A small amount of baseline drift may be evident as the carrier gas flow rate decreases during the course of a temperature program. The flow through the ECD cell changes, and the baseline therefore drifts corresponding to the temperature program. Using a pressure programmable injector in the constant flow mode to maintain a constant flow throughout the temperature program may help (Section 7.7). Detectors equipped with electronic flow controllers can be programmed to increase the makeup gas flow rate to compensate for a decrease in carrier gas flow rate. This may also helps to minimize the amount of baseline drift.

If the stationary phase contains a species that elicits an ECD response, normal column bleed is evident as a baseline rise. The rise should only occur as the temperature approaches the upper temperature limit of the column. Detector temperature fluctuations results in baseline drift as the temperature changes. A malfunction in the detector heating unit or sensor is the primary cause of a consistent temperature drift.

Water in the injected samples can cause appreciable noise problems which may last for some time due to adsorption effects within the detector. The baseline problem occurs around the solvent front or in the retention time region where water elutes from the column. The water peak is very broad and irregular for many columns with nonpolar ones being the worse. Depending on the detector, sensitivity settings and the amount of water, the disturbance may last only 1–2 minutes or up to 5–10 minutes. A large amount of a halogenated compound may also cause a substantial baseline disturbance.

#### 8.5.8.4 Negative Peaks with an ECD

In most cases, negative peaks are a sign of a highly contaminated detector or a detector that needs reconditioning. If the negative peaks appear after some of the positive peaks in the chromatogram, a dirty cell is the most likely culprit. Usually the size of the negative peak corresponds with the size of the positive peak. If the negative peaks are random or in the same location away from positive peaks, a contaminated cell may not be responsible.

There are electropositive compounds that can contribute electrons to the ECD cell. This increases the current in the cell, thus generating a negative signal. If the negative peak is always in the same location relative to the other peaks and its size corresponds to the amount of sample injected, there may be an electropositive compound in the sample. Compound causing negative peaks with an ECD are relatively rare. In some cases, a high concentration of a compound that does not produce an ECD response can affect the detector. The most common example are hydrocarbons. At very high levels, hydrocarbons in the sample can produce a quenching effect and are sometimes observed as negative peaks.

Negative peaks with an ECD are not unusual. A small negative peak immediately after the solvent front is fairly common. As long as the negative peaks do not interfere with any peaks of interest and do not seem to be related to the age or cleanliness of the ECD, they are not of major concern.

#### 8.5.8.5 ECD Linear Range Problems

It is quite easy to exceed the linear range of an ECD. The $10^3$ range is relatively small, thus the potential concentrations of many samples fall outside the linear range. Dilution of the too concentrated samples and re-injection is sometimes necessary. Calibrations curves with a large number of points will also help to accurately quantitate samples in the nonlinear range of the response curve. Good calibration curves eliminate the need to dilute and re-inject samples.

One of the first signs of a contaminated (dirty) ECD cell is a decrease in the linear range. If frequent or large, multiple point calibration curves are not generated, the loss of linear range may not be noticed until the errors in the quantitative results become quite severe. As ECD cells become older, the linear range starts to decrease. Eventually, it will become too small to use for most applications.

#### 8.5.8.6 Miscellaneous Problems with an ECD

Due to the sensitivity of ECD's to halogens, background contamination can be a problem. High volume use of volatile halogenated solvents such as Freons or methylene chloride in the vicinity of the GC can be detected as higher backgrounds. Sometimes the ECD background signal shifts, depending on the frequency of use, time of the last usage, and the volume used of the solvent. The airborne solvent vapors may contaminate vials, pipets, solvents, or anything else coming in contact with the samples. The high sensitivity of the ECD can respond to even very low levels of contamination. Depending on the column and GC operating conditions, a peak representing the contaminating compound may be seen.

## 8.6 Thermal Conductivity Detector (TCD)

### 8.6.1 TCD Principle of Operation

A simplified drawing of a generic TCD is shown in Figure 8-6. The detector consists of two heated cells. The carrier gas from the column flows through the sample cell; the reference cell is supplied with the same type and flow of gas that flows through the column. For most TCD's, each cell contains two interconnected filaments. The same amount of a controlled current is applied to each filament. Each filament establishes an equilibrium temperature if the current and gas flow rate across the filament remain constant. Gas flowing across the filament cools it to some extent by absorbing some of the heat. The amount of heat loss is a function of the thermal conductivity of the gas flowing through the cell. The ability of the gas to conduct heat away from the filament changes if the composition of the gas changes. The gas flowing through the reference cell does not change, thus

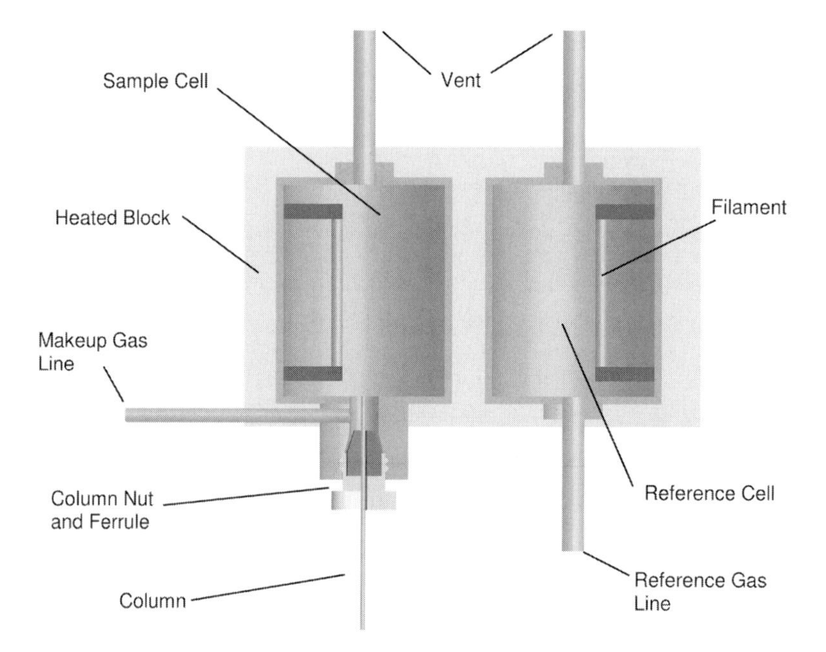


**Figure 8-6** Major components of a dual cell TCD.

the filament temperature remains constant. The gas flowing through the sample cell can change since it may contain compounds eluting from the column. If compounds pass through the sample cell, its filament changes in temperature since the composition of the gas changes. The resulting change in the sample cell filament temperature is compared to that of the reference cell filament temperature. The difference between the filament temperatures is converted into an electronic signal. This signal is processed by the recorder into a corresponding peak in the chromatogram. Greater amounts of the compound generate a larger signal, thus giving a larger peak.

Most TCD's were designed for use with packed columns. Their large cell volumes are not well suited for use with capillary columns. Large volumes of makeup gas are necessary, which reduces the sensitivity. A newer TCD designed for capillary columns utilizes a single cell and filament. The gas flow to the cell is very rapidly switched between the column effluent and reference gas. While the carrier gas from the column is flowing through the cell, it acts as the sample cell; while the reference gas is flowing through the cell, it acts as the reference cell. The difference in the sample and reference periods are measured and compared to generate the final signal. This design allows the use of a very small cell volume, and thus lower total gas flows. This increases sensitivity over the much larger cell volume TCD's.

### 8.6.2 TCD Gases

Makeup gas is usually required for optimal performance even with low volume cell TCD's. Gas flow rates are critical to the sensitivity of a TCD. The best sensitivity occurs at lower flow rates. Excellent results are obtained when the total gas flow is maintained at a minimum of about 5 mL/min. This ensures a rapid sweeping of the outside of the column and eliminates any dead volume problems in the detector. For most TCD's, flow rates below 5 mL/min result in tailing or broad peaks. Usually, a total flow rate of 5–20 mL/min for the carrier plus the makeup gas is used.

Hydrogen carrier gas provides the best sensitivities; however, its use may result in sensitivity changes as the filament ages. There is also the possibility of reactivity problems of the compounds with the hot filament in a hydrogen environment. Helium is the recommended carrier and makeup gas; it has sensitivities close to hydrogen without the possible reactivity problems. Nitrogen gives lower responses, but is useful if helium or hydrogen is being analyzed. A split peak may be obtained when trying to analyze hydrogen using helium as the carrier gas. A special mixture of 8.5% hydrogen in helium is recommended for analyzing hydrogen and minimizes the nonlinear response of TCD's at low levels of hydrogen.

### 8.6.3 Column Position in a TCD

The position of the column relative to the detector cell is important for optimal sensitivity and peak shape. The end of the capillary column must be at the entrance of the sample cell. Placement of the column end in the cell results in poor peak shapes and poor sensitivities. Placement of the column end too far below the cell entrance may result in broad and tailing peaks (dead volume) and/or the loss (absorption) of active sample components. The GC's instruction manual is the best source for the proper positioning of the column in the detector.

### 8.6.4 TCD Temperature

The temperature of the detector is very important. Sensitivity increases as the temperature difference between the cell and the filament increases. It is advantageous to keep the cell temperature as low as possible. It must be hot enough to minimize temperature fluctuations from the outside air and to prevent condensation of high boiling sample components. Temperatures in the range of 150–250 °C are normally used. Lower filament currents require the use of lower detector temperatures to maintain good sensitivity.

### 8.6.5 TCD Selectivity

TCD's are universal in their response. Any compound in the carrier gas elicits a response from a TCD. For this reason, the TCD is used in applications where a FID is not suitable. Analysis of permanent gases ($O_2$, $H_2$, $N_2$, He, Ar, Xe, etc.), oxides of nitrogen (NO, $NO_2$, $N_2O$, $N_2O_2$), sulfur compounds ($H_2S$, $SO_2$, $CS_2$, COS, etc.), $CO_2$, CO, and water are typical applications for a TCD.

### 8.6.6 TCD Sensitivity and Linear Range

The minimum detectable quantity is 5–50 ng of compound. The primary drawback of a TCD is its lower sensitivity. The FID has about 10–100 times greater overall sensitivity, a greater linear range and more reliable signal for quantitation purposes. For these reasons, the FID is a more popular detector. The TCD responds to compounds that are not detectable with a FID (water, $CO_2$, CO, some sulfur compounds, oxides of nitrogen, and permanent gases). The $C_1$–$C_3$ hydrocarbons may give a better response with a TCD, and especially the low cell volume designs.

Sensitivity is increased as the temperature difference between the filament and the cell becomes larger. There are two methods to control the temperature difference. The easiest and most reliable method is to increase the filament current. The additional current increases the temperature of the filament. If the current is too high, noise may develop and there is the risk of premature burning out of the filament. The second method is to decrease the temperature of the cell (detector). The temperature must be hot enough to minimize temperature fluctuations caused by changes in the outside air. Also, the temperature should not be so low that high boiling sample components condense in the detector. Usually the filament current is altered since it is the easiest and quickest method to alter the temperature difference between the filament and cell. Experimentation with the amount of current to maximize sensitivity is suggested, but sensitivity criteria must be balanced against reduced filament lifetime and high noise at high current levels. The GC instruction manual provided some current settings as a starting point.

The TCD is a concentration dependent detector meaning sensitivity decreased at higher flow rates of carrier and makeup gases. Lower gas flows are preferred to maximize sensitivity, but the flow needs to be high enough to effectively sweep the cell. In many cases, optimal sensitivity and minimal peak broadening can not be achieved at the same flow rates; therefore, a compromise must be reached. Higher flow rates that sacrifice some sensitivity are used to maintain narrow peak widths. Another cause of reduced sensitivity is the lower filament temperature caused by high gas flows. This reduces the temperature difference between the filament and cell which reduces sensitivity.

The linear range of most TCD's is about $10^5$. This is similar to many detectors, but smaller than the FID. The TCD may be better for some sulfur compound analyses than some sulfur selective detectors. Specifically, the FPD (Section 9.7) is very selective and sensitive for sulfur compounds, but it is not linear in the sulfur mode. When samples are over a wide concentration range and detectable by a TCD, the TCD may be a better detector. The nonlinear response of the FPD makes quantitation difficult and requires the use of many calibration standards.

### 8.6.7
### TCD Maintenance

The primary maintenance for a TCD involves changing the filament. Most procedures involve improving filament life or keep the filament from becoming damaged or contaminated. A constant presence of oxygen can permanently damage filaments through oxidative processes. The most common source of oxygen is elevated levels in the carrier or makeup gas or a leak near the detector. Oxygen traps are recommended for the carrier and makeup gases to reduce oxygen levels. Proper column installation techniques and regular leak checks (especially after column installation) help to keep leak problems to a minimum. The damage caused by oxygen is more severe at high filament currents. Chemically active sample components such as acids and halogenated compounds may attack and damage the filaments. Avoiding these compounds when possible increases filament life. Turning off or substantially reducing the filament current when the TCD is not in use prolongs filament life.

Increased filament lifetime will result if the following start up scheme is used. Purge the detector with carrier and makeup gas for 10–15 minutes before turning on the filaments. This prevents oxidation of the filaments due to the presence of oxygen (air) that has diffused into the cell under no flow conditions. Turn on the filaments at the lowest possible current setting then increase the filament current in several increments to the desired value. This reduces the power surge upon current introduction and prolongs filament lifetime.

Cell contamination is a problem when lower a detector temperature is used to improve sensitivity. Also, low filament currents promote contamination since the filament temperature are lower at reduced currents. If the cell becomes contaminated, a solvent flush of the detector may help to remove the condensed material. It is important to check with the GC manufacturer to confirm whether the TCD can tolerate a solvent flush. Cool the cell to room temperature and remove the column. Place a septum in a nut or fitting assembly that fits onto the detector entrance. Place the nut or assembly on the detector fitting and tighten. Verify the presence of makeup gas flow. Inject 20–100 μL volumes of toluene or benzene into the detector through the septum. Inject a total volume of at least 1 mL of solvent. Do not inject halogenated solvents such as methylene chloride and chloroform into the detector. After the final injection, allow makeup gas to flow for 10 minutes or more. Slowly raise the temperature of the cell to 20–30 °C above the normal operating temperature. After 30 minutes, decrease the temperature to the normal value and install the column as usual.

## 8.6.8 Common Problems with a TCD

### 8.6.8.1 Change in TCD Sensitivity

Sensitivity changes can usually be attributed to the filament. Any event or situation that changes the temperature of the filament affects sensitivity. On the most basic level, a change in the current setting changes the sensitivity (higher current = higher sensitivity). If the filament has be altered, a different current may be necessary to maintain a specific sensitivity. As filaments become older, the current necessary to obtain a specific sensitivity will change. Contaminated or damaged filaments also affect sensitivity. Contamination is a greater problem with low filament currents and detector temperatures. Obviously, injection of dirty samples or ones containing high molecular weight compounds will contaminate the cell or filament much quicker. High oxygen levels or samples containing high levels of halogenated or acidic compounds damage the filament.

Temperature and flow changes also affect sensitivity. TCD response changes with detector temperature. A lower detector temperature increases sensitivity. Changes in the makeup gas flows cause a noticeable shift in the sensitivity. Changes in the carrier gas flow have a smaller impact. Lower flows increase sensitivity, but broader peaks are usually evident at lower flow rates.

Leaks in the area of the detector can cause flow fluctuations and temperature variations. The change in the temperature and/or flow alters sensitivity. Excessive baseline noise or short filament life usually accompanies a leak in the detector.

#### 8.6.8.2 Peak Shape Problems Attributed to the TCD

Improper column installation or too low makeup gas flow are the most common source of broad or tailing peaks caused by the TCD. If the column is positioned too low (i.e., below the cell entrance), there is excessive dead volume. If a column is forced too far into the detector, a broken column end may result. It is possible to insert the column into the cell for some TCD's. Having the column a small distance into the cell usually does not cause any peak shape problems; however, a sensitivity problem may occur. For all of these column position problems, tailing or broad peaks may result with the problem being more severe for very volatile compounds.

Too low or the lack of makeup gas flows may cause tailing or broad peaks. Most TCD cells have large volumes, and thus require high makeup gas flows to minimize the dead volume. Usually a large increase in sensitivity occurs with excessively low gas flows.

#### 8.6.8.3 TCD Baseline Problems

A small amount of baseline drift may be evident as the carrier gas flow rate decreases during the course of a temperature program. The flow through the TCD cell changes, and the baseline usually drifts corresponding to the temperature program. Using a pressure programmable injector in the constant flow mode to maintain a constant flow throughout the temperature program may help (Section 7.7). Detectors equipped with electronic flow controllers can be programmed to increase the makeup gas flow rate to compensate for a decrease in carrier gas flow rate. This may also helps to minimize the amount of baseline drift.

Depending on the sensitivity setting and the column, normal column bleed is evident as a baseline rise. The rise should only occur as the temperature approaches the upper temperature limit of the column. Detector temperature fluctuations result in baseline drift as the temperature changes. A malfunction in the detector heating unit or sensor is the primary cause of an inconsistent temperature drift.

Another cause of baseline drift may be the condensation of high boiling compounds on the filaments. This is a greater problem for samples containing high molecular weight compounds or when using low detector temperatures or filament currents. Temporarily increasing the filament current to volatilize the contaminants may help; however, excessive times at high currents decrease filament life.

If the baseline suddenly goes off scale and the signal is high, regardless of the instrument and recorder sensitivity settings, the filament is probably burnt out and needs to be replaced. If the signal gradually drifts too high, a burnt out filament is not the problem. A contamination, temperature or flow problem is the probable cause.

#### 8.6.8.4 Negative Peaks with a TCD

When mixed with the carrier gas, some compounds decrease the thermal conductivity of the gas flowing through the TCD cell. This increases the heat of the filament and result in a negative signal. A negative instead of positive peak occurs. Negative peaks for a TCD is not unusual. If all of the peaks are negative, check the polarity of the detector and recorder to ensure proper connection.

#### 8.6.8.5 Short TCD Filament Lifetimes

Short filament life is normally caused by excessively high filament currents. Decreasing the current prolongs filament life, but decreases sensitivity. If very high filament currents are necessary to detect the sample compounds, placing more sample on the column (if possible) or switching to a more sensitive detector may be necessary.

Oxygen damages TCD filaments especially at high filament currents. A leak in the GC system is the most common source of oxygen. Acids and halogenated compounds may attack and damage the filaments. Using hydrogen as the carrier gas usually accelerates chemical or oxygen damage.

## 8.7 Flame Photometric Detector (FPD)

### 8.7.1 FPD Principle of Operation

A simplified drawing of a FPD is shown in Figure 8-7. Carrier gas exiting the column is mixed with oxygen (air) then introduced into a hydrogen flame via a jet. Compounds in the carrier gas are burned in the flame. The sulfur compounds form $S_2$ and phosphorus compounds form HPO. Under the conditions of the combustion flame, $S_2$ emits light with a maximum around 394 nm and HPO with a maximum of around 526 nm. In the sulfur mode, a filter is used to screen out all wavelengths except 394 nm. In the phosphorus mode, a filter is used to screen out all wavelengths except 526 nm. Only the specific wavelength corresponding to the sulfur or phosphorus species reaches the photomultiplier tube (PMT). Some FPD's contain two filters for the simultaneous monitoring of both wavelengths. Upon stimulation by the transmitted light, a current is generated by the PMT. This signal is processed by the recorder into a corresponding peak in the chromatogram. Greater amounts of the compound generates a larger signal, thus giving a larger peak.

There are dual burner FPD's that use two flames. The first flame is for decomposing the compounds and the second flame for formation of the light emitting species. Two different air supplies (flows) are used to maintain the flames. This designs minimizes the quenching problem of FPD's (Section 8.7.6). Also, this design prevents large amounts of solvent from extinguishing the flame (solvent flameout) that occurs with some single burner FPD's. The filter and PMT portions of a dual burner FPD are the same as a single burner design.

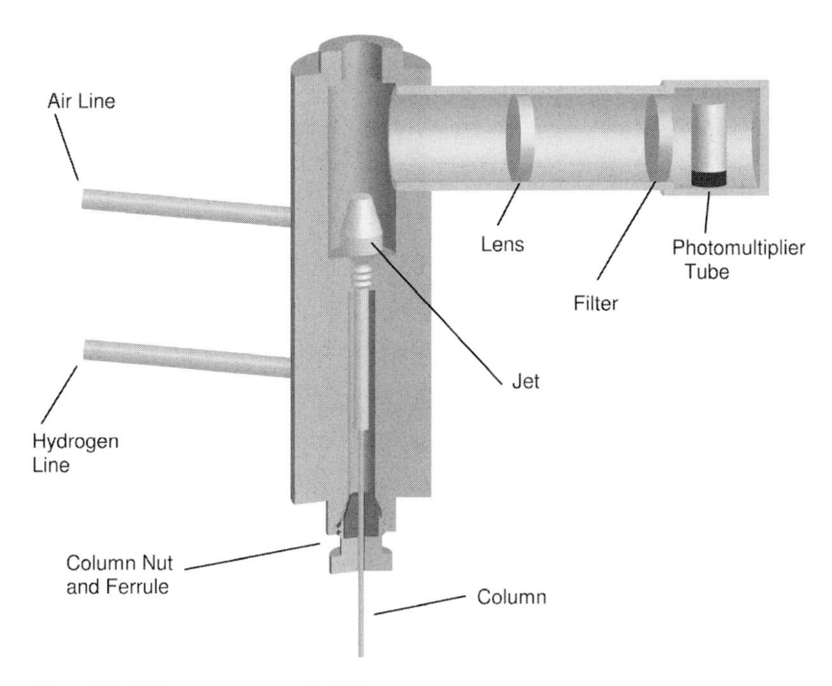


**Figure 8-7** Major components of a FPD.

### 8.7.2 FPD Gases

Hydrogen and air are used as the combustion gases. There are different detector gas flows for the sulfur and phosphorus modes. The actual flows are found in the GC instruction manual. They may vary between the different manufacturers due to differences in the designs. For the sulfur mode, hydrogen flows of 60–80 mL/min and air flows of 80–100 mL/min are common. The flows are usually set so that the ratio of oxygen (in the air) to hydrogen is 0.2–0.3. For the phosphorus mode, hydrogen flows of 120–170 mL/min and air flows of 100–150 mL/min are common. Sensitivity changes may result if the hydrogen and air ratios are changed.

Makeup gas of nitrogen enhances the sensitivity of the FPD especially in the sulfur mode. The nitrogen gas flow should be 50–100 mL/min, but this value is not extremely critical. Actual flow values can be found in the GC's instruction manual. Makeup gas flow rate affects sensitivity, thus a correct and consistent flow is important for the best performance.

The choice of carrier gas is not critical. Using hydrogen or nitrogen as the carrier for wide diameter columns may result in slight response changes or baseline drift during a temperature program. Usually the changes are not noticeable.

### 8.7.3
### Column Position in a FPD

The position of the column within the detector is critical for FPD's. Most sulfur and phosphorus containing compounds are very active and are readily absorbed by non-inert surfaces. The end of the column should be at the tip of the jet to minimize dead volume and compound contact with active surfaces. If the column end is below the tip of the jet, broad or tailing peaks may result due to dead volume effects. Adsorption of active compounds may also occur with tailing or reduced size peaks. It is easy with some FPD's to insert the column into the flame. The column end will burn in the flame and excessive noise and baseline problems occur.

### 8.7.4
### FPD Temperature

The PMT is positioned close to the burner assembly. This limits the detector temperature to 150–275 °C. PMT noise increases with increasing detector temperature and sensitivity decreases with increasing detector temperature especially in the sulfur mode. For these reasons, lower detector temperatures are normally used. High molecular weight compounds may condense in the detector at the lower temperatures, thus creating a problem with detector contamination. Areas such as the filter are particularly susceptible to contamination and have a significant impact on sensitivity. A balance between sensitivity and detector contamination has to be reached; however, it only becomes an issue when analyzing higher molecular weight compounds. Some FPD's have separate temperature control of the detector body and base. The temperature of the detector base is less critical. Even for the analysis of very volatile compounds, a detector body temperature of at least 125 °C is recommended so that water condensation does not occur.

### 8.7.5
### FPD Selectivity

The selectivity of a FPD for sulfur and phosphorus over hydrocarbons is $10^4$–$10^5$. Due to the overlap in the emission wavelengths, the selectivity of phosphorus over sulfur is in the range of 5–10. This results in the possible detection of high concentrations of sulfur compounds in the phosphorus mode. The selectivity of sulfur over phosphorus is $10^3$–$10^4$, thus there is practically no direct interference from phosphorus compounds in the sulfur mode.

### 8.7.6
### FPD Sensitivity and Linear Range

The minimum detectable quantity is 10–100 pg of sulfur compounds and 1–10 pg for phosphorus compounds. Increased sensitivity occurs at lower detector temperatures; however, contamination of the detector increases at lower temperatures. Changes to the hydrogen, air or makeup gas flow rates affect the sensitivity. The gas, amount and direction of the flow change, and the compound determine the amount of the sensitivity shift. If hydrogen is used as the carrier gas, a large change in its flow affects sensitivity. Hydrogen flow to the flame changes, thus the sensitivity also changes.

The presence of non-sulfur and non-phosphorus compounds in the sample can present a problem. If one of these compound even partially co-elutes with a sulfur or phosphorus compound, the response of the compound can be reduced. This effect is called quenching. While any compound can cause quenching, hydrocarbons are particularly effective at reducing FPD response. Sulfur compounds seem to be more affected than phosphorus compounds. Usually a 10–50 fold excess of the quenching compound significantly reduces the response. The greater the concentration, the greater the amount of quenching. In many cases, there is a complete loss of FPD response with complete co-elution of a quenching compound.

The linear range of a FPD for phosphorus is $10^3$–$10^5$. FPD responses are nonlinear for sulfur. Values up to $10^2$–$10^3$ sometimes are reported, but this range is very dependent on the operating conditions, detector design and especially the compound. Calibration curves for FPD's are usually obtained by plotting two times the log of the response versus the log of the sulfur concentration. Errors can still be introduced with this approach. In any case, multiple point calibration curves over the entire possible concentration range of each compound are needed.

### 8.7.7
### Verifying Flame Ignition of a FPD

The signal output as displayed on the data signal or GC display can be used to verify flame ignition. Check the signal prior to lighting the FPD. The signal should be higher when the flame is burning. By comparing the two values, it is easy to determine if the flame is lit. Different GC's and data systems report detector output in different units or values. This is not a problem since only a comparison between the lit and unlit detectors are needed.

To physically verify that the flame is lit, place a cool, shiny object (e.g. wrench, knife blade, etc.) over the vent of the detector. After several seconds, condensation should be visible on the shiny object after it is removed from the vent. If no condensation is visible, the detector flame may not be lit.

### 8.7.8
### FPD Maintenance

FPD's require minimal maintenance to keep them performing at satisfactory levels. The hydrogen, air and makeup gas flows should be occasionally measured. They can drift over time or be unintentionally changed without knowledge of it occurring. Each gas flow should be measured independently to obtain the most accurate values.

The FPD requires periodic cleaning. In most cases, this only involves the jet and less frequently, the filter. Cleaning kits are available containing brushes and wires that simplify the cleaning of all of the detector parts. The brushes are used to dislodge particulates clinging the metal surfaces. A fine wire is used to clean the jet opening of particles. Do not force too large a wire or probe into the jet opening or the opening may become distorted. A loss of sensitivity, poor peak shape and/or lighting difficulties may result if the opening is deformed. The filter or any of the window parts should be handled gently. Scratches or other surface deformities reduce the amount of light passing through the filter, thus reducing response. The filter and related parts should be clean and free from fingerprints.

The PMT need periodic replacement. High detector temperatures reduces the PMT life. When not in use, turn off the PMT to maximize its usable lifetime. Some PMT's may have a shelf life and should not be stored for prolonged periods before use.

## 8.7.9 Common Problems with a FPD

### 8.7.9.1 Change in FPD Sensitivity

A change in detector temperature affects sensitivity. FPD response decreases with an increase in the detector temperature. Sensitivity is dependent on the flow rates of hydrogen, air and the makeup gas. Changes in the makeup gas flows cause a noticeable shift in the sensitivity; changes in the carrier gas flow have a smaller impact. Lower makeup gas flows increase sensitivity, but broader peaks are usually evident at lower flow rates. Leaks in the detector can cause flow fluctuations and temperature variations. The change in the temperature and/or flow alters sensitivity.

Quenching is another cause of sensitivity problems. FPD response can be severely reduced if a non-sulfur or non-phosphorus compound partially or fully co-elutes with the analyte of interest. Eliminating or reducing the co-elution increases the peak response. Changing the temperature program or column may remove the co-elution, thus drastically improving sensitivity.

Contamination of the filter window reduces FPD sensitivity. Cleaning or replacement of the window is necessary to correct this problem. As the PMT ages, sensitivity decreases either as reduced peak size or increased noise.

### 8.7.9.2 Peak Shape Problems Attributed to the FPD

Improper column installation in the detector or too low makeup gas flows are the most common source of broad or tailing peaks caused by the detector. If a column is forced too far into the detector, a broken column end may result. The broken column end and resulting debris disrupt the flow at the end of the column. This turbulent flow may cause broad or tailing peaks. When a column is too far away from the tip of the jet, there will be excessive dead volume. Tailing or broad peaks may result with the problem being more severe for very volatile compounds. Too low or nonexistent makeup gas flows often cause tailing or broad peaks. Significant sensitivity loss usually accompanies insufficient makeup gas flows.

#### 8.7.9.3 Loss of FPD Linear Range

The most common cause of linear range problems is a change in the detector gases with the makeup gas being the most influential. The linear range in the sulfur mode is very small. It depends on the operating conditions and the compound. Multiple point calibration curves are essential for FPD's.

#### 8.7.9.4 FPD Flame Frequently Goes Out

Large sample volumes (usually above 2 μL injected) may extinguish the detector flame. This usually occurs when the sample solvent passes through the detector. There is little that can be done to eliminate the problem except to inject a smaller volume. Probably the largest factor concerning solvent flameout is the design of the FPD.

#### 8.7.9.5 Miscellaneous Problems with a FPD

Some FPD's pop loudly upon flame ignition and this is normal. This is due to the high hydrogen flow rates that are used and the large volume of FPD's. A buildup of hydrogen in the detector occurs very rapidly at these high flow rates. Ignition of the flame immediately after the hydrogen gas is turned on reduces the frequency and severity of the popping.

## 8.8 Mass Spectrometers (MS)

### 8.8.1 MS Principle of Operation

A very simplified drawing of a representative quadrupole mass spectrometer (MS) typically used as a GC detector is shown in Figure 8-8. The entire mass spectrometer is maintained under a vacuum of $10^{-4}$–$10^{-6}$ torr by use of diffusion or turbomolecular pumps. A roughing pump may be used to initially lower the pressure from ambient to aid the diffusion or turbomolecular pump. Carrier gas exiting from the column is introduced into the ionization chamber or source via an interface. The source is heated to 150 – 300 °C to prevent condensation of the sample. A filament outside of the source is used to produce a supply of electrons. The electrons enter the source through a small hole near the filament. A negatively charged shield behind the filament directs the electrons towards the hole in the source. Magnets are often used to focus the electrons into a narrow beam. The potential difference between the filament and source controls the energy of the electrons entering the source. Larger potential differences increases the energy of the electrons. A potential difference that imparts an electron energy of 70 eV is used in most cases. When the electrons enter the source, some of them collide

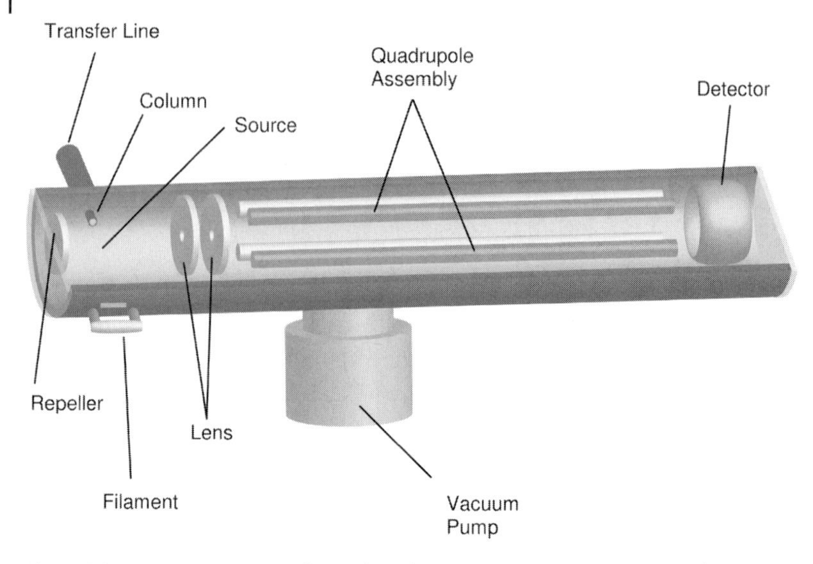


**Figure 8-8** Major components of a quadrupole mass spectrometer. (Not to scale).

with the molecules eluting from the column. The collision imparts energy throughout the molecule which then becomes ionized. If a sufficient amount of energy accumulates in a single bond or group of bonds, the molecule dissociates into smaller, positive ion fragments. The fragments produced (daughter ions) are primarily dependent on the structure of the original molecule (parent ion) and the energy of the electrons. The positively charged fragment ions are ejected from the source by being repulsed by a repeller held at a slightly positive potential. The ions travel through a series of focusing plates which accelerate the ions and focus them in a narrow beam. The focused beam of ion fragments then enters the quadrupole analyzer. The quadrupole is four accurately machined rods set in a square array. Inside of the array a complex field is generated by potentials applied to the rods. The ion fragments entering the quadrupole are set into sinusoidal oscillations. At a particular field, only ions with a specific mass-to-charge (m/z) ratio can pass completely through the quadrupole array without colliding with one of the rods. Any ion that collides with the rods does not reach the detector at the end of the quadrupole array. The field is continuously swept at a very fast rate (usually 2–4 scans per second) to allow ions within a range of m/z ratios to pass through the quadrupole. This allows the quadrupole to act as a mass filter. The detector is at a negative potential so to attract the ions from the quadrupole. Most GC/MS systems use an electron multiplier as a detector which acts to amplify the signal resulting from the impact of the ions. As each ion impacts the detector, its m/z ratio is extrapolated from the value of the potentials applied to the quadrupole rods. The amount of each ion and its m/z ratio measured during one scan is stored in the data system. All of the scans during a run are sequentially combined. The total intensity (abundance) of the ions detected during one scan is plotted versus the start time of the scan to produce a total ion chromatogram (TIC). A background or baseline signal is always present due to the presence of the carrier gas impurities, system contaminants and normal column bleed. Introduction of a compound into the source produces a higher level of ions, thus giving a detector signal above the background. A peak is generated whose size corresponds to the amount of the compound.

## 8.8.2
## Mass Spectral Data

Figure 8-9 shows a TIC. Retention time and peak area data are obtained from a TIC in the same manner as any other chromatogram. The additional information available is the m/z ratio and intensity (abundance) of all the ions detected during one scan anywhere in the chromatogram. The plot of this data is called a mass spectrum (Figure 8-9). The mass spectrum obtained from the peak is for the compound represented by that peak. Each line in the spectrum represents a fragment of the parent molecule. In many cases, the parent molecule does not completely fragment and retains its original structure. This ion is called the molecular ion.

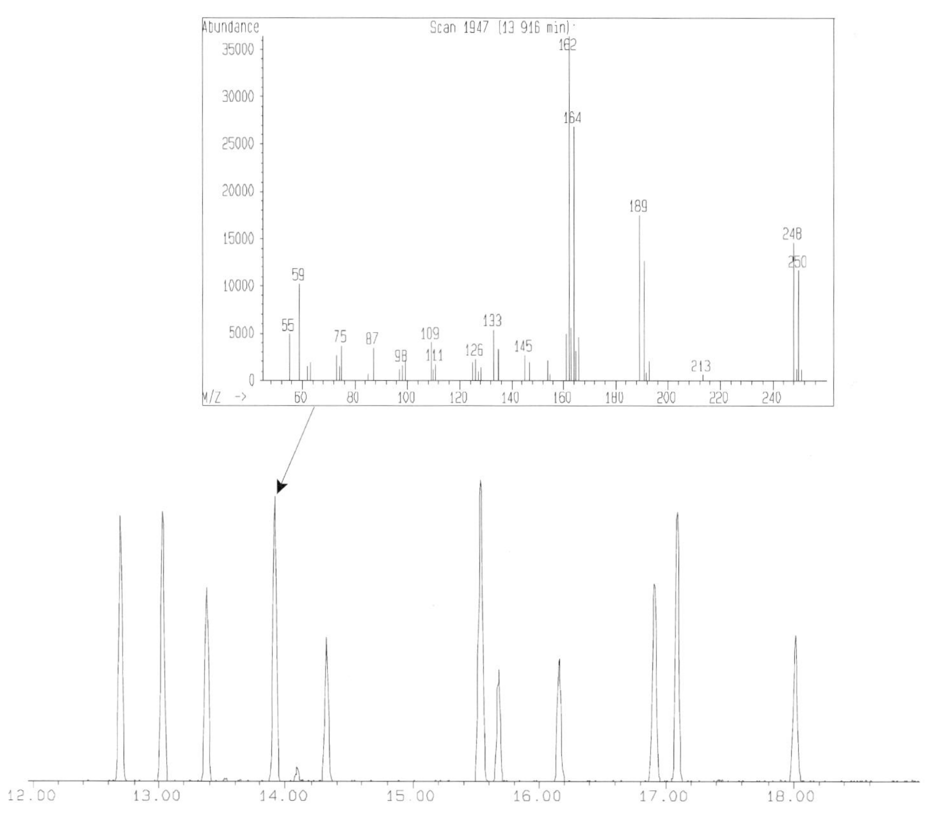


**Figure 8-9** TIC and mass spectrum.

The fragmentation pattern of a given molecule is the same each time providing the electron energy used for fragmentation is the same. The mass spectrum from a peak in the chromatogram can be compared to known mass spectra stored in the computer library or produced from the analysis of analytical standards. If the patterns match, the identity of the compound is confirmed with a high degree of certainty. Along with the retention time data, very high confidence identification of compounds is possible.

The mass spectrum from a single peak representing two co-eluting compounds is a composite of the two spectra. The identification of compounds from their mass spectra is difficult and uncertain for co-eluting peaks. Fortunately, good mass spectra can be obtained from partially co-eluting peaks. The mass spectrum for each compound is obtained from the portion of the peaks that do not overlap.

The practical mass limit for most quadrupole GC/MS systems is 600–800 amu (mass units). Many systems allow scanning at the higher mass ranges, but a loss of many of the higher mass fragments occurs. One of the features of the more expensive quadrupole systems is their higher mass range.

### 8.8.3 Other Ionization, Detection and Mass Filtering Modes

Ionization and fragmentation of molecules using electrons is called election impact or ionization (EI). Another ionization technique is called chemical ionization (CI). Instead of electrons, small gas molecules are used. Methane, butane and ammonia are the most common gases used for CI. CI is considered a softer fragmentation technique. Less fragmentation occurs, thus fewer fragments are produced. Some molecules undergo massive fragmentation with EI and no molecular ions are formed. The large number of low abundance fragments and the lack of a molecular ion makes identification more difficult and uncertain. Some isomeric molecules have very similar EI spectra, but the CI spectra are unique allowing unequivocal identification. Most GC/MS systems require some amount of dismantling and downtime to switch between the two ionization modes. CI is usually an option on most systems, but at additional cost. Coupled with the fact that EI works for most compounds and is simpler, EI is the most common ionization mode.

Instead of an electron multiplier, a few GC/MS systems use a conversion dynode and a photomultiplier tube (PMT). The ions impact the diode and generate secondary electrons that collide with a phosphor disk. This creates photons which are detected and amplified by the PMT. The number of photons hitting the PMT is directly related to the number of ions leaving the quadrupole. PMT's have some advantages over electron multipliers as MS detectors. In general, they provide more sensitivity and much longer lifetimes.

Another design of a benchtop MS system is called the ion trap detector (ITD). The primary difference is the mass filter portion of the unit. Instead of the ions traveling in a semi-linear pattern (i.e., through the quadrupole array), the ions in an ion trap travel in a circular pattern. Ions with the incorrect m/z ratio are ejected from the circular pattern and do not reach the detector. The ion trap collects and detects more of the generated ions than a quadrupole. This makes ion traps about 10–20 times more sensitive than many quadrupole GC/MS systems. Recent improvements in quadrupole design has increased the sensitivity into the same range as an ITD. The higher cost and complexity, sometimes unique EI spectra and concentration dependent spectra makes ion traps less common than quadrupole based GC/MS systems.

### 8.8.4 MS Selectivity

MS selectivity is based on the mass of the fragments. The masses detected and measured are selected by the analyst. If the mass spectrometer scans for all the masses over a broad range, this is called the full scan mode. All of the ions that impact the detector are collected. A full mass spectrum is available for each peak. If the most abundant and unique mass fragments of a compound are known, selected ion monitoring (SIM) can be used. The SIM mode does not scan all of the masses, but only a select few. Only the selected ions are collected and all of the others are ignored. Only the compounds that have one or more of the selected mass fragments are represented by a peak in the chromatogram. Since only a few ions are monitored, obtaining a full mass spectrum or information on non-selected ions is not possible. Most systems allow the selected ions to be changed at pre-selected times during the run. This improves selectivity by limiting response only to the ions expected for the peaks eluting within a particular retention time region.

Usually 2–3 of the most abundant or unique ions for a compound are monitored in the SIM mode. If the masses are carefully chosen, very few compounds have the same mass fragments especially at the same relative abundances. For a known set of compounds, the identity of each compound can often be determined by the presence and abundance of its unique set of ions. Co-eluting peaks present less of a problem in the SIM mode providing each compound has an unique set of ions. By measuring only the ions unique to each compound, the presence and amount of each compound can be determined.

The SIM mode is often used to screen samples. If it appears that a certain compound is present based on the SIM mode information, the sample is re-run using the full scan mode. The complete mass spectrum is more definitive as a means of compound identification. Many isomers and homologs have the same major ions, but the fine detail available only in their full mass spectra can be used to distinguish which compound is actually present.

### 8.8.5 MS Sensitivity and Linear Range

MS sensitivity depends primarily on the scan mode. The sensitivity in the full scan mode is 0.1–1 ng of compound. The sensitivity in the SIM mode is 0.1–10 pg of each compound. Better sensitivity is obtained when the selected ions are substantially more abundant than the non-selected ions. More of these ions are created, thus boosting the number reaching the detector. Extensive fragmentation results in numerous ions, but all of them are at low abundances which lowers the intensity of their detector signal. If too many ions are selected in the SIM mode, the sensitivity advantages are reduced. For analyses with a large number of compounds, the selected ions are changed during the run to the ones corresponding to the peaks expected within a particular retention time range. This minimizes the number of ions selected at any one time and helps to maintain the high sensitivity available with SIM.

Better response is obtained if the electron multiplier voltage is increased. This also increases the noise and no real increase of sensitivity is obtained. Higher voltages are required as the multiplier gets older. Eventually the multiplier needs replacing since the noise becomes too high.

Better sensitivity is obtained at lower carrier gas flow rates. Using lower linear velocities or smaller diameter columns reduces the amount of carrier gas reaching the source. Longer run times and the possibility of an efficiency loss occur when reducing the linear velocity. The major sacrifice with smaller diameter columns is a lower sample capacity.

The linear range of most quadrupole MS systems is $10^4$–$10^5$. The linear range is primarily influenced by the electron multiplier and its operating voltage. The linear range will shrink as the multiplier ages. Linear range is often called the dynamic range for MS systems equipped with electron multipliers.

### 8.8.6
### MS Temperatures

The transfer line, source and quadrupole analyzer are maintained at higher temperatures. This is mainly to prevent compounds from condensing on their surfaces. The transfer line is heated to keep the sample rapidly moving towards the source. Some systems use the heat of the transfer line to conductively heat the source. The source temperature remains 100–150 °C below the transfer line temperature. The transfer line and source temperatures can not be independently controlled, but this rarely causes a problem. Other systems use a separate source heater, thus permitting independent temperature control. Transfer line and source temperatures of 200–300 °C are the most common. Source temperatures does not have a significant affect on sensitivity. It just needs to be high enough to prevent condensation of contaminants on the source. Most systems do not directly heat the quadrupole analyzer. The quadrupole temperature is normally not directly controllable and its temperature is a function of the source temperature.

Systems that use a single heater require longer times to equilibrate after a temperature change. The transfer of heat to the various regions requires more time to occur and stabilize than if the areas were directly heated. Longer equilibration times are needed after periods of non-heating such as during column installation and system maintenance. Excessive background noise, baseline drift and calibration (tuning) difficulties are the usual symptoms of an unequilibrated system.

### 8.8.7 Column Position in a MS

There are several different designs of the interface between the MS source and the capillary column. Regardless of the design, the column effluent is directed into the interior of the source. In some systems, the column is fed through a transfer line until it is positioned in or adjacent to the source. Other systems use a transfer line usually containing a length of deactivated fused silica tubing. The column is connected to the transfer line which directs the column effluent into the source. Either design directs the column effluent into the source.

A column or transfer line that is not properly positioned creates problems. A large amount of dead volume occurs if the column or line end is to far away from the source. Tailing or broad peaks may result due to the dead volume. A loss of peak size may occur due to adsorption of active compounds on hot metal or glass surfaces. It is possible with some interfaces to push the column or line too far into the source. Excessive noise, sensitivity loss or complete failure of the system usually occurs in these situations. Sometimes only a degradation of peak shape is seen.

### 8.8.8 Carrier Gas Flow Rate Considerations for MS Detectors

Higher carrier gas flows reduce the sensitivity of MS systems. The carrier gas inhibits the ionization and fragmentation of the compounds. Larger volumes of carrier gas decrease the sensitivity by reducing the number of ions reaching the detector. The flow limit of most benchtop GC/MS systems is around 2 mL/min. Higher flows can be used, but tremendous sensitivity losses occur. Some newer GC/MS systems are capable of tolerating much higher flows due to larger and improved pump designs.

One method to reduce carrier gas flow is to operate at lower linear velocities. A lower volume of gas reaches the source, but a loss of efficiency may occur and retention times will increase. The best method to reduce carrier gas flow rates is by using smaller diameter columns (0.18–0.25 mm I.D.). Small diameter columns operate at high linear velocities, but low flow rates. An added bonus is the higher efficiency of smaller diameter columns, but longer retention times and lower sample capacities also occur.

If large diameter columns and their high carrier gas flow rates are required with MS detection, a jet separator is usually needed. The column effluent passes through the jet separator before it enters the MS. The jet separator acts to remove the lower molecular weight compounds. Since the carrier gas is the lowest molecular weight compound in the column effluent, large amounts of the carrier gas are removed. A much lower amount of the heavier compounds are removed from the column effluent. This reduces the amount of carrier gas reaching the source without substantially reducing the amount of the compounds. A greater number of light compounds than heavier compounds are lost via the jet separator; thus the jet separator discriminates against the lower molecular weight compounds. Substantial losses of volatile compounds can occur when using a jet separator. Jet separators require a vacuum pump to function. The jet separator and pump adds thousands of dollars to the cost of the GC/MS system in addition to increased maintenance requirements.

### 8.8.9
### MS Maintenance

The GC/MS instruction manual is an essential resource for maintenance schedules and procedures. It contains specific information about the exact model of GC/MS in use. The main areas that need routine maintenance or attention are the vacuum pumps, source, and the tuning parameters.

Most GC/MS system have a roughing or foreline pump. The oil level should be checked every 1–2 weeks. If it is low, fill the pump with the oil recommended by the manufacturer. It is recommended to change the oil about every 6 months. If there are any traps associated with the pump, the trap should be replaced or refilled whenever the pump oil is changed. The main pump on most GC/MS system is a diffusion or turbomolecular pump. If these have any oil or fluids, they need to be checked every 6–12 months. It is advisable to replace the oil or fluid every 12 months.

The ion source becomes fouled with use. High molecular weight materials from the samples and other sources slowly accumulate on the ion source. It requires occasional cleaning as necessary. The manual has precise instructions on removing, disassembly and cleaning the source. The instructions need to be followed exactly or performance problems and/or damage to the MS will occur. The general procedure for source cleaning is polishing of the metal parts with an abrasive slurry. The parts are then rinsed and ultrasonicated in a series of solvents. After thorough drying, the ion source is reassembled. Extra care should be taken to ensure none of the parts have any residues of solvent, abrasive materials, fingerprints, etc. The frequency of source cleaning depends on the samples injected into the GC/MS. Dirtier samples increase the frequency of source cleaning.

Tuning is the process of adjusting the MS parameters to obtain the best performance. Tuning is preformed by introducing a specific compound with well understood properties into the MS. Perfluorotributylamine (PFTBA) is a common tune compound for the EI mode. While the compound is present in the MS, various parameters are adjusted until predetermined performance criteria are met. The MS can be tuned for good performance over a wide mass range for general purpose use or over a limited mass range for specific purposes. The primary performance criteria for most tuning parameters are sensitivity, relative ion abundances and mass assignments. The tune compound and criteria are determined by the analyst, stated in a particular method or as recommended for the GC/MS system. Most GC/MS systems have a variety of built in tuning programs commonly called autotune programs. They vary mainly in the amount of automation and tuning criteria (e.g., high sensitivity, mid-mass). The frequency of tuning depends on a number of factors. Any time a major change is made in the system it will need tuning. A maximum of one week between tunings is usually suggested. Many methods require tuning procedures to be performed more frequently. There is no harm in tuning a MS too much except for the time wasted making unnecessary tuning adjustments.

The filaments and electron multiplier need occasional replacement. Filament life depends on its treatment. In general, having the filament on when large amounts of solvents or oxygen are in the source reduce its life. Electron multipliers have a finite life of 6–24 months. Dirty samples foul the multiplier and unnecessarily high voltage reduce its life.

### 8.8.10 Common Problems with a MS

Due to the greater complexity of mass spectrometers, there are more possible problems that can occur than with standard GC detectors. Also, there are problems unique to the MS detector. The instruction manual is the best source of troubleshooting information. Most of the data systems have diagnosis programs and provide error messages. All of these are useful when trying to diagnosis and solve problems.

#### 8.8.10.1 Change in MS Sensitivity

The source of most sensitivity shifts is the electron multiplier. Either it is dirty or the voltage is too low. Increase the multiplier voltage to determine if the sensitivity increases. If the sensitivity does not increase, look for another source of the problem. As electron multipliers age, they require more voltage to maintain a reasonable sensitivity. Eventually, they reach the maximum voltage allowed and can not be made more sensitive. The multipliers need to be replaced if they reach the maximum operating voltage. Since electron multipliers rarely look dirty, visual inspection can not be used to determine multiplier performance.

A dirty ion source reduces sensitivity. The residues or debris reduces the ionization efficiency. This decreases the number of ions reaching the detector, thus reducing sensitivity. The ion source has to be cleaned using the procedure in the instruction manual. An old or damaged filament decreases sensitivity. Insufficient electrons are produced to ionize and fragment the compound molecules. Replacing the filament or switching to the other filament (if available) is required.

If the vacuum pressure is too high, poor sensitivity results. Usually a leak is the cause of excessively high vacuum pressure. Another cause of high vacuum pressure is when the carrier gas flow is too high. In either case, the number of ions reaching the detector is reduced, thus explaining the loss of sensitivity.

Other possible causes of sensitivity changes are different source temperatures and an incorrect tune. Check both to determine whether they may a source of the sensitivity problem.

#### 8.8.10.2 Excessive Noise or High Background in a MS

Excessive noise or high background can be due to an old electron multiplier. Higher voltages are needed to obtain the desired sensitivity. The higher voltages increase the noise levels. Usually the noise levels increase gradually as the multiplier ages. If the noise level suddenly increased or the multiplier is fairly new, look for other sources of the noise. Usually it will be from an air leak, contamination or a damaged or old filament.

Contamination of the MS or the GC can cause high background levels. Contaminants in the GC can be difficult to diagnose and identify. Most of them originate from the samples and can vary tremendously, thus the masses rarely help to identity the source of the contamination. Sometimes the masses can be used to help identify the possible contaminants. Table 8-2 lists some of the more common masses and sources. If the ion source was recently cleaned, it is common to see residual solvent masses especially if the parts were not thoroughly dried. Diffusion pump fluid usually enters the MS from a sudden or improper venting of the vacuum. Vapors of diffusion pump fluid can enter the MS if there is no carrier gas flow for prolonged periods. Hydrocarbons originate from fingerprints on ion source parts (poor cleaning technique) or foreline/roughing pump oil (foreline pump on with no carrier gas flow and the diffusion pump off). Silicone contamination can originate from many sources. One of the most common sources are the septum and silicon based stationary phases. Many consumer products (e.g., soaps, lotions, deodorants, cleaners) contain a large amount of silicone based materials. Improper handling of columns, injector and MS parts, and column nuts and ferrules with bare or unclean hands often results in silicone or hydrocarbon contamination of GC/MS systems.

The MS adjusts voltages or currents to maintain the desired electron supply from the filament. As filaments age, they have more difficulty in producing a

**Table 8-2** Common contamination masses.

| Substance | m/z |
|---|---|
| Solvents: | |
| • Methanol | 31 |
| • Acetone | 43, 58 |
| • Hexane | 43, 57, 71 |
| • Toluene | 91, 92 |
| Silicones | 73, 147, 207, 221, 281, 295, 355, 429 |
| Diffusion pump fluid | 170, 262, 354, 446 |
| Phthalate plasticizers | 149 |
| Hydrocarbons | masses spaced 14 apart<br>(fingerprints, foreline or roughing pump oil) |

good supply of electrons. Excessive noise may occur when the filament becomes old or if it is damaged.

Small particles of graphite can enter the MS if 100% graphite ferrules are used at the MS interface or transfer line. Even a small amount of graphite entering the MS can cause a large amount of noise and sensitivity loss. Graphite/Vespel ferrules (40/60, 20/80 or 15/85 mixtures) are recommended for GC/MS systems to avoid particulate problems.

#### 8.8.10.3 **Leaks in the MS**

Leaks are normally due to vacuum seals that are not properly seated or damaged. The column nut at the MS interface is one of the most common leak sites. Graphite/Vespel ferrules tend to loosen, especially when new, and develop leaks. Re-tightening the ferrule at the interface or transfer line is almost always necessary since leaks develop after the ferrule becomes heated.

The various seals and O-rings around the MS are the other probable leak locations. Leaks are evident from several different symptoms. The most common are higher than normal vacuum pressure, high background, poor sensitivity and masses characteristic of air which are m/z 18, 28, 32, 40 and 44 (water, nitrogen, oxygen, argon and carbon dioxide, respectively). The most reliable method is to check the abundances of these ions. The abundance of m/z 28 should be less than the abundance of m/z 18. If an air leak is present, the ratio of m/z 28 to m/z 32 is about 5 : 1.

One method to locate a leak is to use an inert gas such as argon. Obtain a small lecture bottle of argon and install a flow regulator. Set up the MS to look for m/z 40. Spray the argon over the areas where a leak may be possible. After spraying each area, wait for 10–20 seconds. If a sudden increase at m/z 40 occurs, the leak is located in the area where the argon was sprayed. Repair the leak and check the other areas in case there are multiple leaks. Tightening the appropriate fittings or replacing the seal or O-ring are the usual repair techniques for leaks.

# 9
# Column Installation

## 9.1
## Importance of a Properly Installed Column

The importance of a properly installed capillary column is under-appreciated in most cases. A poorly cut or installed column can lead to inferior chromatographic results. Also, permanent damage to the column may result if the proper installation precautions are not taken before column conditioning and use. Any gas chromatograph is limited by the weakest portion of the system – column installation is a part of the system just as much as the column or the instrument itself.

## 9.2
## Installing Fused Silica Capillary Columns

### 9.2.1
### Column Installation Steps

The basic column installation steps are the same for practically every column and GC system. Every precaution applies to all columns and situations. Some of the later steps are not necessary to properly install a column; however, they are verification of a proper installation and often prove to be very useful in the event that troubleshooting is necessary at a later date. The basic steps are:

1. Cutting the column
2. Installation in the injector
3. Turning on and verifying the carrier gas flow
4. Installation in the detector
5. Verifying column installation and detector operation
6. Conditioning the column
7. Setting the linear velocity
8. Performing a bleed test
9. Injecting a column test mixture

*The Troubleshooting and Maintenance Guide for Gas Chromatographers, Fourth Edition.* Dean Rood

ISBN: 978-3-527-31373-0

### 9.2.2 Cutting Fused Silica Capillary Columns

For the best performance, it is critical for both ends of a fused silica capillary column to be even and without any burrs, chips or uneven areas. This is accomplished by using the proper technique and tools when cutting the column. A carbide or diamond tipped pencil, a sapphire cleaving tool or ceramic wafer is needed to cut the column. Scissors, files or similar tools are not acceptable cutting tools. The choice of a cutting tool is primarily a personal preference and perhaps a cost issue. The cutting tool should only be used to cut columns and not other items or materials. Regardless of the tool, the basic technique is the same.

The tip of a pencil or the edge of a wafer or sapphire tool are used as the cutting edge. Lightly scribe the polyimide coating with the cutting edge. The goal is to create a weak spot in the polyimide coating that exposes the fused silica tubing. Attempting to cut through the fused silica tubing with the cutting tool results in an uneven and unsatisfactory cut since the column is crushed in this area. Place a short length of the column against a finger or palm. This makes it much easier to feel how much pressure is being applied to the tubing by the cutting tool. The tubing flexes if it is not supported which makes it difficult to control the amount of applied pressure. Lightly drag the cutting tool edge across the column in the desired location. Gasp the column 1–2 cm below the scribe mark. Gently bend the part of the column above the scribe away from the scribe mark. If properly scribed, the column breaks easily and evenly. If the column does not break, re-scribe the column about 1 cm below the previous scribe. Attempt to break the column again making sure it breaks at the point of the new scribe and not at the first one. A small amount of practice is required to consistently cut a fused silica column. After a while it becomes easy and routine; however, the cut should always be inspected regardless of one's experience and skill. Sometimes the column breaks unevenly due to no fault or error in the cutting technique or method.

It is recommended to inspect the column ends with a magnifier. After cutting, the ends of the column should be examined with a 10–20X magnifier to verify a clean and straight cut has been obtained. If the column end is jagged or uneven, re-cutting is necessary. The column needs to be re-cut until a satisfactory end is obtained. A poorly cut column may result in peak tailing or adsorption. Small pieces of the jagged edge of the column can fall into the column and cause severe peak tailing or blockage of the tubing. Small pieces may also enter the detector and cause it to malfunction. Regardless of the analysis, column or GC system, a good column cut is essential to obtain the best performance.

### 9.2.3 Column Placement in the GC Oven

Most GC's have a column hanger in the oven. Some hanger designs only allow one column to be placed on the hanger. If a second column is installed in the same GC oven, another approach has to be taken. Sometimes the second column can be hung directly on the cage of the first column. If this is not possible, there are usually several holes in the top of the GC oven where small pieces of wire or paper clips can be attached. The wire can be fashioned into a suitable hanger and the second column cage can be hung from the pieces of wire.

Unwind enough of the tubing on both ends of the column so there is ample tubing free from the cage. The column should be placed in the GC oven so that there is no contact between the tubing and other objects in the GC oven such as column identification tags, oven walls, column hangers, etc. The forced air currents inside the oven cause movement of the column and these contact points may abrade the column. This may lead to breakage at these abrasion points since the polyimide has been weakened in these areas.

Avoid tight or sharp bends in the tubing. These sharp bends place an excessive amount of stress on the tubing, thus increasing the possibility of breakage. The avoidance of sharp bends becomes more critical as column diameter increases. Large diameter tubing can not tolerate sharp turns or bends as well as small diameter tubing; therefore, it is more susceptible to breakage due to stress. Regardless of tubing diameter, it is much better to remove several centimeters of column length instead of subjecting the column to an extreme bend or turn. The loss of several centimeters has no affect on column performance.

### 9.2.4
### Column Installation in the Injector

After placing the column in the GC oven, the next step is to install the column in the injector. Place the column nut and ferrule on the tubing making sure of the proper alignment of each piece. Consult the GC manual if necessary. In most cases, the flat surface of the ferrule should touch the flat surface in the fitting or column nut; the tapered region should touch a tapered region in the fitting or nut.

Any time a ferrule is placed on a column, 2–3 cm from the end of the column must be cut off. There are no exceptions! Upon sliding the ferrule on the column, very small fragments are shaved from the inner surface of the ferrule by the sharp edge of the column end. These ferrule particles can enter the column and severely interfere with the chromatography. Peak tailing and adsorption are the most common symptoms of ferrule particles in the column. Sometimes, the particles migrate long distances into the column and permanently affect its performance. This is a greater problem with larger diameter columns.

The insertion distance of the column into the injector is an important measurement. The GC manual provides this measurement. Installing the column at the incorrect depth may lead to efficiency losses, changes in injector discrimination or peak shape problems. In some cases, an improvement may occur. The distance is usually stated from the end of column to a reference point such as the top of the ferrule or the bottom of the column nut. It is very easy to move the column from this distance during installation. Section 9.9.5 describes some methods to ensure the proper installation depth.

The carrier gas can be on or off during column installation. If it is on, this is insurance that the carrier gas is on during the later stages of installation where carrier gas flow is essential. Also, if the carrier gas is on, the injector fittings can be checked for leaks after the column is installed.

It is not necessary to cool the injector to install a column. As long as hot fittings can be tolerated, it may be best to leave the injector at a higher temperature. Fittings have tendency to leak when subjected to large changes in temperature. If a column is installed while the injector is cool, it is recommended to slightly tighten the column nut after the injector has reached the set temperature. The use of liquid leak detection fluids require the injector fittings to be below 100 °C.

### 9.2.5
### Turning On and Verifying the Carrier Gas Flow

Heating of a column without carrier gas flow results in serious and usually permanent damage of the column. Before the oven is heated after column installation, it is necessary to verify that there is a reasonable flow of carrier gas through the column. If there is a reasonable pressure on the gauge, it is often assumed that there is carrier gas flow in the column. This is not a completely reliable method. Flow controller or pressure regulator failures, or blockage in the gas lines or column can all create a pressure reading on the gauge; however, there is little to no carrier gas flow through the column. Damage to the column occurs upon subsequent heating. Worse is the damage to any replacement columns installed in the same GC with the same problem.

The best method to verify column flow is very quick and simple. Place the end of the column in a small vial containing a light solvent (e.g., hexane, methanol). If there is carrier gas flow through the column, a steady stream of bubbles is visible. An absence or a very slow stream of bubbles indicates a problem. Check the GC system and gas supply, and make any adjustments or repairs before proceeding with column installation.

### 9.2.6
### Column Installation in the Detector

Column installation in the detector follows all of the same guidelines and precautions as for installation in the injector (Section 9.2.4). The proper alignment of the nut and ferrule, a good column cut, the proper insertion distance and leak free connection are all important. Proper column installation in the detector may not be as critical as for the injector, but great care should be taken anyway.

One of the carryovers from packed GC column usage is leaving the detector end of the column disconnected until the column is conditioned. While this is not absolutely necessary for capillary columns and does not result in any harm to the column or GC, it is still practiced. This technique is discussed in Section 9.2.8.4.

### 9.2.7
### Verifying Proper Column Installation and Detector Operation

A leak allowing oxygen (air) into the column or no carrier gas flow while the column is at higher temperatures (~50 °C or higher) can rapidly damage a column. After the column is installed in the GC, the next step is conditioning the column at a high temperature. Before heating the column, it is best to verify the existence of carrier gas flow (even though it was done previously). The easiest and most secure method is to inject a non-retained compound into the GC while the oven is at 40–50 °C. A pressure on the column head pressure gauge is not always reliable. Do not heat the column above 50 °C before carrier gas flow is verified. Before injecting the non-retained compound, the detector needs to be on and the column head pressure set at an appropriate value. Table 9-1 lists the approximate head pressure ranges for some common column dimensions and carrier gases so that the linear velocity is set close to the recommended values. The head pressure (linear velocity) does not need to be exactly adjusted and set at this time; this is done later when setting the average linear velocity.

**Table 9-1** Approximate head pressures for column conditioning.

| Diameter (mm) | Length (m) | Head Pressure (psig) | | |
|---|---|---|---|---|
| | | Nitrogen | Helium | Hydrogen |
| 0.10 | 5 | | | |
| | 10 | | | |
| 0.18–0.20 | 10–12 | 3–5 | 8–12 | 8–10 |
| | 20–25 | 6–10 | 16–24 | 16–20 |
| | 40–50 | 12–20 | 32–48 | 32–40 |
| 0.25 | 15 | 2–4 | 6–9 | 6–8 |
| | 30 | 4–8 | 12–18 | 12–16 |
| | 60 | 8–12 | 24–36 | 24–32 |
| 0.32 | 15 | 1–2 | 4–6 | 4–5 |
| | 30 | 2–4 | 8–12 | 8–10 |
| | 60 | 4–8 | 16–24 | 16–20 |
| 0.53 | 15 | 0.5–1 | 1–2 | 1–2 |
| | 30 | 1–2 | 2–4 | 2–4 |
| | 60 | 2–4 | 4–8 | 4–8 |

The compounds used to verify carrier gas flow are the same ones used to set the average linear velocity (Section 6.4.1). Table 9-2 lists some recommended compounds. One advantage of injecting a non-retained compound at this time is the additional information that is gained. The presence of a peak for the non-retained compound confirms that the detector is on and that the data system is connected; the shape of the peak helps to confirm proper column installation. Avoid large volume injections or high concentrations since column overload can be misinterpreted as poor peak shape. Using a split injection works the best. If a peak does not appear, there is no carrier flow, the head pressure is set too low, not enough of the non-retained compound was injected, the detector is not functioning properly, or the recorder is not on or properly connected. It is absolutely necessary that a peak be obtained before proceeding. The peak obtained must be extremely sharp and without any tailing. If a broad or tailing peak is obtained, the following items should be inspected or checked:

1. Injector liner (breakage)
2. Position of the column in the injector and detector
3. Leaks in the injector and detector fittings
4. Leaks in the septum
5. Quality of column ends (i.e., poor cuts)
6. Ferrule or septum debris in the column
7. Split ratio (too low)
8. Inadequate makeup gas flow

**Table 9-2** Suggested non-retained compounds.

| Detector | Non-retained compounds |
|---|---|
| FID | methane, butane[1] |
| ECD | methylene chloride[2),3)], dichlorodifluoromethane[2),3)] |
| NPD | acetonitrile[2),4)] |
| TCD, MS | methane, butane, argon, air |

1) A disposable lighter is a possible source of butane.
Place the syringe needle into the flame outlet of the lighter.
Depress the small button allowing the gas to escape.
Pull up 1–2 μL of the gas.
2) Headspace or diluted in solvent.
3) Use a column temperature of 40 °C or greater.
4) Use a column temperature of 90 °C or greater.

*Headspace:* Fill an autosampler vial with about 10 drops of solvent.
Tightly seal the vial with a cap. Shake the vial for several seconds.
Pierce the septum with the syringe needle, but do not insert the needle into the liquid.
Pull up 1–2 μL of the headspace above the liquid.

*Solvent dilution:* Add 25–50 μL of the appropriate solvent to 10 mL hexane or iso-octane and thoroughly mix. Inject 0.1–0.2 μL.
Further dilution may be necessary depending on the sensitivity of the detector.

## 9.2.8
## Column Conditioning

### 9.2.8.1 What is Column Conditioning?

All columns have a small amount of volatile contaminants originating from their handling, installation and storage. Column conditioning involves heating the column to a high temperature to rapidly remove these contaminants. Conditioning is necessary any time a column is installed even if it has been previously conditioned. Failure to condition a column results in baseline problems. The baseline severely rises as the column is heated and numerous peaks, humps and blobs may be visible. While these contaminants do not foul or harm a detector, they are interferences in the chromatogram.

### 9.2.8.2 Conditioning Temperatures

There are two approaches concerning the best temperature to condition a column. One is at the isothermal temperature limit of the column or 20–25 °C above the highest oven temperature that will be used without exceeding the upper temperature limit. Using the upper temperature limit conditions the column faster, but it may shorten column lifetime by exposing the column to unnecessarily high temperatures. If the highest operating temperature is more than 50 °C lower than the column's upper temperature limit, conditioning at 20–25 °C above that operating temperature may be better. The drawback to conditioning at lower temperatures occurs if a higher column temperature needs to be used at a later date. The column usually has to be re-conditioned at a higher temperature to accommodate the new temperature conditions. Failure to re-condition the column may result in a rising and erratic baseline when the column temperature exceeds the previous conditioning temperature.

The rate at which a column is heated has no impact on column conditioning. Rapidly heating a column does not damage it, thus slowly heating the column to the conditioning temperature is not necessary. A temperature program is not needed nor are repeated cycles of heating and cooling. The GC oven can be directly set at the conditioning temperature and the column heated at whatever rate the oven reaches this temperature. There is no evidence that temperature programming or cycling between temperatures improves conditioning or increases column lifetime or performance.

### 9.2.8.3 Conditioning the Column While Connected to the Detector

Occasionally there is a fear that the contaminants eluting from a column during conditioning will contaminate or foul the detector. While this is a valid concern with packed GC columns, it is not for capillary columns with a few exceptions (e.g., very thick film, polar stationary phases; PLOT columns). The amount of stationary phase and contaminants in capillary columns is quite small. Also, most column manufacturers precondition the columns for prolonged periods as part of the production process. There are several advantages to conditioning a column while it is attached to a functioning detector. Conditioning a column while it is disconnected from the detector is not a problem; however, some information is lost and a short conditioning step is still needed. Use of this conditioning technique is further described in Section 9.2.8.4.

It is best to monitor the progress of column conditioning so that problems can be immediately noticed before column damage can occur. The detector should be already functioning due to the carrier gas flow verification step (Section 9.2.7). Set the recorder or data system with a 120 minute run (stop) time and the attenuation or sensitivity setting used for the analyses. Input the conditioning temperature into the GC and start the recording device. The baseline should start to sharply rise after several minutes and continue to rise beyond the time when the GC oven reaches the set temperature. After 5–15 minutes at the conditioning temperature, the baseline should start to rapidly drop. After 30–90 minutes at the conditioning temperature, a steady baseline or detector signal should be obtained. As soon as the baseline is stable for more than several minutes, conditioning should be stopped. Usually the entire plot looks like a large blob with some peaks riding on top (Figure 9-1). Further conditioning reduces the baseline by only a very small amount and usually is a waste of time. More polar stationary phases, thick film columns and greater sensitivity settings usually require longer conditioning times. For very low level analyses (sub ppb), 4–8 hours of column conditioning may be required.

If the baseline is unstable or excessively high after 60–90 minutes, column conditioning should be stopped. Most likely there is a leak in the GC or a contamination problem. If there is a leak, the baseline remains at a high level since damage to the stationary phase is occurring. The high baseline is due to the elution of the degradation products of the stationary phase. Additional heating only further damages the column, thus conditioning should be stopped. The most common location of leaks is around the injector. Check the septum, column nut

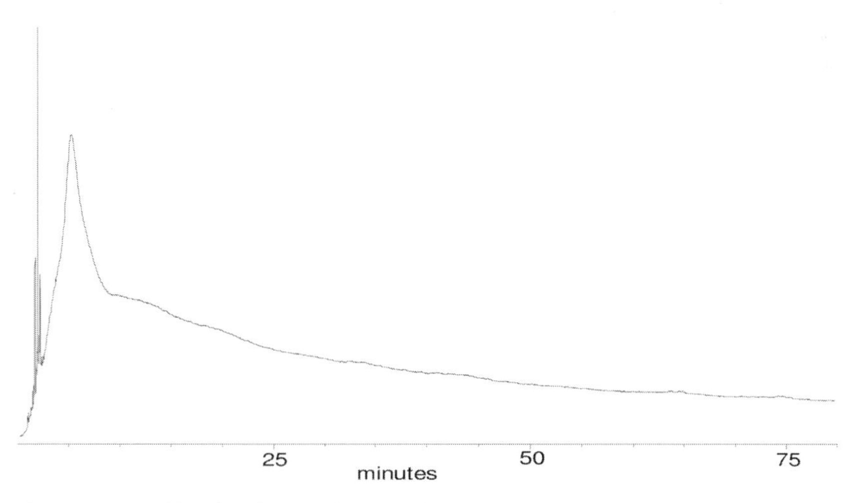


**Figure 9-1** Typical baseline during column conditioning.

and all of the injector fittings for leaks. Be sure that the recording device is not set at one of its most sensitive levels. Normal but small deviations in the detector signal appear to be very large at these very sensitive settings.

Contamination is primarily visible as an erratic baseline and/or the presence of various size and shape peaks. The GC is usually not suitable for use while contaminated, thus conditioning should be stopped so that the GC can be cleaned. A dirty injector or gas lines are the most likely source of any contamination. In some cases, installation of a new column into a contaminated injector results in a contaminated column. If this contaminated column is transferred to a different, contaminant free GC for troubleshooting purposes, the same problem arises. The incorrect conclusion that the column is defective may be reached when in fact the column was contaminated by the first GC system. A new column should not exhibit high bleed. High quality columns do not self destruct upon shipping or storage, and are tested for bleed levels during the manufacturer's testing process.

Some detectors require several hours to stabilize especially if they have been off for more than 1–2 hours. The most common detectors that exhibit this tendency are the ECD, NPD and MS. Upon monitoring the baseline, it may appear that the baseline does not stabilize and continues to drift downward. This drift may be due to detector stabilization. The column maybe fully conditioned, but the detector requires additional time to stabilize. It is difficult to determine whether the drift is column or detector related. Experience with a particular detector usually helps in determining the source of the baseline drift.

#### 9.2.8.4 Conditioning the Column While Disconnected from the Detector

Fears of detector fouling by contaminants eluting from the column during conditioning leads some chromatographers to leave the column disconnected from the detector. Fouling of the detector by the column is a very rare occurrence. Only PLOT columns and very thick film stationary phases have enough material to foul a detector.

Columns can be conditioned while they are disconnected from the detector without any problems. The baseline or background signal can not be monitored during conditioning. This makes it difficult to determine whether the column is fully conditioned. Also, ongoing damage to the column due to a leak at high temperatures can not be detected until significant damage has occurred.

If a column is conditioned while disconnected from the detector, 10–20 cm of the free end needs to be cut off before installation in the detector. A small amount of air enters the end of the column and damages the stationary phase in this region. After the column is installed in the detector, a small amount of additional conditioning is still necessary. The process of handling and installing the column slightly contaminates the end of the column. This requires the column to be held for about 15 minutes at the conditioning temperature or until a stable baseline is obtained. Some detectors may requires hours before they stabilize and a flat baseline is obtained.

### 9.2.9 Setting the Carrier Gas Average Linear Velocity

The best time to accurately set the linear velocity is after the column is properly conditioned. The linear velocity can be set before conditioning; however, time is required for the column to stabilize (condition) at the temperature used to set the linear velocity. Since carrier gas flow has been previously verified, there is no need to set the linear velocity prior to conditioning.

The retention time of a non-retained compound is used to set the average linear velocity. A list of recommended compounds can be found in Table 9-2. The desired average linear velocity and column length is used in Equation 9-1 to determine the required retention time for the non-retained compound. The column head pressure is adjusted downward to increase the retention time and adjusted upward to decrease the retention time until the desired one is obtained. A retention time of ±0.05 minutes from the calculated value is sufficient in most cases. A table of recommended average linear velocities can be found in Table 6-3.

$$t_{\mathrm{m}} = \frac{L}{60\,\overline{u}}$$

$t_{\mathrm{m}}$ = retention time of a non-retained compound (min)
$L$ = column length (cm)
$\overline{u}$ = desired average linear velocity (cm/sec)

**Equation 9-1** Non-retained peak retention time to obtain a specific average linear velocity.

The average linear velocity is dependent on the column temperature. It is important to set the linear velocity at the same temperature for a given analysis. Using a higher temperature results in a slower velocity and a lower temperature results in a faster velocity than desired. Changes in retention times and efficiency occur with the measurement error. Usually the initial temperature of a program is the most convenient. In practice, any column temperature can be used; however, a consistent temperature of measurement is the most important consideration.

After the linear velocity is set, the column is ready to use. There are several more steps that are not necessary for the best performance of the GC system; however, they may prove to be very useful when troubleshooting problems.

### 9.2.10 Bleed Test

After the column is conditioned, a bleed test is one method to measure the amount of column bleed. As a column is used, the bleed level slowly increases. Eventually, it reaches a level that renders the column unsuitable for most uses. By having a bleed level measurement for the column when it was new, there is a reference point for future comparisons.

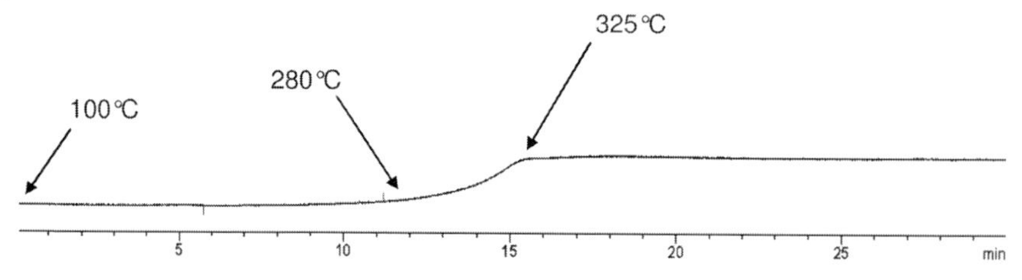


**Figure 9-2** Typical bleed profile.
*Column:* DB-5ms, 30 m × 0.25 mm, 0.25 μm
*Oven:* 100–325 °C at 15°/min, 325 °C for 15 min

There are two methods most commonly used to determine the amount of column bleed. The detector signal at the column's conditioning temperature is used as a measurement. The highest background signal should be obtained at the conditioning temperature or isothermal temperature limit, whichever is the highest. The signal can be obtained from the data system/recorder or detector output signal. It is better to compare this signal to the background signal at a lower column temperature (usually between 50–100 °C). Contributions from sources other than the column are now factored out of the measurement. The difference between the two signals is used as the bleed level measurement.

Collecting a bleed profile is probably the best method to measure the bleed level. A bleed profile is generated by running a temperature program without an injection (i.e., blank run). The temperature program should follow these guidelines: initial temperature of 50–100 °C; ramp rate of 10–20 °C; final temperature equal to the isothermal upper temperature limit of the column or the conditioning temperature (whichever is lower); and a final hold time of 10 minutes. The appearance should be similar to that in Figure 9-2. If the appearance is significantly different, it may be a sign of a column or GC problem. The size of the rise in the baseline is the important feature. As long as the same scale is used to draw both chromatogram (before and current), the rise in a current bleed profile can be compared to the one generated when the column was installed.

### 9.2.11
### Injecting Column Test Sample

After the column has been conditioned and a bleed test run, some type of column test mixture should be injected and analyzed. There are two types of test samples.

One is the same sample used by the column manufacturer to test a newly made column. The other is a representative or actual sample like the ones analyzed with the GC. Perhaps both can be used to test different aspects of the GC's performance.

Nearly every capillary column comes with a performance chromatogram generated using a specific test mixture. If the same conditions and sample are used, the chromatogram generated should be nearly identical to the one sent with the column. Large differences are indicative of a problem with the gas chromatographic system or perhaps the incorrect test conditions. When a good chromatogram is obtained, it should be stored for future reference. If a problem should arise in the future, the test mixture can be injected and the chromatogram compared to the one generated with the properly working system. The diagnosis of many problems can be accomplished with the proper interpretation of the results of the test mixture chromatogram. Each of the compounds in the test sample provides a specific bit of information about the column. This information is very useful when trying to troubleshooting column problems. Detailed information about test mixture compounds can be found in Chapter 10.

Instead of a manufacturer's test sample, a representative or actual sample can be used to test system performance. In some cases, this sample is a better test of the performance of the gas chromatographic system. Some analyses are less demanding of the column than the column test mixture. A column that appears to be bad with the column test mixture performs satisfactorily for the samples of interest. It is probably unreasonable to discard that column based on the results of the column test mixture when it adequately performs the desired analyses. Some analyses are more demanding of the column than the column test mixture. A column that appears to be good with the column test mixture performs less than satisfactorily for the samples of interest. While this is fairly uncommon, a different sample than the column test mixture is needed to adequately judge the performance of the column. Usually there are only a few compounds or peaks that place high demands on the column. The test sample should contain these compounds; the other sample compounds may not be necessary especially if they are not as demanding on the column.

One simple method to monitor system performance is to make an overhead transparency of a reference test chromatogram. The transparency can be placed over a corresponding test chromatogram generated using the same conditions and printed in the same size as the one depicted on the transparency. Acceptable performance variations can be noted on the transparency. Factors such as peak retention, shape and size, and baseline rise can all be easily noted and measured even the most inexperienced GC user. For example, a box can be drawn around a reference peak on the transparency and the corresponding peak in the generated test chromatogram has to reside within the box boundaries to be acceptable. The need for system maintenance can be easily discovered before a major problem develops. Also, this helps to prevent the use of incorrect conditions or methods since the chromatograms will not match.

## 9.3
## Column Ferrules

Graphite and graphite/vespel are used to make capillary column ferrules. Ferrules made of both materials are available for some GC's while others may be limited to only one. Essentially, ferrule material is often a personal preference; however, a few models of GC's require a specific type of column ferrule. In these cases, the exact ferrule is required since non standard fittings are often used for the injector and detector.

Graphite ferrules are very soft which makes them easy to seal. This makes it easier to get a leak free connection. The softness of graphite also creates a few problems. Upon over tightening, graphite ferrules sometimes generate small flakes of graphite. These small flakes can contaminate injectors, detectors, fittings and columns. The soft graphite flows into the threads and openings of the injector and detector fittings. In severe over tightening cases, graphite can be extruded into the injector or detector. Removal of this material is necessary before the next column is installed. Graphite ferrules are not recommended for use on the detector fitting for GC/MS and ECD systems. Small pieces of the graphite can contaminate these particulate sensitive detectors. GC/MS systems require hours of work to clean out particulate matter. The vacuum in a MS system also makes it easy to suck ferrule particles into the source. ECD cells are sealed, thus ferrule particles in the cell necessitate return of the detector to manufacturer and days of downtime. Over tightening a graphite ferrule makes it difficult to re-use the ferrule multiple times. A graphite ferrule can be used 10–15 times providing it is not deformed by repeated over tightening. Techniques to avoid over tightening fittings can be found in Section 9.4.

Graphite/vespel are much harder than graphite ferrules. This makes their properties the opposite of graphite ferrules. A leak free seal is more difficult to obtain with graphite/vespel ferrules since they are much harder. It is much harder to over tighten graphite/vespel ferrules. Graphite/vespel ferrules do not flake and are better suited for GC/MS and ECD systems. Graphite/vespel has the tendency to creep at higher temperatures. The ferrule changes shape upon heating. A leak free fitting at one temperature can leak at a different temperature. If the fitting temperature is changed by more the 25–30 °C, it is recommended to check the fitting for any leakage after it has reached the new temperature. Upon over tightening especially at higher temperatures, graphite/vespel ferrules usually stick to fused silica tubing. Often the tubing has to be cut since the ferrule does not come off of the tubing.

Ferrule size is determined by the GC fittings and column diameter. The GC fittings governing the outer diameter and height of the ferrule. The most common outer diameter is 1/8″ with a few 1/4″ and metric sizes. This size does not refer to the outer diameter of the ferrule, but to the size of the fitting. The inner diameter is governed by the column diameter. There are three different ferrule inner diameters in common use. Table 9-3 lists the ferrule inner diameter need for the various column diameters.

**Table 9-3** Column size vs. ferrule inner diameters.

| Column i.d. (mm) | Ferrule i.d. (mm) |
|---|---|
| 0.05–0.25 | 0.4 |
| 0.32 | 0.5 |
| 0.45 | 0.8 |
| 0.53 | 0.8 |

## 9.4 Tightening Fittings

Column nuts are often over tightened. This is due to all of the cautions about avoiding leaks and the damage they cause. It is tempting to use a lot of force when tightening the column nut. While this does not cause any real damage, most ferrules can not be re-used if they are seriously over tightened. This can be quite expensive over the course of a year and a number of column installations. There are several guidelines that can be used to avoid excessive tightening of fittings.

Graphite ferrules are very easy to seal, thus little tightening of the fitting is needed. When using a new graphite ferrule, after tightening the column nut by hand, the column nut only needs to be turned about 1/4 of a turn using a wrench. Any more is using excessive force. A used graphite ferrule often seals in less than 1/4 of a wrench turn. It is easy to tell if the column nut is tight enough. If the column can not be pulled from the fitting, it will not leak. If the column moves during the pull test, make sure it is re-positioned in the proper location. Upon removal of the fitting, if the graphite ferrule is highly deformed or a portion has been forced through the opening in the fitting or column nut, the fitting was over tightened.

Graphite/vespel ferrules are hard, thus more difficult to seal than graphite. When using a new graphite/vespel ferrule, after tightening the column nut by hand, the column nut only needs to be turned about 1/2 of a turn using a wrench. Column nuts on GC/MS transfer lines may require slightly more than 1/2 of a turn. Any more is using excessive force. A used graphite/vespel ferrule often seals in less than 1/2 of a wrench turn. If the column can not be pulled from the fitting, it will not leak. It is more difficult to over tighten a graphite/vespel ferrule. If the ferrules permanently stick to the fused silica tubing on a consistent basis, the fitting was probably over tightened.

## 9.5 Techniques for Measuring Column Insertion Distances

The proper position of the column in the injector and detector is important. The position of the column is usually controlled by maintaining a distance between the end of the column and a reference point on the column nut/ferrule assembly. It is not easy to maintain this distance while inserting the column into the fitting and tightening the column nut. As little as 2 mm of slippage can result in poor performance for some GC's. There are several easy techniques to ensure the column remains in the proper position throughout the installation process.

The most familiar procedure is to use liquid correction fluid ("White-Out") to mark the column. The column is positioned as noted in the instruction manual, the correction fluid is applied immediately below the column nut. The top of the mark aligns with the bottom of the column nut (Figure 9-3a, left). After the correction fluid is applied, make sure the ferrule does not drop or slip into wet fluid. Only a small amount of fluid is necessary to mark the column. The presence of excess material in or on the ferrule may prevent a leak free seal from forming between the column and the ferrule, and may also cause contamination. If the ferrule slips into the wet fluid, remove the ferrule, cut off the marked portion of the column, and start the process again.

Another method of measuring column insertion distances utilizes a used disk-type GC septum. The column is inserted through the hole in the septum made by previous injections. The column nut and ferrule are then placed on the column. The septum is position so that the column nut and ferrule are in the proper location (Figure 9-3b, right). The septum fits snugly on the column and holds the column nut and ferrule in the proper place during installation. The septum stays on the column during use, but it does not affect the column in any manner.

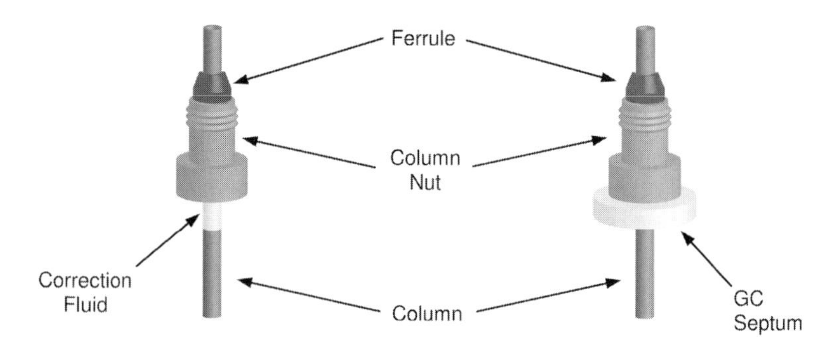


**Figure 9-3** Marking column insertion distances.

## 9.6 Leak Detection

Detecting leaks can be a difficult task. There are many places where leaks can develop with the injector fittings being the most common area. Liquid leak detectors are often used to detect leaks, but they do not work in many applications. Using an electronic leak detector is the most convenient, but they are very expensive. They work on hot fittings which is not possible with liquid leak detection fluids.

Liquid leak detection fluids are easy to use and inexpensive. The most common fluid is referred by the brand name of Snoop. When squirted onto a fitting, a leak becomes visible by the continuous formation of bubbles in the liquid. The absence of bubbles indicates that there is no leak present. A 50/50 mixture of iso-propanol and water can also be used as a leak detection fluid.

Liquid leak detection fluids do not work on hot fittings. Most detection fluids have an upper temperature limit of 90–100 °C. Most injector and detector fittings are maintained at temperatures well above 100 °C. To check the fitting for leaks, it has to be cooled below 100 °C before the leak detector fluid can be used. Fittings often develop a leak upon heating especially if equipped with a graphite/vespel ferrule. If the fitting is leak checked at a lower temperature, tighten the fitting by about 1/16–1/8 of a turn after it has reached the higher temperature.

One method to determine whether an injector or the gas lines leading into it are leaking is to use a static leak check. This involves pressurizing the injector and gas lines and checking the column head pressure gauge for a loss of pressure. The details on this method can be found in Section 12.4.5.

# 10
# Column Test Mixtures

## 10.1
## Column Performance Testing

One of the best means to evaluate the performance of a capillary column is by the injection of a column test mixture. Test mixtures contain compounds that have a range of chromatographic behaviors. Each compound in the test mixture usually provides a piece of information about the column. By evaluating the shape, height, width and retention of the test compound peaks, a substantial amount of information is obtained about the column and possibly the chromatographic system.

Practically every capillary column manufacturer individually tests each column using a test mixture and this test chromatogram is included with the column. The test sample used to evaluate the column can be very useful in the laboratory. It is useful in determining column performance and in many troubleshooting situations. It is important to save the test chromatogram and the accompanying information for future reference. This test chromatogram often provides additional information about the column such as the testing conditions, temperature limits, actual dimensions and serial numbers.

Depending on the laboratory and the analyses being performance, a test sample other than the column manufacturer's may be better. The manufacturer's test mixtures and conditions are designed to test the column and not the entire GC system. Test samples can be designed to evaluate the performance of the GC system. This is usually more desirable since it is the performance of the entire system that influences the final results. The manufacturer's test mixture and conditions may prove to be more valuable as a troubleshooting and column evaluation tool. Regardless of the situation or goals, some type of column or system test sample should be used.

*The Troubleshooting and Maintenance Guide for Gas Chromatographers, Fourth Edition.* Dean Rood

ISBN: 978-3-527-31373-0

## 10.2 Column Test Mixture Compounds

### 10.2.1 Hydrocarbons

Hydrocarbon peaks are the standard against which all other peaks are compared. Due to the lack of functional groups, hydrocarbons interact only with the column's stationary phase. Hydrocarbons peaks should be sharp and symmetrical without any tailing. Malformed hydrocarbon peaks are due to carrier gas flow problems in the injector or detector, solid debris in the column, very poor injection techniques, an extremely contaminated column or an extensively damaged stationary phase. Carrier gas flow problems include dead volumes or leaks in the injector, poor positioning of the column in the injector or detector, a broken or obstructed injector liner, insufficient make up gas flow, too low a split ratio or a poorly cut column end. If the amount of tailing decreases as peak retention increases, the problem is usually in the injector area.

Since hydrocarbons are not susceptible to activity problems, they are used to measure a number of column characteristics. The broadening of a hydrocarbon peak is solely due to the behavior of the hydrocarbon in the stationary phase, thus the hydrocarbon peak is used to measure the uniformity and thickness of the stationary phase. Hydrocarbons are used to measure the number of theoretical plates and utilization of theoretical efficiency since these values are affected by how well the stationary phase is coated in the tubing. The retention factor ($k$) of a hydrocarbon peak is used to measure retention, and thus the film thickness of the column. Straight chain or normal hydrocarbons are used in the test mixtures with $C_{10}$–$C_{16}$ being the most common ones.

### 10.2.2 Alcohols

Compounds with hydroxyl groups can interact with many materials through hydrogen bonding. Ideally the only interaction occurring inside the column is with the stationary phase. If this occurs, a symmetrical peak for the hydroxyl containing compound is obtained. In cases where an interaction between a hydroxyl containing compound and the tubing surface or contaminants in the column occurs, a tailing peak is obtained. A damaged stationary phase can also cause a tailing peak for a hydroxyl containing compound.

Straight chain alcohols are used in column test mixtures as sensitive indicators of column activity. Alcohols with chain lengths between $C_8$ and $C_{12}$ are most often used. Diols are sometimes used in the mixtures as a more sensitive indicator of activity, especially for hydroxyl containing stationary phases. The most common diol used for this purpose is 2,3-butanediol.

Activity may also be caused by the presence of non-volatile or semi-volatile contaminants in the column or injector. The most common source of these materials are previous sample injections. Hydroxyl groups often interact with these residues, thus tailing peaks for alcohols and related compounds may not always be due to column activity. Contaminants causing activity problems can often be removed by solvent rinsing the column. If the alcohol peak tailing remains after properly solvent rinsing the column, either the tailing is caused by column activity or some of the contaminants are insoluble in the rinse solvents. In either case, the column is probably not suitable for use unless only inactive compounds are being analyzed.

### 10.2.3 Acids and Bases

Extremely acidic and basic compounds are among the most difficult to chromatograph since many of them are very active. Most column test mixtures contain both acidic and basic compounds to measure column activity and a column's pH behavior characteristics. Hydroxyl containing compounds are more sensitive to column activity problems than most acids and bases. In additional to column activity, acidic compound peaks tail if the column is too basic and the basic compound peak tails if the column is too acidic. In some cases, the acid or base peaks may be grossly distorted.

Adsorption is another problem with acidic and basic compounds. One of the differences between equivalent columns from different manufacturer's is the pH character. One manufacturer's column is better for acidic compounds because it is more acidic than another manufacturer's; therefore, the other manufacturer's column is better for basic compounds. To measure adsorption, the heights of the acid and base peaks are compared to the height of a hydrocarbon or FAME peak. Hydrocarbons and FAME's are not susceptible to adsorption, thus they are a good measure of the case where there is no adsorption. Height ratios below the established values indicate adsorption, thus a reduction in the peak height for the acid or base compound peak.

Dimethyl and chloro substituted phenols and anilines are the most common acid and base test mixture compounds. Dicyclohexylamine and 2-ethylhexanoic acid are often used when more sensitive basic and acidic compounds are desired. The peaks for these two compounds exhibit tailing with most columns.

### 10.2.4 FAMEs

Fatty acid methyl esters (FAMEs) are often included in column test mixtures. FAMEs exhibit no appreciable activity which makes them suitable for many performance measurements. They are often used to calculate the number of theoretical plates, resolution and coating efficiency for columns with very polar stationary phases. Also, their retention indices are routinely used to determine whether the stationary phase selectivity and polarity are correct. Most column test mixture use the $C_{10}$–$C_{14}$ methyl esters.

### 10.2.5
### Other Compounds

Occasionally, there are some additional or different compounds in the column test mixture. Naphthalene or related compounds are found in some column test mixtures. They are primarily used as a retention index compound. Additional compounds are sometimes added to the column test mixture for specialty columns. The compounds are often ones specified as test compounds in regulated methods or standardized testing protocols. These compounds are added to the column test mixture to ensure the proper performance (e.g., response factors, peak shape, retention) of the column for the specific method. Sometimes the special compounds are in a different sample, thus the columns are tested separately with this test mixture in addition to the standard column test mixture.

## 10.3
## Column Testing Conditions

### 10.3.1
### Injectors

Column manufacturer's are more interesting in testing column performance. The test conditions used for quality checking new columns are designed to minimize the amount of influence the GC system has on the final test chromatogram. Column test mixtures are usually run using a split injector at a high split ratio. High split ratios minimize any injector related efficiency losses. Split ratios of 1 : 50 to 1 : 250 are often used for most tests. Other injectors can be used if necessary, such as for low concentration samples or when exact analysis conditions have to be duplicated.

### 10.3.2
### Detectors

Column manufacturer's use a flame ionization detector (FID) for their tests since it is very reliable and nearly universal in its response. This makes the test mixtures suitable for use with MS and TCD systems also. The use of other detectors poses a problem since many of the typical test compounds do not respond with a NPD, ECD or FPD. The presence of a responsive functional group on the test compound is required for selective detector use. Addition of halogens rarely changes the activity or important properties of compounds such as hydrocarbons or alcohols. The addition of nitrogen, phosphorus or sulfur often changes the activity of many compounds. This make its more difficult to find suitable test compounds for some of the more selective detectors.

Some specialty columns may be tested using a selective detector such as an electron capture detector (ECD) for a chlorinated pesticides column or a MS for low bleed columns. These tests are often run in addition to the standard column test mixture. In any case, it should be possible to duplicate these tests in the laboratory.

### 10.3.3 Column Temperature

Isothermal temperature conditions provide the best overall test of the column. A compound spends the same amount of time in every portion of the column's stationary phase. Every portion of the column has an equal opportunity to interact with each test compound. This is not the case with a temperature program. A compound spends more time in the front portion of the column than in the back half. The actual time spent in each portion depends on the temperature program (primarily the hold times and ramp rates). Problems in the front portion of the column have a greater influence on the chromatography than problems in the back portion. This is not equal opportunity testing of the column and it can create false impressions about its performance.

Temperature programming reduces the amount of visible peak tailing and creates peaks that are much narrower than isothermal conditions. Peak tailing is more evident using isothermal conditions. The primary problem with isothermal conditions is the need for test compounds that elute within a reasonable length of time. To obtain the best measure of the column, the isothermal test temperature should be greater than 90 °C. If only the column is being evaluated, isothermal temperature conditions provide the best and most complete information about the column. If the purpose of the testing is to determine column or system performance for a specific analysis or set of conditions, a temperature program is satisfactory. A few column manufacturers use temperature programs while most use isothermal conditions in their column tests.

### 10.3.4 Test Sample Concentration

The concentration of a test sample is dependent on the injector, detector and intent of the test. Using high split ratios with a split injector requires the use of fairly concentrated test samples. Injectors that introduce a large amount of the sample into the column require lower concentration test samples. More sensitive detectors such as a NPD or ECD require the use of lower concentration test samples than less sensitive detectors such as a FID or TCD.

In general, lower concentration samples are more difficult to analyze. Peak shape problems, adsorption and baseline interferences are much greater problems when the GC system operates in a very sensitive mode. Many chromatographic problems are hidden when using concentrated samples. Peak tailing and small interferences are not as visible in the chromatogram since high attenuation values are used for highly concentrated samples. Problems are more visible at lower attenuation values. If a test sample that represents actual samples or conditions is being used, the sample concentration should be close to the concentration of the lowest level, actual sample to be run. If the system is working properly for the lower level samples, it will perform well for the higher level samples. If the test samples are at a higher concentration level, the system check reveals no problems; however, problems are experienced during the analyses (when it is too late) for the lower level samples.

## 10.4 Grob Test

The Grob test mixture is familiar to many gas chromatographers. It was first introduced around 1979 as an universal and demanding test for capillary columns. It is still used by some column manufacturer's for their column performance tests. If differs in several ways from the column specific test mixtures and conditions used by many column manufacturers.

Due to their widely different retention on most columns, a temperature program is needed to analyze the compounds in the Grob mixture. The Grob mixture uses the same compounds and temperature program conditions for every column. While having an universal sample and conditions is more convenient than individual test samples for each stationary phase, if any of the compound co-elute on a particular stationary phase, the information provided by those compounds is lost. A different temperature program can be attempted to better resolve the co-eluting compounds, but usually only small improvements are obtained.

The specific compounds in the Grob mixture are listed in Table 10-1. The Grob mixture compounds provide the same type of information as the isothermal column test mixtures; however, there are few differences. The *n*-decylamine, 2-ethylhexanoic acid and 2,3-butanediol are more active than most of the compounds used in other column test mixtures. Some non-Grob test mixtures contain some or all of these very active compounds. Slight tailing is visible for these peaks for nearly every capillary column. Also, the heights of the alcohol peaks are compared

**Table 10-1** Grob mixture compounds.

| **Acids** | **Hydroxyl containing** | **Hydrocarbons** |
|---|---|---|
| 2-Ethylhexanoic acid | 1-Octanol | *n*-Decane |
| 2,6-Dimethylphenol | 2,3-Butanediol | *n*-Undecane |
| **Bases** | **FAMEs** | **Aldehydes** |
| 2,6-Dimethylaniline | Methyl decanoate | 1-Nonanal |
| Dicyclohexylamine | Methyl undecanoate | |
| | Methyl dodecanoate | |

to a hydrocarbon or FAME; this value is not measured in most isothermal column test mixtures. The value of the trennzahl (TZ) is calculated using the FAME's in the Grob mixture. The TZ value is very dependent on column temperature, thus comparisons must use exactly the same temperature conditions. The trennzahl number is rarely calculated for the isothermal column test mixtures. Since the Grob mixture requires the use of a temperature program, retention indices are normally not calculated. Besides, retention indices are not as accurate for temperature programs as for isothermal temperature conditions.

The use of a temperature program can create a false impression of column performance. Temperature programs generate narrower peaks than isothermal conditions. Sometimes this creates the impression than one column is more efficient than another when the only real difference lies in the temperature conditions. Also, temperature program can conceal small amounts of peak tailing. It is more difficult to detect peak tailing for narrow peaks. Using a high chart speed for the chromatogram print out is suggested to better observe the active compound peaks for tailing behavior.

## 10.5
## Own Test Mixture

The best indicator of column and system performance is a sample containing some or all of the target compounds in the actual samples to be analyzed. The most relevant information is obtained in this manner. For example, if only hydrocarbons are being analyzed, the fact that the alcohols in the column test tail rarely affects the results of the actual hydrocarbon analysis. A test mixture containing only the important hydrocarbons is all that is necessary. Even though the column is not performing in the best possible manner (i.e., tailing alcohol peak), it would be premature to dispose of the column in this case. For this type of situation, the column test mixture is too rigorous for the actual analyses.

Some compounds are notoriously difficult to chromatograph and perfect chromatography should not be expected. A column that appears to be perfect with a column test mixture may exhibit tailing or adsorption for an actual sample. Perfect peak shapes and no adsorption are very difficult to achieve for low molecular weight primary and secondary amines, carboxylic acids (especially dicarboxylic acids), alcoholic amines, and some diols and dienes. Attempting to achieve perfect chromatographic performance with these types of compounds is usually a futile exercise. If these types of compounds are being analyzed, a test sample containing these compounds provides the most realistic and accurate measure of system performance.

Column test mixtures usually only provide efficiency measurements. Efficiency is not always a good measure of the amount of resolution for some closely eluting peaks. Slight differences in carrier linear velocities or other conditions can affect the resolution, but have a much smaller impact on the measured column efficiency. An actual test sample provides this type of important information.

The most difficult to resolve compounds can be included in the test mixture to make sure that critical resolution requirements are met. Other compounds such as those very susceptible to adsorption or tailing can be also included in the test mixture. These types of compounds often exhibit problems before the compounds in the manufacturer's test mixture. Routine checks of the system with an easy test mixture may not detect problems with the more difficult to chromatograph compounds.

Each analysis and laboratory probably uses a different set of performance criteria. Using an actual test sample to measure performance allows each method or laboratory to have its own set of testing protocols. For example, some methods may require high resolution, another needs good peak shapes while another needs sufficient peak response. The areas of the greatest importance or concern can be selected for monitoring while other unimportant areas can be virtually ignored. The areas of interest have to be selected based on previous analysis knowledge and experience. However, the areas must be realistically selected and reasonable performance levels have to be set. This is to ensure that problems do not go undetected while normal variations within a method or technique are allowed.

## 10.6 When to Test a Column

It is recommended to test a column after installation, whether the column is new or has been previously used. Not only does this test the GC system before use, it also provides a performance reference point in the event of future problems. Upon encountering a problem situation, the test mixture can be injected, analyzed and compared to the original chromatogram for clues to the source of the problem. This makes troubleshooting much easier, especially if a column manufacturer's test mixture and conditions are used.

Some laboratories inject a test sample at regular intervals to insure the continued and ongoing performance of their GC systems. This verifies the proper operation of the GC system and increases confidence in the results. Whether this occurs before each set of samples, on a daily basis, or on some other schedule, is a decision that can only be made by the chromatographer or perhaps mandated by a regulatory agency.

# 11
# Causes and Prevention of Column Damage

## 11.1
## Causes of Column Damage and Performance Degradation

There are numerous misconceptions concerning the factors that can damage or affect the lifetime of fused silica capillary columns. Capillary columns are durable and stable, but they still require proper care to maintain their high levels of performance. Some types of column damage are permanent while others may be of a temporary nature. Breakage, overheating and constant exposure to high levels of oxygen are the three primary causes of column damage. The most common cause of column performance degradation is contamination of the column with sample residues.

## 11.2
## Column Breakage

### 11.2.1
### Causes of Column Breakage

Fused silica columns do not need to be handled with extreme caution and surgeon-like care. Fused silica tubing is flexible which allows it to be wound onto cages and easily installed in gas chromatographs. Rough handling of the column can result in breakage usually at a future date and not at the time of handling.

If the polyimide coating is abraded or scratched due to rough handling or contact with a sharp edge, the column is weakened at that location. Vibrations from the oven fan, continuous heating and cooling of the column or future handling may result in breakage of the column. Contact with identification tags, fan shrouds, metal edges in the oven or careless handling are all causes of premature column breakage.

It is a rare occurrence for a column to break without some sort of physical abrasion. The process of column manufacturing subjects the fused silica tubing to a series of harsh environments such as high temperatures, corrosive and reactive solutions, and tremendous amounts of physical manipulations. If a piece of tubing has a weak point, it usually breaks during the manufacturing process.

*The Troubleshooting and Maintenance Guide for Gas Chromatographers, Fourth Edition.* Dean Rood

ISBN: 978-3-527-31373-0

### 11.2.2
### Symptoms of Column Breakage

The chromatographic symptom of a broken column is the lack of peaks after an injection. While other problems can cause the absence of peaks, it requires only a few seconds to visually inspect the column for breaks. In nearly every case, one or two column ends are readily visible. It is rare, but not impossible, for a fused silica column to develop a hole and not break.

Depending on the location of the break, a drop in the column head pressure may be evident. The closer the break is to the injector, the greater the drop in the head pressure. Obviously, there are other problems or situations that can cause a drop in the head pressure.

### 11.2.3
### Prevention of Column Breakage

The column should be placed in the GC oven so that the tubing is not subjected to excessive bending. Bends approaching 180° or greater subject the tubing to tremendous stresses at the bend points. These areas are stressed and are more likely to break than other portions of the column. The avoidance of excessive bends becomes more critical as column diameter increases. Large diameter tubing can not tolerate sharp turns or bends as well as small diameter tubing. Columns are slightly stressed from being wound on cages and the amount of stress on a column is dependent on the diameter of the column cage. Columns wound on small cages are more likely to break.

Winding an 0.53 mm i.d. column on a cage below seven inches in diameter may result in premature column breakage. Small winding diameters places tremendous stress on wider bore tubing. Narrow bore (≤ 0.25 mm i.d.) fused silica tubing can be wound onto cages of 3–4 inches without any excessive breakage problems. Regardless of the tubing diameter, using less than seven inch cage requires extreme care to minimize any other possible sources of tubing stress or damage (Section 3.5).

It is important that any points of contact between the fused silica tubing and column identification tags, oven walls, column hangers, etc. are eliminated. The forced air currents inside the oven causes movement of the column and these contact points may abrade the column. This leads to breakage at these abrasion points since this has the same overall effect as scribing the column for cutting.

### 11.2.4
### Recovery from Column Breakage

If a column breaks, it may not be rendered completely useless. A loss of 3–4 loops of tubing from the column does not significantly reduce its resolution capabilities. As long as a small decrease in the retention times is acceptable, install the longer length of the column back in the GC. If the same retention times are desired or needed, use an union to join the two pieces. If the break occurs further into the column, use a union to repair the break. Column performance is not sacrificed if the union is properly installed. No more than two unions should be installed in a column to repair breaks.

If the column breaks while in use and is subjected to a high temperature for a prolonged period, part of the column may become damaged. There is no carrier gas in the length of column past the breakage point, thus it has been exposed to oxygen at an elevated temperature. This usually result in significant damage to the column within 1–2 hours. There is carrier gas flow in the part of the column before the breakage point, thus it does not suffer from damage due to tubing breakage.

## 11.3 Thermal Damage

### 11.3.1 Causes of Thermal Damage

Exceeding a column's upper temperature limit for prolonged periods results in damage to the stationary phase. Short term exposure or temperatures slightly above the upper limit do not cause significant damage. There is more damage at higher temperatures above the column's upper temperature limit and the longer the time of high temperature exposure. Thermal damage most frequently occurs when two columns with different upper temperature limits are in the same column oven. One column is used without noticing that the analysis temperatures exceed the upper temperature limit of the other column. The second column is damaged since it is maintained at excessively high temperatures for prolonged periods.

Thermal damage also occurs when the column is heated without carrier gas flow. This usually occurs after changing a column or gas cylinder. The lack of carrier gas at temperatures above 50 °C can result in damage to the stationary phase of some columns. In general, the more polar the stationary phase, the lower the temperature where significant thermal damage occurs.

### 11.3.2 Symptoms of Thermal Damage

Usually the first symptom of a thermally damaged stationary phase is excessive column bleed. The baseline rises to a very high level near the upper temperature limit of the column. The amount of bleed increases with greater thermal damage to the column. Peak tailing for active compounds occurs shortly after or simultaneously with the increase in column bleed. If the thermal damage is severe, all peaks tail with severe shape distortion being common. In some cases, a loss of efficiency and resolution occurs upon thermal damage. This is usually due to the distortion of the peak shapes. In most cases, excessive column bleed and activity problems render the column before efficiency losses become too severe.

### 11.3.3
### Prevention of Thermal Damage

Thermal damage is avoided by not exceeding the column's upper temperature limit. The easiest method to avoid overheating a column is to set the oven maximum temperature on the GC to 5 °C above the upper limit of the column. GC ovens sometimes exceed the set temperature by 1–2 °C during rapid heating. If the oven maximum is not set slightly above the final temperature, the GC may shut off if only slight overheating occurs. Finally, the upper temperature limit can be prominently displayed on the GC as a reminder. Regular checks of the carrier gas cylinders and regulators insures that there is a constant supply of carrier gas.

### 11.3.4
### Recovery from Thermal Damage

Most of the damage caused by overheating a column is permanent. If the overheating is not too severe, very little damage is noticed. In many less severe cases, only a small increase in column bleed occurs along with a small reduction in column life. If a more severe thermal damage situation occurs, a bake out may return some of the column's performance. Disconnect the detector end of the column and supply carrier gas to the column. Heat the column to its isothermal temperature limit and leave it at this temperature for 12–24 hours. After the bake out, cut off 10–20 cm from the free end. Install the column as usual and evaluate. The column will not return to the pre-damage performance level; however, it often is still useable. If the column is not suitable for use, it should be discarded since nothing else can be done to repair the damage.

## 11.4
## Oxygen Damage

### 11.4.1
### Causes of Oxygen Damage

At elevated temperatures, oxygen damages a stationary phase. The introduction of air into the carrier gas through a leak is the usual source of oxygen. Higher oxygen levels and higher column temperatures accelerate oxygen damage. Lower oxygen levels and column temperatures slowly degrades the stationary phase while higher oxygen levels and column temperatures rapidly degrade the stationary phase. Polar phases, especially polyethylene glycol (PEG) based stationary phases, are particularly sensitive to oxygen damage. A small introduction of oxygen such as with air injections do not harm the stationary phase. It is constant exposure to oxygen at elevated temperatures that causes the real damage to the stationary phase.

### 11.4.2
### Symptoms of Oxygen Damage

The symptoms of an oxygen damaged stationary phase are the same as for thermal damage. Usually the first symptom is excessive column bleed. The baseline rises to a very high level near the upper temperature limit of the column. The amount of bleed increases with greater oxygen damage to the column. Peak tailing for active compounds occurs shortly after or simultaneously with the increase in column bleed. If the oxygen damage is severe, all peaks tail, with severe shape distortion being common. In some cases, a loss of efficiency and resolution occurs upon oxygen damage. This is usually due to the distortion of the peak shapes. In most cases, excessive column bleed and activity problems render the column useless before efficiency losses become too severe.

### 11.4.3
### Prevention of Oxygen Damage

Avoiding leaks is the best method to prevent oxygen damage. Frequent and thorough leak checks are recommended especially after injector maintenance and column installation. High purity carrier gases are recommended since they contain very low levels of oxygen. The use of oxygen traps is often recommended especially when using lower grades of carrier gas. The cost of the traps and the increased lifetime of the columns have to be compared to determine whether oxygen traps are worth their expense and extra maintenance. The cost of the traps may be justified in the savings realized by reduced column replacement when using lower grades of carrier gases.

### 11.4.4
### Recovery from Oxygen Damage

The damage to stationary phases caused by exposure to oxygen is permanent. A small amount of damage may not render the column useless; however, the damage is not reversible. Sometimes, the high bleed level caused by the oxygen damage slowly drops with column use, but this bleed decrease is usually small. In most cases, the column needs to be replaced.

## 11.5
## Chemical Damage

### 11.5.1
### Causes of Chemical Damage

There are relatively few compounds that actually damage a stationary phase. Non-volatile compounds contaminate a column and can negatively affect the chromatography, but they do not physically attack and destroy the stationary phase. Some solvent/compounds combinations may not be compatible with a stationary phase and poor peak shapes are obtained. These solvents can not be used in these cases because of the peak shape problem and not due to any damage to the stationary phase.

#### 11.5.1.1 **Bases**

Inorganic or mineral bases are particular damaging to stationary phases (Table 11-1). Substantial stationary phase destruction results if these types of compounds are introduced into the column. Damage occurs even if the base is neutralized by the addition of an acid prior to injection. There is always enough of the free base present in a neutralized sample to cause appreciable damage to the stationary phase. A low number of injections does not cause a significant amount of damage. Large numbers of injections, high pH and high concentrations of the base compound speed up the onset of damage related problems.

Organic bases such as alkyl amines do not harm the stationary phase since they a possess sufficient volatility to elute from the column. The inorganic bases are not volatile, thus they deposit in the column and remain for prolonged, if not indefinite, periods of time. This long residence time along with the strength of inorganic bases distinguishes inorganic from the organic bases in terms of the detrimental effects on column lifetime. PEG stationary phases may take on the respective pH characteristic of the base in the sample. Subsequent use of the column for acidic compounds often results in poor peak shapes.

#### 11.5.1.2 **Acids**

Inorganic or mineral acids cause damage to stationary phases (Table 11-1). Substantial stationary phase destruction results if these types of compounds are introduced into the column. Damage occurs even if the acid is neutralized by the addition of a base prior to injection. There is always enough of the free acid present in a neutralized sample to cause appreciable damage to the stationary phase. A low number of injections does not cause a significant amount of damage.

**Table 11-1** Compounds causing stationary phase damage.

| Inorganic or mineral acids | Inorganic or mineral bases | Perfluoro acids |
|---|---|---|
| • Hydrochloric acid ($HCl$)[1]<br>• Sulfuric acid ($H_2SO_4$)<br>• Phosphoric acid ($H_3PO_4$)<br>• Nitric acid ($HNO_3$)<br>• Chromic acid ($CrO_3$)<br>• Hydrofluoric acid (HF)<br>• Perchloric acid ($ClHO_4$) | • Sodium hydroxide (NaOH)<br>• Potassium hydroxide (KOH)<br>• Ammonium hydroxide ($NH_4OH$)[1] | • Trifluoroacetic acid (TFAA)<br>• Pentafluoropropanoic acid (PFPA)<br>• Heptafluorobutanoic acid (HFBA) |

[1] Condition dependent.

Large numbers of injections, low pH and high concentration of the acid compound speed up the onset of damage related problems.

Organic acids do not cause damage to the phase since they possess sufficient volatility to elute from the column. High concentrations (> 10%) of organic acids may cause slight phase damage especially if low column temperatures are employed. PEG stationary phases may take on the respective pH characteristic of the acid in the sample. Subsequent use of the column for basic compounds often results in poor peak shapes.

Perfluoro organic acids are particularly strenuous on stationary phases. They can damage a stationary phase when present in high concentrations. Their usual source is as the by-product of the use of perfluoro acid anhydride derivatization reagents. Removal or reaction of the excess acid by-product is recommend to prolong column life and maintain performance.

#### 11.5.1.3 **HCl and $NH_4OH$**

Hydrochloric acid (HCl) and ammonium hydroxide ($NH_4OH$) can be safely used under certain conditions. Hydrogen chloride is a gas at room temperature; however, it is usually dissolved in water as hydrochloric acid. Ammonium hydroxide is ammonia in water. Ammonia and hydrochloric acid are gases, thus they immediately elutes from most columns. Since HCl and $NH_3$ have little to no residence time in the stationary phase, damage does not occur. If water is present in the sample, HCl and $NH_3$ dissolve in the water. The HCl and $NH_3$ elute with the water. At column temperatures above 100 °C, the water, HCl and $NH_3$ rapidly elute from the column, thus no damage occurs to the stationary phase. At column temperature below 100 °C, the water condenses in the column. This traps the HCl and $NH_3$ in a concentrated form in the column. The long residence time of HCl and $NH_3$ results in damage to the stationary phase. Temperature programs starting below 100 °C, but passing 100 °C fairly rapidly are not a problem since HCl and $NH_3$ remain in the column for only a short time period. Isothermal column temperatures below 100 °C or temperature programs remaining below 100 °C for a prolonged time present ample opportunity for HCl and $NH_3$ to damage the stationary phase.

#### 11.5.1.4 **Organic Solvents and Water**

Bonded and cross-linked stationary phases are not damaged by water or organic solvents. Even in the splitless and on-column injection modes where solvent actually condenses in the column, there is no damage to the stationary phase. Poor chromatography may result with some solvent, compound and stationary phase combinations, but no damage to the stationary phase occurs. Polyethylene glycol (PEG) based stationary phases may exhibit a small amount of degradation in the first 10–20 cm of the column after several hundred injections of a sample with water or an alcohol (e.g., methanol) as the sample solvent.

Non-bonded and cross-linked phases exhibit some minor damage to the front part of the column after prolonged use in the splitless or on-column injection modes. About 10–20 cm of the front portion of a non-bonded and cross-linked column has to be removed periodically to maintain satisfactory and consistent performance. For these types of columns, the use of a guard column is strongly recommended.

### 11.5.2 Symptoms of Chemical Damage

The first symptom of stationary phase damage by mineral acids or bases is peak shape problems. The stationary phase is destroyed in the front portion of the column where sample introduction occurs. The sample band is distorted which leads to peak shape problems. Also, active sites are formed and tailing peaks for active compound is common. In some cases, smaller peaks occurs due to adsorption. Since the stationary phase degradation only occur over a small region of column, the increase in column bleed is small. The bleed usually increases as the damage become more widespread. Loss of efficiency occurs if the damage is severe. The efficiency loss is primarily due to the peak shape distortion and not just a broadening in the peak width.

If the column has become damaged by acids or bases, there is probably a large amount of the acid or base compound in the injector. The acids and bases causing damage are non-volatile or only slightly volatile. This means that an appreciable amount remains in the injector. Liner deactivation is lost by the action of the acid or base on the glass. If damage to the column has occurred from an acid or base, the injector should be cleaned and the liner replaced.

### 11.5.3 Prevention of Chemical Damage

The best method to prevent damage by acids or bases is to avoid injecting them into the GC. Often this is not possible due to sample preparation requirements, unknown sample origins and difficulties involved in removing the problem compounds. Attempting to neutralize a high or low pH sample does not prevent column damage. The problem acid or base is still present in solution. Also, the free acid or base often forms upon injection into the GC. Sometimes the acid or base added to neutralize the sample is as bad or worse for the column as the original problem compound. Ideally, the acid or base can be removed from the sample using some type of sample preparation method. Ion-exchange SPE cartridges are ideal for this purpose.

If a problem acid or base is present, there are several methods to reduce the extent of possible column damage. A guard column provides some protection against the action of the acid or base. Instead of collecting in the front of the analytical column, the acid or base collects in the deactivated fused silica tubing of the guard column. The very short resident time of the sample compounds in the guard column greatly reduces the contact between the compounds and the acid or base. By keeping the acid or base away from the stationary phase, much of the damage is avoided. Eventually, the guard column needs to be replaced as it becomes saturated with the acid or base compound. Peak shape problems or adsorption are usually the first signs that the guard needs replacement.

If a sample has to be pH modified for stability or to keep the compound in the proper form, use an organic acid or base, if possible. Acetic acid works well to acidify samples. Triethylamine is a good alkaline modifier. Similar compounds can be used if preferred. Peaks for acetic acid, triethylamine and similar compounds may be observed with some columns or conditions.

### 11.5.4 Recovery from Chemical Damage

Since stationary phase damage caused by chemical means is localized in the front of the column, the best method to repair the problem is to remove 0.5–1 meter of the front of the column. This removes the area of greatest damage and is especially beneficial since the front portion of the column has the greatest impact on the chromatography. Solvent rinsing the column is of little help. While it removes the materials causing the damage, any existing damage is not repaired by the solvent rinsing procedure. At least further column damage is prevented upon removal of the acid or base compound by the solvent rinse.

## 11.6 Column Contamination

### 11.6.1 Causes of Column Contamination

The most common source of chromatographic problems is column contamination. These contaminants are materials that do not elute from the column within a reasonable amount of time. They accumulate in the column and cause a series of different problems. Contamination problems mimic almost every other GC problem. This can make it difficult to positively diagnose the problem. Contaminants can be classified into two groups: non-volatile and semi-volatile. It is possible to have both types of contaminants present at the same time. The symptoms of each type of contamination are similar, but there are a few differences.

Non-volatile residues are any portion of the injected sample that does not elute from the column. They accumulate in the column and injector, and it is this build up of materials from numerous injections that leads to the loss of performance. Anywhere from several to hundreds of injections may be made before the build up of residues interferes with the chromatographic process. The non-volatile residues deposit in the front portion of the column since they are not volatile enough to be transported by the carrier gas very far down the column. The majority of compound residence time in the stationary phase occurs in the front section of the column; therefore, residue contamination can have a very large influence on the quality of the chromatogram. This layer of residue causes several problems. It may interfere with the proper and efficient partitioning of the compounds between the carrier gas and the stationary phase. The contaminants may also prevent a sample band from properly forming at the front of the column. Finally, the contaminating residues can interact with the sample. In all of these cases, peak shape problems occur. Non-volatile residues should not cause any baseline problems since these materials are not eluting from the column.

Semi-volatile residues are contaminating materials that eventually elute from the column; however, it may be hours or days before they exit the column and cause any problems. These semi-volatile residues elute from the column and contribute to baseline disturbances and the presence of ghost peaks. They also cause peak shape problems for the same reasons as for non-volatile contaminants. Semi-volatile residues can contaminate substantial lengths of a column since they are mobile and can travel down the column. Some semi-volatile residues may originate from the thermal breakdown of non-volatile residues present in the injector or column. The high temperatures of the injector and column, and the repetitive cycling to high column temperatures are conducive to the breakdown of non-volatile residues.

The most common source of column contamination is the injected samples. Even samples that have been subjected to rigorous extraction procedures contain traces of semi- and non-volatile residues. With enough injections of a clean sample, a build up of contaminating residues can occur. A colorless or clear sample does not mean that there are no harmful residues in the sample. The brown materials that sometimes discolor injector liners are non-volatile materials. Obviously, some samples have larger quantities of contaminating residues than others. The use of selective detectors can contribute to contamination problems. There may be a large amount of compounds in the sample, but only those eliciting a detector response show up as a peak in the chromatogram. This can create the false impression that the samples are cleaner or less complex than in reality. Rapid contamination of the column can occur. If some of these non-detected residues elute late in the chromatogram, the GC run may be terminated before all of these compounds elute from the column. This leaves the contaminants in the column, thus they can interfere with subsequent analyses.

There can be a large number of compounds that do not gas chromatograph. These substances are introduced into the GC and accumulate until they cause a problem. Every compound in a sample does not have to be represented by a peak in a GC chromatogram. Just because there are six peaks in the chromatogram does not mean there are only six compounds in the sample. It means that there are only six compounds that are suitable for GC analysis in the sample.

Some types of samples are infamous for contaminating GC systems. Biological fluids and tissues, soils, sludge and waste water all contain high levels of non-volatile residues and can rapidly foul a column. Some samples like extracted drinking water, distilled solvents and petroleum distillate products contain very low amounts of non-volatile residues. These samples rarely contaminate columns. Other sources of contaminants can be non-sample related. These include residues or contaminants in sample vials and caps, solvents, pipets, septa and ferrule particles, and other non-sample related sources. Contamination sources of this type can be difficult to find and remedy.

### 11.6.2
### Symptoms of Column Contamination

Column contamination problems mimic about every common GC problem. The most frequent symptoms of contamination are baseline disturbances (noise, drift, humps, blobs and peaks), peak shape problems (tailing, split, broad, and misshapen), and adsorption. Retention time shifts, inconsistent peak sizes and non-linear response can also be caused by column contamination.

It is often difficult to distinguish between column contamination and other problems. The use of the column test mixture is often helpful. The alcohols in the test mixture often exhibit obvious peak tailing. Sometimes the acid or base compound also exhibit peak tailing. The later eluting alcohol peak (if present) should exhibit more tailing than the earlier one. The hydrocarbons, FAME's or similar compounds are not affected by contamination unless it is fairly severe. If the alcohol peaks do not exhibit any tailing, contamination is probably not the problem.

If a temperature program is being used or can be used for testing, there are two quick contamination tests. If the presence of non-volatile contaminants is suspected, cut off 0.5–1 meter from the front of the column. Since most of the contaminants are located in this region, the most problematic area is removed. If the problem is substantially reduced, contamination is the probable cause of the problem. In some cases, removing 0.5–1 meter of the column completely eliminates the problem. Another test is to reverse the direction of the column. The contaminants are now located at the end of the column which is the least influential portion. If some improvement is noted, column contamination is a possibility. Care should be taken when using this technique. If there are significant amounts of semi-volatile contaminants at the front of the column, the reversal moves them very close to the detector. They may elute from the column and cause baseline disturbances which may complicate the evaluation results.

There is an easy method to determine whether an appreciable amount of non-volatile residues is present in a sample. Deposit about 10 μL of the sample on a microscope slide or small watchglass. Carefully place the slide or watchglass on the injector port or a hot plate. The glass becomes heated to the approximate temperature that the sample experiences inside of the injector port. The solvent and any volatile sample components evaporate and the non-volatile portion is left on the glass surface. This residue is representative of the materials that may deposit and accumulate in the injector and front section of the column. If a large amount of residue is present, some type of sample clean-up should be undertaken to remove these residue before injecting the sample into the gas chromatograph.

### 11.6.3
### Prevention of Column Contamination

The best defense against column contamination is to avoid having the contaminants in the sample. Good sample preparation techniques remove most of the problem substances. Even the best procedures and techniques do not remove every trace of each problem substance, thus contamination may still occur regardless of the prevention efforts. Some samples require extensive preparation or there are situations when sample preparation is not possible or reasonable. If proper sample preparation requires substantial labor, high cost, or unavailable equipment or expertise, tolerating column contamination may be the only choice. There are cases where it makes sense (economic and effort-wise) to frequently replace or clean columns instead of properly preparing the samples.

If there are unavoidable contaminants in the sample, the use of a guard column (Section 11.8) or packed injector liner (Section 11.9) may help. Both require periodic maintenance and can create other problems if not properly installed, used and maintained.

### 11.6.4
### Recovery from Contamination

Cutting 0.25–1 meter from the front of the column often reduces or eliminates contamination symptoms, especially if the contaminants are of the non-volatile variety. If the non-volatile contaminants extend more the 2–3 meters into the column, it may be best to cut off 0.5–1 meter and then to reverse the column (only if a temperature program is being used). This can only be done once since reversing the column a second time only returns it to the original direction.

Solvent rinsing is the best method to remove contaminants from the column. Solvent rinsing involves flushing the interior of the column with a series of solvents; it is not injecting large volumes of solvent while the column is still installed in the gas chromatograph. It requires several hours of time; however, many columns return to their original performance levels. Also, it can be done numerous times without any harm or change to the column. Not all columns can be solvent rinsed, thus this needs to be determined before attempting any rinsing. If solvent rinsing columns has been repeatedly successful, it is well worth the time and effort. Two columns of the same description can be used for an analysis. As one becomes contaminated, it is replaced with the other one. The contaminated column is then solvent rinsed and used to replace the second column upon its contamination. The two can be used and rinsed many times until they fail from old age and long term use.

After solvent rinsing, it might be necessary to discard a column if it does not return to a suitable performance level. Before doing that, it may be possible to salvage half of the column. Much of the permanently contaminated or damaged area is in the front of the column. Cut the column in half and evaluate the back portion. Often the back half of the column performs very well. It is the front half that is responsible for the unacceptable performance. The short length may be satisfactory for the original analysis since excessively long columns are often used. If the half column is too short, the column may prove to be suitable for a different analysis. Perhaps, two of the shorter lengths can be joined with a union. This joined column may be a good backup in cases of emergency, a methods development column, or a column used to run very dirty or potential column damaging samples.

Baking out (i.e., keeping the column at a very high temperature) a contaminated column for 8 or more hours is often used to remove the contaminants. This is a common practice with packed GC columns. Capillary columns should not be baked out for any more than about 2 hours. Exposing the capillary column to continuous, elevated temperatures may cause some of the residues to polymerize or react in addition to shortening column lifetime. Then the only means to remove the polymerized materials is by removing the section of column containing the residues. In some cases, this means that up to five meters of column has to be discarded before chromatographic performance is be restored. In some cases, the column is no longer useable after a prolonged bake out.

## 11.7
## Solvent Rinsing Columns

Capillary columns eventually become contaminated with residues after repeated injection of any type of sample. Often these residues can only be removed by solvent rinsing the column. Solvents are forced through the column to wash out the contaminating materials. Only bonded and cross-linked stationary phases can be safely solvent rinsed. Do not rinse a non-bonded stationary phase or damage to the column will occur. Some column can be rinsed, but there are some solvents that should be avoided. Check with the manufacturer of the column if there is any uncertainty about solvent rinsing the column.

### 11.7.1
### Solvent Rinse Kits

There are several methods to solvent rinse a column. The easiest method is to use a column rinse kit. They are available from nearly every capillary column manufacturer or a kit can be made for single use or temporary purposes. The cost of the pre-made kits is low and they are very easy to use, thus it is simpler to buy one. The investment is easily recovered if only one capillary column is saved from unnecessary disposal.

Figure 11-1a shows the simple design and concept of a typical rinse kit. The detector end of the column is placed in the rinse kit vial through the straight arm of the tee-fitting. The column end should be located only 2–3 mm above the bottom of the vial. Some type of seal around the column is needed to prevent solvent or gas from escaping and to hold the column in place. Solvent is added to the vial

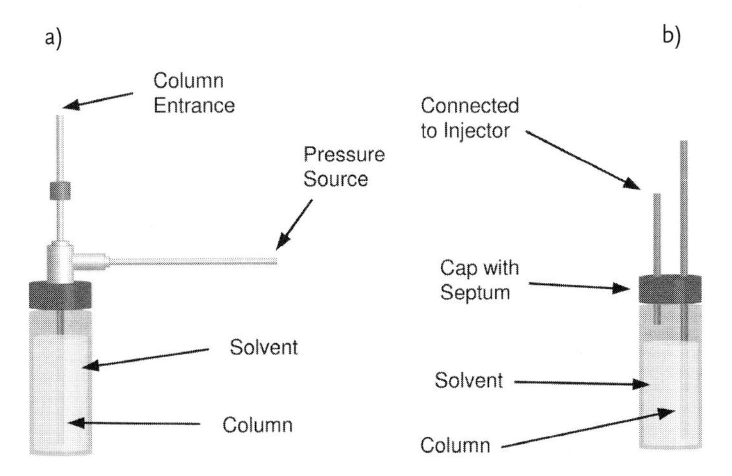


**Figure 11-1** Solvent rinse kits for capillary columns.

and an inert gas is supplied via the other arm of the tee-fitting. Nitrogen, helium or argon can be used; air is suitable if other gases are not available. A pressure of 5–20 psi (30–140 kPa) is recommended. Longer and smaller diameter columns require high pressures. It is not advisable to exceed 20 psi since most rinse kits have some glass components. The pressure generated in the sealed vial forces the solvent to pass through the column. A beaker or similar container is used to collect the solvent as it leaves the column.

A home-made rinse kit can be constructed with readily available laboratory supplies (Figure 11-1b). All that is needed is a 10–20 mL screw cap vial, the corresponding open hole cap and septa, and about a 1 meter piece of fused silica tubing or clean, discarded column. Punch two small holes near the center of the vial septum. Fill the vial with the appropriate solvent and securely screw the septum and cap onto the vial. In the same manner as a regular column, install the piece of fused silica tubing or old column in the GC's injector. Insert the free end of the tubing through the septum so that it protrudes several millimeters into the vial, but not into the solvent. Insert the detector end of the column through the other hole in the vial septum. The end of the column should be 2–3 mm above the bottom of the vial. Using the flow or pressure controller of the GC, supply about 10 psi of pressure to the vial and adjust to maintain the proper solvent flow rate. This home-made kit works better for larger diameter and short columns since lower pressures are required. If the pressurizing gas tends to leak from the holes in the septum, either use smaller holes or use two septa in the cap.

A HPLC pump can be used to push solvent through the column. The column can be connected to the HPLC tubing using a metal union. A standard, graphite/vespel capillary column ferrule works in most metal unions. The desired solvents can be pumped through the column using the flow rate control of the HPLC. Keep the pressure below 50 psi for 0.53 mm i.d. columns and 100 psi for smaller diameter columns.

Another rinsing technique is to use a vacuum source to pull solvent from a reservoir through the column. Care must be taken not to pull solvents in the vacuum source. A trap or reservoir is needed to collect the rinse solvents. Low boiling point solvents such as methylene chloride should be avoided when using a vacuum source. The generated vacuum reduces the boiling points of the rinse solvents and large amounts of solvent vapors can be produced. These vapors are more difficult to trap and can travel directly to the vacuum source. This is often damaging to vacuum pumps or contaminates the flowing water source used in a water aspirator.

### 11.7.2 Solvent Selection, Volumes and Flow Rates

The structure or identity of the contaminants in a column is usually unknown. For this reason, there are no exact guidelines to which rinse solvents to use for a particular situation. More than one solvent is often needed to remove the contaminating materials. A good general purpose rinse solvent selection uses methanol, methylene chloride and hexane. These solvents have proven to work in most situations. Other solvents can be substituted; however, the new solvent should be similar in polarity to the one being eliminated.

There is a set of general recommendations concerning the rinse solvents. Using a range of solvent polarities is more successful. Try to use a polar, an intermediate polarity and a nonpolar solvent as part of the rinse scheme. If the previously injected samples were at a polarity extreme, the opposite polarity solvent can be eliminated. For example, methanol can be eliminated if petroleum samples (i.e., very non-polar) were analyzed. The best column rinsing occurs when each successive solvent is miscible with the previous one. For example, hexane should not be used immediately after methanol since these two solvents are immiscible. Another solvent such as acetone or methylene chloride should be used between the methanol and hexane since they are miscible with both of the other solvents. Another good choice is the sample or injection solvent. The contaminants are reasonably soluble in this solvent; thus they will probably be thoroughly rinsed from the column. Avoid solvents with boiling points above 100 °C especially as the last rinse solvent. Since rinsed columns are heated to remove the last traces of solvent, a longer time is needed to condition the column than when using lower boiling point solvents. As a final rinse solvent, avoid halogenated solvents such as methylene chloride and chloroform if the column is going to be used with an ECD; avoid nitrogen containing solvents such as acetonitrile and DMF if the column is going to be used with an NPD. Trace amounts of these solvents create a high response with the specific detector. Very long column conditioning and detector stabilization times are required in these situations. If aqueous based or extracted samples have been injected (biological extracts, soils, waste water, etc.), include water as one of the rinse solvents. It is usually best to use it as the first solvent. Water and alcohols (methanol, ethanol, iso-propanol) should not be used to rinse PEG stationary phases. A small amount

**Table 11-2** Recommended rinse solvent volumes.

| Column diameter (mm) | Solvent volume (mL)[1)] |
|---|---|
| 0.10 | 1–1.5 |
| 0.18–0.20 | 3–5 |
| 0.25 | 5–7 |
| 0.32 | 7–10 |
| 0.53 | 10–15 |

1) Use the lower end of the range for 10–15 meter columns and the higher end in the range for 50–60 meter columns.

of stationary phase loss occurs especially if the column has been in service for several months.

The rinse solvent volume is dependent on the diameter of the column. Table 11-2 lists the recommended solvent volumes. Larger volumes can be used without any problems; however, this wastes solvent and time. When using multiple solvents, the previous solvent does not have to completely vacate the column before the next solvent can be started through the column. The next solvent can be added to the rinse kit vial or reservoir any time after the previous solvent is gone.

The flow rate of the solvent through the column should not exceed 1 mL/minute. For most columns, less than 1 mL/minute is delivered even at the highest recommended rinse kit pressure of 20 psi. Megabore (0.53 mm I.D.) columns are the only ones that can easily accommodate flow at greater than 1 mL/minute. Smaller diameter and longer columns require longer times to complete rinsing since lower solvent flow rates are delivered at the maximum rinse kit pressure of 20 psi. Columns with 0.25 mm internal diameters and 60 meters in length may require up to one hour for each solvent to pass through the column. Viscous or heavy solvents such as water or methylene chloride flow at slow rates. More time is needed for heavier solvents to pass through the column than for light solvents such as hexane.

### 11.7.3 Conditioning the Column After Solvent Rinsing

After the column has been rinsed, it must be purged of any remaining solvent. Allow the solvent rinse kit pressurizing gas to flow through the column for about 10 minutes after the last solvent has left the column. Detach the column from the kit and install the column as usual. Do not heat the oven or condition the column at this time. Pass carrier gas through the column for 10–20 minutes.

After the carrier gas purge, verify that the injector is leak free. Condition the column using a temperature program. The initial temperature should be 40–50 °C and the ramp rate at 2–3° per minute. Program up to the usual conditioning temperature for that column. Hold the final temperature until a stable baseline is obtained. Usually, obtaining a stable baseline requires 2–4 times longer after solvent rinsing a column than after a regular (i.e., non-rinsed) column installation.

### 11.7.4 Some Solvent Rinsing Considerations

There is no set limit to the number of times a single column can be solvent rinsed. In general, non-polar stationary phases can be rinsed more times than polar stationary phases. PEG stationary phases can tolerate the fewest number of solvent rinses. In most cases, the stationary phase degrades from normal use before the maximum number of rinses can be reached. PEG columns can experience a small amount of stationary phase loss upon solvent rinsing. This usually occurs when large volumes of polar solvents are used and for older columns.

If the same type of samples are routinely analyzed, the rate of solvent rinsing success should be considered. If three or more columns in succession are not saved by solvent rinsing, it is probably a waste of time and effort to continue rinsing columns in the future. If a moderate to large number of successes are experienced, future solvent rinsing of columns is usually worth the time and effort.

## 11.8 Guard Columns and Retention Gaps

### 11.8.1 Deactivated Fused Silica Tubing

A guard column or retention gap is a 1–5 meter piece of deactivated fused silica tubing attached to the front of the analytical column using some type of union. Shorter or longer pieces of tubing can be used, but 1–5 meters seems to be the most useful length. Guard column and retention gaps are installed for different reasons, but both provide the benefits of the other. Deactivated fused silica tubing is used for guard columns or retention gaps. It is the same tubing that is used to manufacture fused silica capillary columns, but it contains no stationary phase. The inner surface is deactivated to prevent any chromatographic interactions between the sample compounds and the tubing. Undeactivated fused silica tubing is available, but it is should not be used for retention gaps or guard columns. The diameter of the guard column or retention gap tubing is usually the same as that of the analytical column, but this is not mandatory.

## 11.8.2
## Guard Columns

A guard column is used to provide a surface for non-volatile residues to deposit so that they do not accumulate in the analytical column. If the non-volatiles residues reach the analytical column, they usually coat the stationary phase and cause chromatographic problems such as peak tailing, splitting and adsorption. The vaporized portion of the injected sample passes through the guard column unretained since there is no stationary phase present. The residence time of the sample in the guard column is extremely short; therefore, a buildup of residues does not significantly affect the injected compounds. Since the non-volatile residues accumulate in the guard column, it can easily be discarded or a portion cut off when it becomes excessively contaminated. This eliminates the need to cut small lengths from the front of the analytical column as it becomes contaminated with residues.

Guard columns do not prevent semi-volatile residues from reaching the analytical column. The semi-volatile residues can easily move from the guard column into the analytical column since there is no stationary phase to retain the residues. The semi-volatile residues have sufficient volatility to be transported by the carrier gas into the analytical column. Non-volatile residues do not have sufficient volatility for this to occur.

The guard column has to be replaced when it becomes saturated with non-volatile residues. The frequency of replacement depends upon the types and number of samples injected. Usually a peak shape problem similar to those experienced without a guard column occurs. The guard column is saturated with residues; thus some of them may have overflowed into the analytical column. If this has occurred, a small amount of the front of the column needs to be cut off; 10–20 cm is sufficient in most cases. Careful monitoring of the number of samples injected before problems occur allows regular guard column replacement before complete guard column saturation has occurred.

## 11.8.3
## Retention Gaps

A retention gap is used to improve peak shapes in specific situations. A retention gap is necessary for cool on-column injections especially when using secondary cooling since broad or tailing peaks are obtained without such a device. In some cases, narrower peaks are obtained with a retention gap when using splitless and Megabore direct injections. The most benefit occurs when there is a polarity mismatch between the sample solvent and the stationary phase.

If a non-bonded stationary phase is being used for splitless or on-column injections, a retention gap often protects the column. Sample solvent is collected and focused at the front of the column. This disrupts a small amount of the stationary phase in this region since the phase is not bonded. After repeated injections, enough damage occurs to cause poor peak shapes. Solvent focusing occurs in the retention gap if one is installed. A retention gap also acts in the same way as a guard column; thus it may protect the column against non-volatile residue accumulation.

### 11.8.4 Unions

There are a variety of unions for attaching two pieces of fused silica tubing. Stainless steel or glass, press-fit unions are used most often. Each has certain advantages with the primary consideration being the ease of installation. If properly installed, unions do not have any detrimental effects upon the chromatographic results.

The low cost and relative ease of installation are the primary appeal of glass, press-fit unions. Their tendency to leak under certain situations is the main drawback. Proper installation is the best insurance against a leak. The fused silica tubing must have clean and evenly cut ends. The proper cutting tools and technique are required to obtain evenly cut ends (Section 9.2.2). Clean the end of the tubing by wiping with a laboratory wiper wetted with methanol or acetone. Clean the interior of the union by rinsing with methanol or acetone. Make sure the union is dry before using. The end of the tubing is inserted into the union until the tubing fits snugly into the tapered region of the union. The polyimide column coating compresses slightly and acts as the sealing force to hold the union in place and prevent leakage of carrier gas. Close inspection of the junction should reveal a continuous, dark ring around the tubing at the contact point with the union. If the ring is broken, the union and tubing do not have a proper seal and the connection will leak; re-installation of the union is necessary. If the union easily drops off of the tubing, it is not correctly installed. In some cases, the size of the tubing and the union are such that a good seal is not possible. If after several attempts to connect the union does not work, try another union. The normal size variances in unions and column outer diameters occasionally cause a specific union and piece of tubing to be incompatible.

Sometimes upon repeated heating to 200–250 °C or above, a glass union and fused silica tubing permanently fuse together. Fusing does not always occur, thus it can not be relied upon to form a leak free seal. If a glass, press-fit union is used in a GC where the initial and final temperatures in a temperature program are far apart, or to connect a transfer line in a GC/MS system, leaks can develop upon use. A permanent and leak free connection can be made using polyimide sealing resin on the union, but the proper technique is essential.

*Polyimide Sealing Resin Procedure for Glass Unions*

1. Clean the union by rinsing with acetone.
2. Lightly scribe the column 1–2 cm from the end, but do not break the column at this time.
3. Rinse several centimeters of the outer surface of the tubing with acetone while holding the open end of the column upward.

4. Using a laboratory wiper (such as a Kimwipe™), apply a thin layer of polyimide sealing resin around the column over the previously scribed portion.
5. While still holding the column with the open end upward, break the column at the scribe mark.
6. Inspect the cut end with a magnifier. If the cut is unsatisfactory (uneven or jagged) or resin has gotten into the column, repeat the cutting and resin application process.
7. Upon a satisfactory cut, place a small drop of resin on one side of the tubing about 0.5 cm from the end of the column.
8. Insert the tubing into the union and press firmly (but not so hard as to break or chip the column end).
9. On the side where the drop of resin was applied, the gap between the tubing and the union should be filled with resin. If a gap is present, add just enough resin to fill the gap, but do not fill the entire gap around the column.
10. Carefully install the assembled column into a GC and establish carrier gas flow. Carefully examine the union for any slippage or looseness.
11. Heat the tubing for at least 30 minutes at 150 °C to cure and set the resin. Longer times are not necessary, but there are no problems associated with the longer cure times. If the union needs to be removed, the tubing can be cut adjacent to the union.

Stainless steel unions can be used instead of glass, press-fit unions. A better seal is usually obtained when compared to glass unions without polyimide sealing resin. Stainless steel unions can be difficult to properly install. They require the use of very small ferrules and it is practically impossible to visually determine if there is a gap between the tubing ends. A gap contributes to peak broadening or tailing due to the excessive dead volumes especially for small diameter columns. The column has to be installed and tested before any problems can be detected. It is not unusual for several attempts to be needed before a good connection is obtained. One union design uses a tapered glass insert so that the column ends can be positioned with a minimum of dead volume. The high thermal mass of stainless steel unions may pose some problems when rapid and extreme heating or cooling of the column oven is involved. The union does not cool or heat at the same rate as the tubing. Peak shape distortions and retention time reproducibility problems may result. Some type of clamp or holder is usually required to hold the heavy union in place. A metal union that is not held in place may lead to column breakage or pull a portion of the column onto the bottom of the GC oven. A final consideration is that the union ferrules often cost more than a single glass, press-fit union. In some union designs, the ferrules are not easily re-used, thus replacement ferrules are always necessary.

## 11.9
## Packed Injector Liners

If removing the majority of the harmful residues from the samples prior to injection is not possible or requires too much effort, using the injection liner as a trapping device may help to delay the onset of performance problems. A loosely packed plug of silanized glass wool inserted in the injection liner can act as a filter or adhesion site for some of the non-volatile residues and this reduces the amount reaching the column. Packed injector liners should be used with split injections and in high carrier gas flow Megabore direct injections; loss of efficiency often occurs with splitless injections.

The glass wool should be packed tight enough to filter out the residues without causing excessive restriction of the carrier gas flow. Distorted peak shapes or quantitation problems are signs of a too tightly packed plug of glass wool. Do not crush the glass wool into a tiny ball. Gently fold the glass wool fibers into a spherical shape and place the plug in the liner. Do not use regular lab grade glass wool. Injection liner glass wool must be silylated or irreversible adsorption and poor peak shapes occur. It is very difficult to correctly and completely silylate glass wool in the laboratory. It is recommended to purchase the silanized glass wool from a chromatography products supply company to ensure the proper material. Use the highest grade glass wool available such as pesticide grade.

For extremely dirty samples, a different type of packed liner can be used with split injections. A small amount of silylated glass wool is placed in the liner followed by 2–3 mm of packed GC column packing material. A 1–3% loading of OV-1, OV-101 or a similar stationary phase is recommended. Another small plug of silylated glass wool is placed on top of the packing to hold it in place. The packed column packing traps the contaminants The packing has to be routinely changed since it becomes dirty with use. Make sure the injector temperature does not exceed the upper temperature limit of the packing material. It is recommended that a newly packed liner be heated in the injector with carrier gas flow before the column is installed. Small amounts of the stationary phase from the packed column material may bleed off and contaminate the capillary column. Several hours at the injector temperature is usually sufficient to condition the liner packing.

## 11.10
## Gas Impurity Traps

One way to prolong column life is to minimize the amount of oxygen in the carrier gas. The use of high purity gases or the installation of an oxygen trap in the carrier gas line are the easiest methods. Moisture or water traps are also used in many systems since water can contribute to a higher background signal especially when using water sensitive detectors.

If a water trap is being used, it should placed before an oxygen trap. Oxygen is not removed by a water trap; however, water is retained on an oxygen trap. If the oxygen trap is placed first, its capacity can be prematurely reached by the presence of water in the gas. Water traps have very high capacities and are often inexpensively re-filled. Oxygen traps are much more costly than water traps. Oxygen traps have much shorter lifetimes, are not re-fillable, and require more frequent changing. The cost of a water trap is quickly returned by the savings in oxygen traps.

Installation of an indicating oxygen trap is often recommended if routine changing of the oxygen trap does not occur. The indicating oxygen trap is installed after the regular oxygen trap and is intended only as a warning that the regular oxygen trap has expired. An indicating oxygen trap has low capacity and is not intended to remove large quantities of oxygen. It quickly expires if a regular oxygen trap is not in place. The indicating trap can be very important since an expired oxygen trap is worse than no trap at all. An expired trap may leach or dump materials back into the carrier gas that often causes high background signals or contamination peaks.

Glass or metal bodied traps are preferred over plastic bodied traps. Some detectors such as ECD's or mass spectrometers are very sensitive to impurities that may be present in the plastic trap bodies. Some plastics may be slightly permeable to oxygen. Traps are usually used just for the carrier gas. A water and oxygen trap is often recommended for the auxiliary or makeup gas for ECD's. Most other detectors are not significantly affected by the presence of water or oxygen in their gases. Traps should be placed in a vertical position. When placed horizontally, it is possible for some of the packing material to settle downward and form an open channel near the wall of the trap. Most of the gas goes down the channel instead of passing through the packing material. Water or oxygen is not removed when this happens. The traps should be placed as close to the GC as reasonably possible and they need to readily accessible for inspection and changing. If multiple GC's are on the same gas supply line, the traps can be placed before the point where all of the gas lines branch off to the individual systems. Minimize the number of gas line fittings past the traps. Each fitting potentially reduces the quality of the gas and counteracts the activity of the traps.

Hydrocarbon traps are available to remove trace levels of volatile organic compounds from the gas. These traps are packed with activated charcoal, thus they trap more than just hydrocarbons. Hydrocarbons traps are used where problems with organics from improperly cleaned gas lines, low quality gases, and compressor supplied gases are used. Hydrocarbons traps are not necessary unless contamination problems have been experienced.

Some GC's have small, internal traps, usually packed with activated charcoal or molecular sieves. These traps need to be replaced or rejuvenated periodically. The recommended procedures and internal trap locations can usually be found in the instruction manual for the gas chromatograph.

When using the higher quality grades of carrier gases, the use of gas impurity traps is debatable. The levels of oxygen and water are so low in the higher grades of carrier gas, the traps do not improve the quality of the gas. While the traps do last a long time, the cost, maintenance and leak concerns with the extra fittings may not be worth any of the possible gains. Traps are often not used when using the higher grades of carrier gas. The extra cost of the higher grade gas is often offset by the savings in not using traps. Also, the higher maintenance (inspection and changing of the traps) and leak possibility with traps is eliminated. Traps are necessary for the lower grades of gases. The traps reduce the levels of water and oxygen, thus improving column life and, in some cases, detector response. For situations where columns are frequently replaced and discarded, the use of traps may not be economically reasonable. No gain in column life is obtained and, as long as sensitivity is not affected by the absence of traps, installing gas traps is an unnecessary expense.

## 11.11 Column Storage

Columns should be stored such that scratching or abrasion of the fused silica tubing is avoided. The original column box is perfect for storage purposes. The column should be sealed by placing a GC septum over each end. Old, used septa can be used. It has been found that flame sealing or welding the end shut is not necessary and sometimes does more harm than good. Columns do not need to be flushed with carrier gas before storage.

The shelf-life of columns is not completely known. Non-polar stationary phases have exhibited no performance degradation even after five years of storage. There have been isolated incidences of new columns with very polar stationary phases exhibiting slight performance changes after years of storage. The changes are small and often do not cause any problems.

## 11.12 Column Repair

Capillary columns with damaged stationary phases can not be repaired. Columns can not returned to the manufacturer for repair since no column manufacturer accepts columns for repair. The cost and effort of repairing a column (if possible) is nearly the same as producing a new column. It is much better and easier to obtain a new column. Besides, many types of damage are not reversible and can not be fixed. There are some rejuvenation solutions available that claim to return capillary column performance. These solution rarely work and any improvement is very short lived.

# 12
# Troubleshooting Guidelines, Approaches and Tests

## 12.1
## Introduction

Most chromatographic problems can be quickly resolved if a logical and efficient approach to the problem is maintained. Usually any superfluous amount of panic, fitting tightening, system alterations, random button pushing and knob turning only makes the actual problem more difficult to isolate and remedy. It is important not to jump to conclusions or make assumptions based on incomplete data. It is very easy to get fooled if some of the facts are not available or ignored. The best allies when troubleshooting are a good understanding of the basics of gas chromatography, capillary GC columns, the proper operation and characteristics of the particular model of gas chromatograph, the proper tools and common sense. Deficiencies in any of these areas make problem solving a more difficult undertaking.

One difficulty in troubleshooting is the multiple cause and effect nature of many GC problems. A single symptom can be caused by many different and non-related problems. Also, a single problem can have multiple symptoms some which are not evident in all cases. Finding the real problem requires the proper approach and evaluation of the situation.

Most problems are evident in the chromatogram; however, the chromatogram is the last link or operation in the chromatographic process. This means that the source of the problem can be anywhere or anything before the data system or recorder. In some cases, the recorder or data system is the source of the problem. Since many problems are only evident via the chromatogram, logical and simple procedures to find or eliminate possibilities are required to quickly remedy any problem situation.

*The Troubleshooting and Maintenance Guide for Gas Chromatographers, Fourth Edition.* Dean Rood

ISBN: 978-3-527-31373-0

## 12.2 Approaches to Solving GC Problems

### 12.2.1 Systematic Approach

It is often forgotten that the gas chromatograph and capillary column function as a complete system and not as two individual parts. In addition, the gas chromatograph can be viewed as constructed of a number of smaller systems or parts (injector, detector, etc.). Many of the small parts can be broken down even further. A problem or deficiency in any part of the GC including the column often results in some type of chromatographic difficulty. Using a logical and controlled troubleshooting procedure to quickly and accurately identify the source of the problem is the systematic approach. Maintaining the realization that the entire gas chromatographic system is a collection of different, but interrelated parts makes troubleshooting an easier and less complicated enterprise.

### 12.2.2 Checking the Obvious

*Often the most obvious problems or solutions are overlooked because of their simplicity.* Usually the simplest items are the easiest ones to check. Even if a setting or measurement was made while the GC was functioning properly, check them again after a problem occurs. Accidental and inadvertent changes occur without anyone's knowledge. Many of the areas that need to be checked seem trivial, but too many times it is deficiencies, malfunctions or incorrect parameters in these areas that are responsible for the problem. Some areas can be instantly eliminated using common sense. For example, if a chromatogram is being generated, it is very likely that the data system is connected properly and the detector is on. If no peaks are generated with an injection, both of these areas are possible causes of the problem.

The usual areas of first inspection and questions that need to be answered are:

1. *Gas supply:* Are the correct pressures and gas types being used?
2. *Gas flows:* Is the carrier gas average linear velocity correct? Are the flows for the split vent, septum purge, splitless purge, detector combustion and auxiliary gases correct?
3. *Temperatures:* Are the correct temperatures set for the column, injector, detector and transfer line?
4. *Leaks:* Are there loose or bad fittings or gas lines? Has the septum been changed?
5. Is the detector turned on?
6. Are the intended injector and detector being used?
7. *Sample:* Is the injection volume correct? Is the sample concentration and solvent correct? Is the sample old or properly stored?

8. *Column:* Is the correct column being used? Is the column properly installed or broken?
9. *Syringe:* Is the syringe plugged or a different volume or style?
10. *Data system:* Are the correct channel and settings being used. Are the connections and cabling correct?
11. *Gas impurity traps:* Are the traps expired or leaking. Are they properly installed?

### 12.2.3 Looking for Changes

*Be conscious of any changes in any area related to the analysis.* Some changes are very obvious such as a new gas cylinder or column. Other changes can be more difficult to detect or notice. For example, vial or cap supply, solvent brand or bottle, syringes, pipets, extraction columns, GC septa, the person making the injections have all been causes of changes in the chromatographic results. Some of the smallest and most trivial changes can affect the results. Changes occurring outside of the laboratory are often not noticed or reported, thus investigations of the onset of a problem can lead to areas outside of one's control.

### 12.2.4 Looking for Trends, Patterns and Common Characteristics

*A set of patterns and trends can often help to narrow the list of possible problems sources.* Characteristics such as compound structure, volatilities and solubilities can provide clues about the cause of the problem. Some trends or patterns are very easy to detect while others are nearly impossible since so many variables can be involved. Simple items such as sample color, brand of vial cap, time of day, number of previous sample injections are easy to find after a little investigation. Others such as sample exposure to light, position in a sample evaporator, ambient air humidity and contamination by solvent extractions performed in an adjacent laboratory can be very difficult to find and verify.

Sometimes the presence of a common trait or characteristic provides useful troubleshooting information. As an example, there is a noticeable pattern of decreasing peak tailing with increasing peak retention. In general, the earlier the peak retention time, the more volatile the respective compound. Thus, the trend in this example is the more volatile the compound, the worse the peak tailing. This points to a problem in the injector or carrier flow behavior since highly volatile compounds are more susceptible to expansion, backflash and vaporization problems. Activity is probably not the problem since peak tailing becomes worse with increased peak retention. Also, only the active compounds would exhibit peak tailing.

### 12.2.5
### Asking "If ... Then ..." Questions

*Some of the best types of questions (after the obvious ones) are the "if ... then ..." variety.* They are the basis of deductive reasoning. A good foundation in basic GC theory and common sense are very powerful GC troubleshooting tools. Before attacking a problem, asking and answering some questions can often help focus the search for the problem. For example, *if* column activity is the cause of the peak tailing, *then* the alcohol peaks in the column test mix should tail but not the hydrocarbon peaks. If the alcohols or similar compound peaks do not exhibit tailing, column activity is probably not the cause of the peak tailing. If all of the peaks tail, then column contamination is a possibility. Another test needs to be run or a new set of "if ... then ..." questions need to be asked. "If ... then ..." questions are very powerful and can often narrow the possible problem causes. Very effective troubleshooting tests can often be built around "if ... then ..." types of questions.

### 12.2.6
### One Thing at a Time

*It is often best to attack or investigate one area at a time.* There is a very strong and unconscious urge to change more than one parameter or variable at a time. Also, there is the urge to take everything apart before examining and evaluating the situation. Trying to examine and repair too many things at once is often less efficient and more time consuming than using a series of simple and conclusive tests. If numerous variables are simultaneously altered, the corrective action that repaired the problem is not known. While fixing the problem is important and the ultimate goal of troubleshooting, knowledge about the actual source and cause of the problem is better. Armed with this knowledge, preventative measures can be undertaken so that the problem does not re-occur in the future, and if it does occur, the proper actions can be immediately taken to repair the problem.

### 12.2.7
### Moving from the General to the Specific

*Do not try to solve a problem with a single, all-encompassing test, experiment or deduction.* Determine the most general possible causes for a problem and then start to narrow the focus on specific causes. As an example, a problem with a loss of resolution occurs. A loss of resolution is caused by a decrease in peak separation and/or an increase in peak width; thus, the real problem is a separation and peak width issue. First, investigate whether a separation decrease or peak width increase (or both) has occurred since this is the general problem. Next, determine the possible causes of whichever problem was detected – this is focusing on the more specific problems. This approaches helps to focus attention on the causes of the real problem instead of the symptoms. This is also known as root cause analysis. Jumping to conclusions without finding the real problems often wastes time and effort since the wrong problems are often investigated.

### 12.2.8
### Eliminating the Possibilities

*The easiest way to get to the source of a problem is to reduce the number of possible causes.* Instead of trying to find the source of the problem, try to eliminate the sources that can not be the cause. One of the remaining causes is probably the reason behind the problem. Simple experiments or tests are the best. Trying to eliminate too many choices with a single test often confuses the situation. Common sense usually dictates the plan of attack. As an example, some peaks are missing in the chromatogram. Areas such as the detector being on, the presence of carrier gas flow, correct data system connection can be eliminated since some peaks are observed. Problems in these areas would result in the complete absence of all peaks. If the retention times of the peaks are normal, problems in areas such as carrier gas linear velocity and column temperature can be eliminated. Changes in these areas would result in a change in the retention times. If the sample is injected into another suitable GC system, a large amount of information can be gained. If the correct or expected results are obtained, the sample and the syringe can be eliminated as possibilities. This has narrowed the possibilities down to the injector or column. Installation of a verified good column and its proper performance can eliminate the original column as the source of the problem. With a little thought and a few simple tests, many possible problem areas are eliminated. Another method to eliminate possibilities is to divide as much of the GC system into individual components and evaluate each component. By eliminating each component, the one causing the problem should be eventually discovered.

### 12.2.9
### Divide and Conquer

*By dividing the GC system into separate components, the performance of each component can be evaluated.* If a component functions properly, it can usually be eliminated as the problem source. A divide and conquer approach works very well when there are multiple units comprising the GC system. As an example, peak shape problems occur with a GC equipped with a purge and trap sampler. Introduction of a known standard using the vaporization injector instead of the purge and trap sampler divides the system into two sections. If the problem goes away, the source of the problem is associated with the purge and trap sampler; if the problem remains, the problem is probably with the GC or column. By removing the purge and trap sampler from the system ("divide"), it is easy to determine whether it is responsible for the problem ("conquer"). Dividing and conquering is eliminating possibilities in a systematic manner.

There are two approaches to dividing a system. One is to split the system into two equal parts (as much as possible) so that the largest sections are examined.

This is often not possible due to the design and operational requirements of many GC systems. The other approach is to start at the data system and work towards the injector and gas supply. After the data system is investigated, the detector is checked. This continues until the problem is found. This is often necessary in simple GC systems with very few options, accessories or operational possibilities.

## 12.3 Troubleshooting Tools

Maintenance and troubleshooting are easier when the right tools are readily available. Some tools are indispensable for successful troubleshooting and also for maintaining capillary GC systems in the best working order. Every gas chromatograph health care package and first aid kit should contain:

1. *Flow meter:* A digital or manual model is suitable. The volume range of the meter should be compatible with the values being measured. For example, using a 10 mL flow meter to measure 100 mL/min may result in irreproducible and inaccurate flow values. Even with electronic pressure or flow systems, a flow meter is useful to verify flow rates. Be aware some digital flow meters may interfere or alter the actual flow values for electronically controlled flow systems.

2. *Check-out column:* This is a column of known, verified and documented performance and quality. It is used to check the performance of the GC or to eliminate the original column as the problem source (Section 12.4.3).

3. *New syringe:* A clean syringe exclusively reserved for troubleshooting purposes. This helps to determine whether the original syringe is delivering the correct volume or is contaminated.

4. *Non-retained compound:* A non-retained compound is used to measure and verify the linear velocity of the carrier gas.

5. *New septa, seals, O-rings, ferrules and injection liners:* These are used to replace parts that eventually become defective, worn out or dirty with use.

6. *Column test mixture or reference sample:* Test mixtures are used to diagnose select system problems. They are also useful to compare current system performance with previous, satisfactory performance. With the correct interpretation, test mixtures are very powerful diagnostic tools.

7. *Leak detector:* An electronic detector is the best (but most expensive) way to find leaks. Liquid leak detection fluids are satisfactory for detecting leaks in non-heated areas.

8. *Log or record book:* Log books are invaluable for determining the past history of the system (e.g., number and types of samples injected, septa changes,

occurrence of problems, maintenance schedules). A log book makes it easier to spot trends or to determine the frequency of required periodic maintenance. The log book is also a good place to keep performance specifications or requirements. Many of the newer GCs and data systems have electronic logs and error reporting features.

9. *Instrument manuals:* These are not a last resort. Manuals are a very useful resource for determining system performance specifications, proper operating guidelines and instructions on the disassembly of hardware like injectors and detectors. Instrument manuals are a good source of troubleshooting information specific to a particular GC.

## 12.4 Troubleshooting Tests

### 12.4.1 Jumper Tube Test

The jumper tube test is especially helpful in tracking down contamination, baseline drift or noise problems of unknown origin. It is simple, quick and helps to find or eliminate possible problem sources.

There are three steps to a jumper tube test. First, remove the column and cap off the entrance to the detector. Turn on the detector as usual with all of the normal detector gas flows and temperatures. Activate the detector by igniting the flame, turning on currents or whatever else is needed for operation. Give the detector 15–60 minutes to stabilize based on previous experience. Using the same oven temperature conditions and data system settings as when the problem occurred, perform a blank run (i.e., no injection) and collect the chromatogram. If excessive noise, baseline drift or other problems are evident, the detector is probably the source of the problem. If the baseline is flat and there is no evidence of the problem, the detector is not the source of the problem.

The second part of the test is to install a 1–2 meter piece of clean, deactivated fused silica tubing in the GC using the same method and conditions as for a regular column. Do not use an old piece of column. Establish a carrier gas flow similar to the one used with the regular column. Perform the blank run again and collect the chromatogram. If excessive noise, baseline drift or other problems are evident, the source of the problem is most likely the injector, carrier gas or incoming gas lines. The detector has already been verified to be functioning satisfactorily with the first test. Since the column is not installed, it is not responsible for the problem. If the baseline is completely or fairly flat and there is no evidence of the problem, the injector, carrier gas or incoming gas lines are probably not the source of the problem. The problem is most likely associated with the column.

The final part is to install and condition the column as usual. Perform the blank run one more time. If the problem is not evident, the problem is solved.

Sometimes, the process of re-installing a column corrects the problem. If the problem returns after installing the column, the problem is with the column or installation technique. The column may be not damaged, but only contaminated or improperly installed. If an injector or related areas were found to be contaminated, it is likely that the column may be contaminated also. If the contaminants in the injector have also contaminated the column, solvent rinsing the column may be warranted.

### 12.4.2 Condensation Test

The condensation test is used when contamination of the injector, carrier gas or gas lines is suspected. Keep the column at 40 °C for at least several hours; 8–24 hours is better. Afterwards, temperature program the column at 10–20°/min starting at 40 °C and ending at the column conditioning temperature with a final hold of 15–30 minutes. Collect the chromatogram. As soon as the column has cooled down and the GC is ready, immediately start another blank run using the same conditions. Collect the chromatogram and compare it to the first one. If the chromatograms are similar, it is unlikely that the injector, carrier gas or gas lines are severely contaminated. If the first chromatogram (the one kept at the low temperature for 8–24 hours) is worse than the second chromatogram, there is a contamination problem with the injector, carrier gas and/or gas lines, or significant septum bleed.

By allowing the column to stay at a low temperature, many of the contaminants are trapped (retained) by the column. The longer time at the low temperature, allows more of the contaminants to be collected. When the column is temperature programmed, the collected contaminants elute from the column. More contaminants accumulate when the column is left for many hours at the low temperature (first chromatogram), thus the blank run looks worse than the column with minimal time at the low temperature (second chromatogram). Similar first and second chromatograms mean there was only a small amount of contaminants collected in the column even after hours at the low temperature. The first chromatogram often looks slightly worse than the second one. It is common for a small amount of system contamination and septum bleed to be present. It is large differences between the two chromatograms that indicates a contamination problem exists.

### 12.4.3 Check Out Column

A check out column is one that is used to evaluate the performance of the GC. The performance of the check-out column should be thoroughly documented for comparison purposes. Upon removal of the original column and installation of the check out column, the original column is eliminated as a problem cause. If the check out column's performance is normal (as previously verified and documented), the original column or its installation is at fault. If a problem is noticed with the check out column installed, the problem is with the GC (providing the check out column was properly installed and set up).

Ideally, the check out column should be the same size and phase as the regular column. If more than one column is used in the GC, use the column with the most non-polar stationary phase for the check out column. It is not essential that the column be the same as one being used since the GC is being evaluated and not the column. A similar column is still best since its performance, characteristics and behavior are familiar. The check out column should be used exclusively for evaluating system performance and not used for sample analysis. A new column makes a satisfactory check out column; however, its performance in the GC has not been previously documented and adds a bit of uncertainty to the check out procedure.

### 12.4.4 Column Exchange

A column exchange is another method to determine whether a column is the cause of a problem. Another properly functioning column or GC is needed for this test. Ideally, a column with the same description is available. If an installed column is suspect, replace it with the other column. If the problem returns, the GC or the column installation technique is the problem. If the problem does not return, the first column is probably at fault.

Another approach is to install a suspected problem column in a properly working GC. If the problem returns, the column is probably at fault. If the problem does not appear, the original GC is probably at fault. Sometimes the process of re-installing a column fixes the problem, thus re-install the suspected problem column back in the original GC and evaluate before attempting to repair the GC.

### 12.4.5 Static Pressure Check

A static pressure check is probably the fastest and most complete method to determine whether a leak is present in the injector or the carrier gas lines. A column head pressure gauge or in-line pressure gauge in the incoming carrier gas line is necessary to perform a static pressure check. To perform the test, the column is removed from the injector and the column inlet is sealed with a leak free cap. Also, replace the septum with a new one. For a split/splitless injector system, set the split line in the closed position (purge off) so that a leak in the vent solenoid can be detected. Cap off the septum purge line to prevent gas from escaping from this route. Pressurize the injector by supplying carrier gas. After the pressure has stabilized, eliminate the flow of carrier gas to the injector by turning the carrier gas off at its source (i.e. at the cylinder or in-line pressure regulator). Do not use the flow controller or pressure regulator on the GC to stop the gas flow. Some controllers are set so that they do not completely stop carrier gas flow. Also, some controllers may be easily damaged if they are tightened excessively in the off position. The head pressure should not drop more than 1–2 psi over a 10–15 minute time span. A drop in the head pressure indicates that there is a leak in the injector or the carrier gas lines. The most common sources of leaks are loose fittings on the injector or the incoming gas lines.

### 12.4.6
### Column Test Samples

Column test samples can often help to pinpoint problems. The column manufacturer's test sample can measure activity, efficiency and retention characteristics. A self made test sample that mimics the actual samples provides information about the performance of the GC for a particular analysis. For either type of test sample, a lot of information about the GC is obtained. Injecting and evaluating a test sample should be one of the first tests performed after all of the obvious and simple inspections and tests have been completed. Chapter 10 outlines the use of the various types of test samples and the interpretation of the test results.

# 13
# Common Capillary GC Problems and Probable Causes

## 13.1
## Using This Troubleshooting Guide

*This chapter is intended as a starting point or guide through a capillary GC problem. Referencing other book sections, manuals or additional investigative work will be necessary.* It is recommended for novices to GC troubleshooting be familiar with the material in Chapter 12 before attempting in-depth troubleshooting activities.

The chapter is organized by chromatographic problems. Potential causes for each problem are listed in the order of most to least common or likely; however, the order is a loose guideline at best. Where appropriate, the previous book section pertaining to the explanation, prevention or solution to a problem is provided in parentheses.

It is practically impossible for any troubleshooting book or guide to identify, discuss and explain every possible problem and solution. Usually generalizations and simplifications are necessary, and this book is no exception. Also, there seem to be exceptions and contradictions to nearly every problem, cause and symptom.

There are no short-cuts or foolproof methods to easily and quickly diagnosis and fix chromatographic problems. A basic understanding of GC operation and parameters, good data and information, and a deliberate plan are all necessary to effectively and efficiently troubleshoot any GC problem.

## 13.2
## Troubleshooting Checklist and Pre-Work

Preparing a written or mental checklist before working on a GC system is recommended. This generates a more thoughtful and planned approach to solving a problem. It also aids in thinking about the problem before prematurely drawing conclusions and randomly trying solutions. It is very easy to forget or overlook an important detail when starting without a plan. Having the appropriate information and documentation, GC parts and tools before working on the GC is also highly recommended. Sometimes the problem and solution is identified during the process of preparing a checklist, looking up information, and gathering

*The Troubleshooting and Maintenance Guide for Gas Chromatographers, Fourth Edition.* Dean Rood

ISBN: 978-3-527-31373-0

the needed parts and tools. A planned troubleshooting approach nearly always yields a faster and more complete solution to the problem.

The information to obtain and tasks to complete before proceeding with troubleshooting are:

1. *GC parameters:* Obtain the GC analysis method information.
   a) Column description and its use history
   b) Oven temperatures and times
   c) Carrier gas type and velocity (or flow rate)
   d) Injector settings – split flow or ratio, purge times, temperature, liner description, injection volume, liner, etc.
   e) Detector settings – gas types and flows, temperatures, current, mass ranges, etc.
   f) Sample – compounds and their structures, sample solvent, concentration, age, etc.
   g) Other components – operating parameters of other system components such as autosamplers, headspace samplers and purge & trap samplers

2. *Reference chromatograms and performance information:* Obtain a known "good" chromatogram as a reference so the following can be measured or observed:
   a) Peak shape and width
   b) Peak resolution and separation
   c) Size and number of peaks
   d) Bleed
   e) Baseline stability
   f) Signal to noise (sensitivity and selectivity)
   g) Solvent front size and appearance

3. *Tools:* Gather the necessary tools to make measurements, run tests, and disassemble and re-assemble the GC and components (Sections 12.3 and 12.4).

4. *Spare parts:* Gather the consumable items required for periodic replacement or maintenance (Section 12.3).

5. *Check the Obvious:* Check pressures, temperatures and flows settings (Section 12.2.2).

6. *Make note of any changes:* Record anything different from before and after the problem occurrence (Section 12.2.3).

7. *Make note of any patterns or trends:* Record any repetitive, correlated, consistent or coincidental occurrences, patterns or trends (Section 12.2.4).

## 13.3 Baseline Problems

### 13.3.1 Baseline Drift or Wander

Drift or wander is small to moderate upward or downward movement of the chromatogram baseline. It can also be a combination of multiple upward and downward movements within a single chromatogram which may resemble humps or very broad peaks. Make sure not to confuse baseline drift or wander with normal column bleed.

*Primary Root Cause:* A slow addition and/or subtraction to the normal background or detector signal.

1. *GC or column contamination:* Semi-volatile residues are eluting from the column. The residues can be located in the column, injector, gas lines, traps or carrier gas. A condensation test may help in identifying this cause (*Contamination 11.6; Condensation Test 12.4.2*).
2. *Incompletely conditioned column:* The column has not been properly conditioned (i.e., too short of a time) (*Column Conditioning 9.2.8*).
3. *Excessive or severe column bleed:* Usually the rise in the baseline from column bleed should start 30–40 °C below the upper isothermal temperature limit of the column. A severely damaged column may exhibit substantial amounts of bleed and baseline rise below this temperature range (*Column Bleed 4.7*).
4. *A change in the carrier gas flow during a temperature program:* This is normal when maintaining the GC at a constant injector pressure with a concentration dependent or flow sensitive detector. Some amount of baseline drift is normal with these types of detectors; however, the amount of drift may be amplified if there are contaminants in the carrier gas or a flow controller is unstable (i.e., malfunctioning). Operating in the carrier gas constant flow mode often reduces the amount of normal baseline drift (*Detector Sensitivity 8.2.4; Constant Flow Mode 7.7.3*).
5. *The detector has not been allowed to fully equilibrate or condition:* This is most common for ECD, TCD and MS systems. Many of these detectors may require several hours to fully stabilize even though they are operational.
6. *Detector difficulties:* Irregularities or inconsistent temperature, current, filament, bead, lamp or gas flow rates.
7. *Loose, dirty or defective electrical connections:* Usually a problem with the detector or data system cables and circuit boards connections.

### 13.3.2 Noisy Baseline

Baseline noise is always present. It should be small, but visible at more sensitive settings. A reference is needed to determine whether the amount of noise is normal or higher than before. A noise measurement taken during normal and satisfactory operation should be available for comparison purposes. Signal to noise is an excellent measurement for reference purposes (*Detector Sensitivity 8.2.4*).

*Primary Root Cause:* A fairly consistent, and usually permanent, change in the steady state background signal.

1. *Detector problems:* A dirty or defective flame jet, collector or cell. An expired, damaged or dirty bead, filament, electron multiplier or photomultiplier tube. The noise tends to be fairly consistent over short time periods and progressively become worse over longer time periods.
2. *Column or injector contamination:* The noise is usually irregular, changes with time and often accompanied by baseline drift. Sometimes peaks may be visible above the noise (*Contamination 11.6*).
3. *An air leak in an ECD, TCD or MS:* The leak is usually located in the vicinity of the detector (column fitting or gas line).
4. *The column is inserted into the detector flame:* Occurs with a FID, NPD or FPD. Peak tailing is often observed with this cause (*Column Position: FID 8.3.3; NPD 8.4.3; FPD 8.7.3*).
5. *Incorrect air and hydrogen flow rate ratios:* Occurs for a FID, NPD or FPD (*Gases: FID 8.3.2; NPD 8.4.2; FPD 8.7.2*).
6. *Loose, dirty or defective electrical connections:* Usually a problem with the detector or data system cables and circuit boards connections.
7. *Outside interferences:* Electrical or magnetic fields generated by other pieces of equipment in the vicinity of the GC. High volume air currents around the detector.

### 13.3.3 Spikes in the Baseline

Spikes appear as a narrow line and not as a peak (i.e., a single line with no width). They are usually irregularly spaced.

*Primary Root Cause:* An extremely fast increase and decrease in the detector signal.

1. *Particulate matter passing through the detector:* Usually the particles are from a dirty flame detector with the collector being the primary source. Gently tapping on the collector with a non-metallic object will create spikes as particles are dislodged from the collector.

2. *Particles eluting from PLOT columns:* The spikes will be irregularly spaced. The amount of spiking is usually most severe immediately after installing a PLOT column. Poraplot type stationary phases (Q, U, etc.) tend to be the worse than Alumina or Molecular Sieve PLOT columns. Some spiking from PLOT column is not unusual and can not be readily avoided. Installation of a column trap may reduce the amount of spiking (*Porous Layer Stationary Phases 4.2.4*).
3. *Loose connections on cables and circuit boards:* Thermal expansion and contraction of connections may cause the spiking behavior especially with very old or dirty connectors.
4. *Cycling of other equipment in the same room or electrical circuit:* Check the other equipment to determine if the appearance of a spike coincides with a particular event occurring with one of the other pieces of equipment.

## 13.4 Peak Shape Problems

Capillary GC peaks are expected to be symmetrical, narrow and sharp. Deviations from this ideal form are a common problem. Depending on compound structure, column and GC conditions, some compounds peaks are always poorly formed (e.g., tailing, broad, distorted).

### 13.4.1 Tailing Peaks

Peak tailing is caused by many different problems and takes on many different appearances. It is probably one of the most common problems, but it also one of the first signs of a failing column with normal use. Some compounds always exhibit tailing peaks, thus the presence of a tailing peak is not always indicative of a column or GC performance problem. Finding and removing the cause of a peak tailing problem can sometimes be difficult.

*Primary Root Causes:* (1) A condition or event causing some of a solute's molecules to travel down the column at a slower rate than the remaining solute's molecules. (2) A condition or event causing a portion of the sample to enter the column later than the majority of the sample.

1. *Column contamination:* An accumulation of sample residues or solid materials in the column. Active compounds are more susceptible to tailing problems. Severe cases of contamination also cause less or non-active compounds to tail especially if the contaminants are solid particles such as small pieces of ferrule or septa (*Contamination 11.6; Tubing Activity 3.2; Active Compounds 10.2*).
2. *Column activity:* Active compounds are the most susceptible to tailing problems. Inactive compounds do not tail due to column activity. Usually the tailing

is more severe for the later eluting peaks unless the later eluting compounds are significantly less active than the earlier eluting peaks (*Tubing activity 3.2; Active compounds 10.2*).

3. *Injector contamination:* An accumulation of non-volatile sample resides or solid debris in the injector. Tends to be a larger problem with splitless injections.
4. *Liner activity:* An insufficiently or inadequately deactivated liner for the specific analysis. Liners lose their deactivation with use, thus the problem may slowly become worse over time. Tends to be a larger problem with splitless injections.
5. *Dead volume:* The result of poorly installed liners, unions or columns. The peak tailing is usually more severe for the earlier eluting peaks (i.e., more volatile) compounds. A broken liner may also cause the same problem.
6. *Backflash:* Peak tailing is usually more severe for the earlier eluting peaks (i.e., more volatile) compounds (*Backflash 7.2.3*).
7. *Lack of solvent effect or cold trapping:* Occurs for splitless and on-column injections (*Splitless 7.4.3 and 7.4.4; On-column 7.6.2*).
8. *Peak co-elution:* A small peak buried in the base of a larger peak may appear as tailing. Close inspection often reveals a small shoulder or plateau instead of a curved or sloped tail.
9. *Poorly matched stationary phase and sample:* Most problematic when the polarity of the sample compound and solvent is very different from the column, or when there are large differences in the pH of the sample compound and stationary phase.
10. *Sample decomposition in the column:* Usually the peak tail is extended and flatter (plateau-like) compared to a curved or sloped tail for column activity or contamination.
11. *Peak elutes immediate prior to a sample solvent or a very high concentration sample compound:* The compound interacts with the large amount of solvent or other compound adjacent to it in the column.
12. *Too low or no detector makeup gas:* Usually a change in detector sensitivity occurs also (*Detector Makeup Gas 8.2.2*).
13. *Poor injection technique:* An erratic or discontinuous depression of the syringe plunger (*Injection Techniques 7.8*).
14. *Condensation of the vaporized sample on a cooler region in the injector:* The septum, incoming gas lines, transfer lines or a poorly heated region of the injector are the most common areas. The peak tailing is usually more severe for the later eluting peaks (i.e., less volatile) compounds.
15. *Glass wool in the injector liner is old or in the wrong location:* The plug of glass wool is too tightly packed, very low in the injector liner or active. The glass

wool may have moved if the septum nut was removed while the injector was still at a high pressure (*Packed Injector Liners 11.9*).

16. *Overloading of a PLOT column:* PLOT columns exhibit tailing instead of fronting when overloaded (*Column Capacity 2.11*).

### 13.4.2 Fronting Peaks

Fronting peaks are noted by tailing or slanting of the leading edge of the peak. They often appear to be similar to mirror images of tailing peaks.

*Primary Root Causes:* (1) A condition or event causing a portion of the sample to enter the column earlier than the majority of the compound. (2) A condition or event causing some of the solute's molecules to travel down the column at a faster rate than the remaining solute's molecules.

1. *Column position in the injector:* The column is inserted the wrong distance into the injector. The error usually has to be large to have a significant influence on peak shape. Higher boiling point compounds are usually more affected (*Split 7.3.5; Splitless 7.4.7; Direct 7.5.4*).
2. *Compound and solvent/compound polarity mismatch:* This usually occurs when the sample compounds are significantly more soluble in the sample solvent than in the stationary phase.
3. *Column overloading:* More of a shark fin shaped peak instead of a curved front peak (*Column Capacity 2.11*).
4. *Peak elutes immediate after a sample solvent or a very high concentration sample compound:* The compound interacts with the large amount of the solvent or another compound adjacent to it in the column.
5. *Sample decomposition in the column:* Usually the peak front is extended and flatter (plateau-like).
6. *Poor injection technique:* Often caused by having a volatile sample or solvent in the syringe needle when it is inserted into a heated injector. Usually observe greater fronting for the earlier eluting (more volatile) peaks (*Injection Techniques 7.8*).
7. *Lack of a retention gap for on-column injections with secondary cooling* (*On-column Retention Gap 7.6.4*).
8. *Co-eluting peaks:* A small peak buried in the base of a larger peak may appear as fronting. Close inspection often reveals a small shoulder or plateau instead of a curved or sloped leading edge.
9. *Injector is too cold:* Condensation of the sample in the injector (*Injector Temperature 7.2.1*).

### 13.4.3 Extremely Broad or Rounded Peaks

Capillary column peaks should be sharp and pointed at the top. Column efficiency, injector and system conditions all affect peak width, thus they need to be considered before making judgments on any peak broadening situation.

*Primary Root Causes:* (1) Sample entering the column in a broad band without being re-focusing into a narrow band. (2) Sample band broadening while traveling through the column.

1. *Column contamination:* Only occurs when there is severe column contamination (*Contamination 11.6*).
2. *Lack of the solvent effect or cold trapping:* Occurs with splitless or cool on-column injections (*Splitless 7.4.3 and 7.4.4; On-column 7.6.2*).
3. *Compounds eluting before the solvent front when using splitless and on-column injection:* Peaks eluting before the solvent front are often broad or distorted.
4. *Lack of a retention gap for on-column injections using secondary cooling* (*On-column Retention Gap 7.6.4*).
5. *Normal peak broadening of highly retained peaks:* Peaks with very long retention times appear to be much broader than the earlier eluting peaks. This is very pronounced for isothermal temperature conditions.
6. *Normal loss of column efficiency with use:* The loss is gradual unless accelerated by severe thermal or oxygen damage.
7. *Co-eluting peaks:* Two or more co-eluting compounds may appear as a single broad peak.
8. *Extremely high or low carrier gas average linear velocity* (*Recommended Average Linear Velocities 6.6*).
9. *Normal broadening of branched compounds for PLOT columns:* The peak width of branched hydrocarbons (e.g., *i*-butane) are significantly wider than straight chain hydrocarbons (e.g., *n*-butane) for many PLOT columns.
10. *Too low or no detector makeup gas:* Usually a change in detector sensitivity occurs also (*Detector Makeup Gas 8.2.2*).

### 13.4.4 Flat Top Peaks

The larger peaks in a chromatogram may have flat tops since they are beyond the scale of the page or screen. A solvent front is an example of a flat top peak. Changing the scale to get a flat top peak fully on the page or screen may cause some of the smaller peaks to disappear from view. If the peak tops remain flat, adjustments to the detector or data system settings may be necessary. Time programming of the detector range or attenuation settings may be necessary to accommodate a very broad spread of compound concentrations in a single sample.

*Primary Root Causes:* (1) Exceeding the range of the detector or data system. (2) Too small of a scale on the page or screen. (3) An extremely high concentration of a sample compound.

1. *Exceeding the maximum input signal of the data system or recorder:* The full scale signal of the data system or recorder is too low. One or more of the peak signals exceeds the maximum output level of the data system or recorder. The detector output signal (attenuation) can be increased if necessary or if it is an available option.
2. *Exceeding the maximum output signal of the detector:* The maximum output of the detector electronics is too low. The detector range has to be increased to prevent limiting of the signal being sent to the data system. The detector attenuation setting can also be increased. Either adjustment causes a decrease in detector sensitivity.
3. *The full scale or attenuation setting on the display for the data system or recorder is too low:* The peaks are consistently off scale and appear to be flat on the top. In reality, the peaks are well formed, but flat-topped due to the apexes being displayed off the top of the page or screen.

## 13.5 Split Peaks

Most split peaks are caused by some type of sample injection problem or issue. It is rare for a damaged or malfunctioning column to cause split peaks.

*Primary Root Cause:* The sample is introduced into the column in two or more portions.

1. *Poor injection technique:* A jerky or erratic syringe plunger depression or insertion of the needle through the septum (*Injection Techniques 7.8*).
2. *Sample in the syringe needle upon injection:* The portion of the sample in the needle volatilizes and leaves the syringe before the plunger is depressed. More common for very volatile compounds or sample solvents (*Injection techniques 7.8*).
3. *Incorrect column position in the injector:* The column is either substantially too low or high in the injector (*Column position in the injector: Split 7.3.5; Splitless 7.4.7; Direct 7.5.4*).
4. *Sample degradation in the injector:* This occurs when the degradation product peak elutes close to the original compound peak. A separate peak may be observed if the degradation product peak elutes away from the original compound peak.

5. *Sample degradation in the column:* This occurs when the degradation product peak elutes close to the original compound peak. Instead of two distinct peaks, the peaks are often joined by a plateau or curved section that does not extend to the baseline.

6. *Peak co-elution:* Contamination of the sample, syringe or syringe cleaning/ rinsing solvent. It may be a carry-over problem, but one of the peaks should be very broad compared to the other. There may be a change in the sample matrix, thus the peak is in the sample and not an external contaminant.

7. *Severe backflash:* Usually occurs with very large injection volumes, smaller volume liners or very volatile samples. The earlier eluting (i.e., more volatile) peaks are usually more affected (*Injector Backflash 7.2.3*).

8. *Severe detector overload:* The detector becomes "blinded" by the large amount of compound and momentarily decreases in response then returns to normal. The response drop-off creates a small valley which gives the appearance of a split peak.

## 13.6 Negative Peaks

Negative peaks are unusual and often detector or compound related. In a few cases, negative peaks are normal.

*Primary Root Cause:* The detector signal momentarily goes below the baseline value.

1. *Compound structure:* Some compounds generate a negative response with a TCD or ECD (*ECD 8.5; TCD 8.6*).

2. *Dirty or old ECD:* The negative peak usually immediately follows a large, positive peak (*Negative Peaks with an ECD 8.5.8.4*).

3. *Detector quenching:* A reduction of detector response by non-responsive compounds with a selective detector. If no other compound is eluting at the same retention time, a negative peak may occur.

4. *Detector overload:* A high concentration of the sample "blinds" the detector and the response momentarily and rapidly drops and then returns to normal. This creates a negative peak and is most common for element specific detectors such as ECD, NPD and FPD.

5. *Reversal of detector or recorder settings:* If all of the peaks are in the negative direction, check for a reversal of the recorder connections or settings. The polarity setting of the TCD may be reversed from normal.

6. *Detector contamination:* Contaminants may be present in the TCD reference cell gas lines.

## 13.7
## Excessively Broad Solvent Front

Sometimes, a wide solvent front may be normal depending in the injector, column and sample. A reference chromatogram will be necessary to make the determination.

*Primary Root Causes:* (1) A large amount of solvent entering the column.
(2) Transfer of the solvent from the injector into the column occurs over an excessively long time period without later re-focusing of the sample.

1. *Lack of a splitless purge function:* A purge function is required for splitless injections unless a very large solvent front can be tolerated (*Purge activation time 7.4.1 and 7.4.2*).

2. *Too long of a splitless purge activation time:* Times greater than 3–5 minutes result in large solvent fronts (*Purge activation time 7.4.1 and 7.4.2*).

3. *The split ratio is too low:* The minimum value depends on the column diameter being used (*Split ratios 7.3.2*).

4. *Use of very high boiling point sample solvents:* A greater problem at low initial program temperatures or isothermal column temperatures.

5. *Using mixed sample solvents with large differences in their boiling points:* The two solvent elutes over a large temperature and time range, thus creating a broad solvent front.

6. *Excessively large injection volume:* Volumes greater than ~3 μL may result in broad solvent fronts especially for direct, on-column and splitless injections.

7. *Leak in the injector:* A leak at the liner O-ring or seal is the most probable location. This a more common problem when using lower boiling point sample solvents.

8. *Too low of an injector temperature:* This is a greater problem with high boiling point solvents (*Injector Temperatures 7.2.1*).

9. *Poorly installed column in the injector:* The error in the insertion distance has to be large for this problem to occur (*Column Installation in the Injector 9.2.4*).

10. *Very slow depression of the syringe plunger:* This a greater problem with low boiling point sample solvents (*Injection Technique 7.8*).

## 13.8 Loss of Resolution

Resolution is a function of peak width and peak separation. The first step is to determine which has occurred since the causes can be very different.

*Primary Root Causes:* (1) Increase in peak width. (2) Decrease in peak separation.

1. *Column contamination:* Sample residues are present in the front portion of the column. Usually results in increased peak widths, but peak separation may be decreased in more severe cases (*Contamination 11.6*).
2. *Oxygen, thermal and/or chemical induced damage to the stationary phase:* Usually higher column bleed or activity is also experienced. Usually results in increased peak widths, but decreased separation may occur in more severe cases (*Thermal Damage 11.3; Oxygen Damage 11.4; Chemical Damage 11.5*).
3. *Change in the carrier gas average linear velocity:* Retention time shifts without a change in $k$ should be observed (*Retention Factor 2.3.3; Measuring Average Linear Velocities 6.5.1*).
4. *Full or partial co-elution of another compound with one or both compounds:* Usually results in increased peak widths, but a shift in separation may occur.
5. *Normal column degradation with use.*
6. *A large change in the amount of one or both compounds in the sample:* Usually results in increased peak width, but a change in peak retention time may occur with very large concentration changes.
7. *A change in the column temperature.*

## 13.9 Retention Changes

Retention time shifts are very common and caused by numerous factors. Usually a small retention time shift is not an indication of a problem or error. It is critical to determine whether there is also a change in the retention factor ($k$) in addition to a change in retention time, thus calculating the $k$ values before proceeding is recommended.

### 13.9.1 Retention Time ($t_r$) Change Only

Only the retention time shifts, but the retention factor ($k$) remains the same.

*Primary Root Causes:* (1) Change in the carrier gas average linear velocity.
(2) Change in the column length.

1. *Change in the carrier gas head pressure:* Even changes of several tenths of a psi can cause a change in retention times especially for large diameter or short length columns. The change may not be readily visible on an analog pressure gauge.

2. *Leak in the injector:* Retention times shift to lower values if a leak is present. Some injector designs (back pressure regulated) are more prone to this behavior.

3. *Average linear velocity was set at the wrong or different column temperature:* A different column temperature is used than specified in the method or previously used (*Measuring Average Linear Velocities 6.5.1*).

4. *Change in the column length:* Caused by repeated trimming of the column upon installation. The retention times shift to lower values.

### 13.9.2 Retention Factor (*k*) Change

The retention time and retention factor ($k$) shifts with this problem situation.

*Primary Root Causes:* (1) Change in the column. (2) Change in the column temperature.

1. *Difference in column diameter or film thickness:* There is normal variation in diameter and film thickness between columns of the same description from the same manufacturer. This will not occur for the same column.

2. *Change in the sample solvent:* A small $k$ shift for compound peaks eluting near the solvent front may occur. Larger differences in the polarities of the solvent and analytes increases the possibility of this cause.

3. *Large change in amount of compound entering the column:* Usually a 100-fold or higher difference in the compound amounts is required.

4. *Column contamination:* Usually, the amount of contamination has to be severe. Sample compounds are prevented from interacting with the stationary phase at the front of the column resulting in a decrease in $k$. In rare cases, one or more of the compounds are retained by the contaminants, thus causing an increase in $k$. Usually peak shape problems are also evident (*Contamination 11.6*).

5. *Column activity:* Severe activity or long lengths of the column have to be active to cause a shift in $k$. Active compound will show a significant amounts of peak tailing.

6. *Change in column temperature:* All of the sample compounds are not shifted in retention by the same amount, but the shift will be in the same direction. Oven temperatures can drift over time and oven re-calibration is periodically needed.

## 13.10
## Peak Size Problems

Peaks that are too small are the most common peak size problem. Peak size changes can be difficult to solve since there are numerous possible causes.

*Primary Root Causes:* (1) Column activity. (2) Change in the sample. (3) Change in detector response. (4) Change in injector behavior.

### 13.10.1
### No Peaks

The complete absence of all peaks is indicative of a major GC parameter or setup problem. Checking the obvious (see Section 12.2.2) is the best place to start.

1. *Detector is not activated or turned on:* Flame is off, no current to the filament, etc.
2. *Plugged syringe:* This is more common with an autosampler.
3. *Insufficient sample in the autosampler vial:* The vial is empty or the sample volume is too low.
4. *Extremely low or no carrier gas flow through the column* (*Turning On and Verifying the Carrier Gas Flow 9.2.5*).
5. *Broken column* (*Column Breakage 11.2*).
6. *Sample was injected into the wrong injector.*
7. *Column is installed in the wrong injector or detector* (*Verifying Proper Column Installation and Detector Operation 9.2.7*).
8. *Detector is not properly connected to the data system or recorder.*

### 13.10.2
### All Peaks Change in Size

The peak size change may not be equal for all peaks; however, the direction of the change is usually the same (i.e., all larger or smaller). A combination of larger and smaller peaks is often indicative of multiple problems.

*Larger, smaller, or a combination of larger and smaller peaks.*

1. *Change in the detector response:* Usually each compound peak size changes by a different amount and not always in the same direction. The most common causes are incorrect type or flow of detector gases, temperatures, settings or detector contamination.
2. *Change in the syringe technique:* For example, the volume of sample contained in the syringe needle may be different for various models of syringes, thus

the injection volume is different; failure to rinse any residual solvent in the syringe from a previous injection or cleaning (*Injection Techniques 7.8*).

3. *Change in the sample:* For example, an error in sample preparation would result in a sudden change in the peak sizes or a loss of sample solvent due to evaporation would result in larger peaks.
4. *Change in the split ratio* (*Split Ratios 7.3.2*).
5. *Change of the purge on time for a splitless injection* (*Purge Activation Times 7.4.2*).
6. *Change in the septum purge flow:* Earlier eluting (i.e., more volatile compounds) peaks are usually more affected (*Septum purge flows: Split 7.3.3; Splitless 7.4.5; Direct 7.5.3*).

*Smaller peaks only.*

7. *Leak in the injector:* Earlier eluting peaks (i.e., more volatile compounds) are usually more affected and often accompanied by peak tailing (Leak *Detection 9.6; Static Pressure Check 12.4.5*).
8. *Severe sample backflash in the injector:* Earlier eluting peaks (i.e., more volatile compounds) are usually more affected and often accompanied by peak tailing. Sometimes loss of later eluting peaks (i.e., less volatile compounds) occurs upon condensation of the sample compounds on a cool spot (e.g., carrier gas line) (*Injector Backflash 7.2.3*).
9. *Partially blocked or clogged syringe:* The relative peak size change is the same for each compound.
10. *Too low of an injector temperature:* Later eluting peaks (i.e., less volatile compounds) are usually more affected. The peaks may also be broad or rounded (*Injector Temperatures 7.2.1*).
11. *GC was started while the baseline was dropping or drifting downward:* Only the tops of the peaks are visible, thus giving the appearance of decreased peak size. Some peaks may not be visible at all. The baseline looks exceptionally noise-free.

### 13.10.3
### Some Peaks Change in Size or Missing Peaks

A missing peak may not be completely gone. Expanding (magnifying) the baseline in the region of the missing peak may show the peak being present, but very small.

1. *Column or liner activity:* Adsorption of active compounds by the column or injector liner. Peak tailing may also be evident. The loss of inactive compounds is indicative of a different problem such as an injector, detector or sample deficiency (*Tubing Activity 3.2; Active Compounds 10.2*).

2. *Column or liner contamination:* Adsorption of active compounds by contaminants in the column or injector liner. Peak tailing may also be evident. The loss of inactive compounds is indicative of a different problem such as an injector, detector or sample deficiency (*Contamination 11.6*).
3. *Sample compounds degraded or reacted:* Sometimes one or more extra peaks from the degradation or reaction products are evident in the chromatogram.
4. *Error in sample preparation or extraction:* For example, a compound was not added to the standard or an internal standard was not added to the sample.
5. *pH change or shift in the sample:* A pH shift may create a salt or ionized form of an acid or base that does not chromatograph properly, thus not appear in the chromatogram.
6. *Co-eluting peaks:* A change in retention or resolution results in one or more peaks to fully or partially co-elute. It may appear that one peak disappears while an adjacent one increases in size.

## 13.11
## Extra or Ghost Peaks (Carryover)

A damaged or malfunctioning column can not generate distinct peaks. A column can not generate peaks without the injection or introduction of compounds into the column. In rare cases, short-term baseline drift may appear as small, broad peaks especially near the upper temperature limit of a column. The occasional small peak may appear with even the cleanest and best maintained GC system.

If the ghost peaks are similar in width to the ones in the injected sample, the contaminants probably entered the column with that injection. If the ghost peaks are significantly wider, the contaminants probably entered the column during a previous injection or were carried into the column with the carrier gas before the injection.

*Primary Root Cause:* GC, column or sample contamination.

1. *Contamination of the injector:* The appearance and size of the extra peaks is often inconsistent since the peaks are caused by semi-volatile residues. Usually the size and number of peaks increase the longer the GC is left at lower column temperatures. The extra peaks are often visible in a blank run or solvent injection (*Contamination 11.6; Jumper tube test 12.4.1; Condensation test 12.4.2*).
2. *Contamination of the column:* The extra peaks are broad especially compared to ghost peaks resulting from injector contamination. The appearance and size of the extra peaks is often inconsistent. Injector contamination usually resulting in column contamination (*Column Contamination 11.6*).
3. *Contamination of the carrier gas:* The number and size of the ghost peaks usually becomes worse the longer the GC is left at lower column temperatures. Peaks are usually visible in a blank run. A wandering or drifting baseline sometimes

occurs with carrier gas contamination (*Contamination 11.6; Gas impurity traps 11.10; Condensation test 12.4.2*).

4. *Sample contamination or degradation:* Impurities in the compounds or solvents used to make the samples or standards may appear as peaks in the chromatograms. The number and size of the ghost peaks should be fairly consistent. No peaks should be visible in a blank run. Sample degradation tends to get worse over time, thus the extra peaks become larger.
5. *Septum bleed* (*Septum bleed 7.10.3; Condensation test 12.4.2*).
6. *Sample carryover in the syringe:* The peak sizes and appearance will vary. Peaks should appear with injection of clean solvent, but not in a blank run.

## 13.12
## Rapid Column Deterioration

Column lifetimes are dependent on the analysis conditions, samples and integrity of the GC system. Columns with different stationary phases have different lifetimes. It is difficult to determine normal column lifetime without sampling and use of a large number of identical description columns.

*Primary Root Cause:* Introduction of damaging materials or compounds into the column.

1. *Column contamination by high molecular weight materials:* These contaminants polymerize or bond to the column due to exposure to high temperatures or are insoluble in typical column rinsing solvents. (*Contamination 11.6; Solvent Rinsing 11.7*).
2. *Oxygen damage:* The column was exposed to oxygen especially at high temperatures. The most common reasons for the presence of oxygen in the carrier gas are entry of air through a leak or use of poor quality carrier gas (*Gas Purity 6.8; Oxygen Damage 11.4; Gas Impurity Traps 11.10*).
3. *Chemical damage:* Injection of mineral acids and bases rapidly damage stationary phases. Perfluoro organic acids may also damage stationary phases (*Chemical Damage 11.5*).
4. *Thermal damage:* The column was subjected to temperatures above its upper temperature limit for a prolonged period. Longer exposure times, higher temperatures and higher oxygen levels increase the speed and severity of the damage (*Thermal Damage 11.3*).
5. *Tubing breakage:* Excessive abrasion or scratching of the polyimide coating due to rough handling. Tight bends in the tubing also increases the probability of tubing breakage. Breakage is more common for larger diameter columns (*Column Breakage 11.5*).
6. *Polar stationary phases have much shorter lifetimes than non-polar phases.*

## 13.13 Quantitation Problems

Quantitation problems can be some of the most complex and difficult problems to solve. Nearly every part of the GC and external factors such as samples and standards are involved. This section can only point to the most likely areas. Usually further and thorough investigation is needed to find the root cause of the problem.

Many of the problems causing peak size and shape changes can lead to quantitation difficulties. Fixing these problems occur eliminates the quantitation problem.

1. *Column or liner activity:* Adsorption of active compounds by the column or injector liner. Lower level compounds are more affected (*Tubing Activity 3.2; Active Compounds 10.2*).
2. *Column or liner contamination:* Adsorption of active compounds by the column or injector liner. Lower level compounds are more affected. Peak tailing or shape distortion also occurs, thus altering peak integration. Contaminants eluting from the column cause baseline disturbances such as noise, wander, drift, high background signal, negative peaks or interferences which may interferes with peak integration (*Contamination 11.6*).
3. *Calibration curve generation or use errors* (*Calibration Curves F.5.2*).
4. *Error in preparation of analytical standards* (*Standard Preparation F.7*).
5. *Change in the syringe technique* (*Injection Techniques 7.8*).
6. *Injector discrimination due to a change in injector conditions* (*Injector Discrimination 7.2.4*).
7. *Exceeding the linear range of the detector* (*Detector Linear Range 8.2.6*).
8. *Incorrect detector conditions or settings:* Changes in detector gas flows, temperatures, voltages or currents.
9. *Improper recorder or data system settings:* May lead to incorrect or inconsistent peak integration.
10. *Improper sample handling or storage:* Changes in the sample concentration from evaporation, reaction or compound degradation.

# Appendix A
# Terms

| | |
|---|---|
| $\alpha$ | separation; selectivity |
| $\beta$ | phase ratio |
| $A$ | multi-path flow term |
| $B$ | longitudinal diffusion term |
| $C$ | resistance to mass transfer term |
| CE | coating efficiency (%) |
| $d$ | diameter |
| $d_f$ | film thickness |
| $F$ | flow rate (mL/min) |
| $\bar{F}$ | average flow rate (mL/min) |
| $H$ | height equivalent to a theoretical plate |
| $H_{eff}$ | height equivalent to an effective theoretical plate |
| $I$ | retention index (RI) |
| $I_T$ | retention index, temperature program |
| $K_c$ | distribution constant |
| $k$ | retention factor (partition ratio; capacity) |
| $L$ | column length |
| N | number of theoretical plates |
| $N_{eff}$ | number of effective theoretical plates |
| $r$ | column radius |
| R | resolution |
| $t_m$ | retention time of a non-retained compound |
| $t_r$ | retention time |
| $t'_r$ | adjusted retention time |
| TZ | Trennzahl |
| $\bar{u}$ | average linear velocity (cm/sec) |
| UTE | utilization of theoretical efficiency (%) |
| $W_b$ | peak width at half height (in time units) |
| $W_h$ | peak width at base (in time units) |

*The Troubleshooting and Maintenance Guide for Gas Chromatographers, Fourth Edition.* Dean Rood

ISBN: 978-3-527-31373-0

**For retention index calculations**

| | |
|---|---|
| $x$ | compound of interest (compound $x$) |
| $y$ | number of carbons in normal alkane eluting before compound $x$ |
| $z$ | number of carbons in normal alkane eluting after compound $x$ |

# Appendix B
# Equations

The book section describing the equation is denoted in parenthesis.

### Column and Peak Characteristics

$$W_b = 1.699\ W_h \tag{2.2.1}$$

$$\beta = \frac{r}{2\,d_f} = \frac{d}{4\,d_f} \tag{2.4}$$

$$K_C = \frac{\text{Compound concentration in the stationary phase}}{\text{Compound concentration in the mobile phase}} \tag{2.5}$$

$$K_C = k\,\beta = k\left(\frac{r}{2\,d_f}\right) \tag{2.5}$$

### Retention

$$t_r' = t_r - t_m \tag{2.3.2}$$

$$k = \frac{t_r - t_m}{t_m} \tag{2.3.3}$$

$$t_r = t_m\,(k+1) \tag{2.3.3}$$

$$I = 100\,y + 100\,(z-y)\left[\frac{\log t_r'(x) - \log t_r'(y)}{\log t_r'(z) - \log t_r'(y)}\right] \tag{2.3.4}$$

$$I_T = 100\left[\frac{t_r(x) - t_r(y)}{t_r(z) - t_r(y)}\right] + 100\,y \tag{2.3.4}$$

## Efficiency

$$N = 5.545 \left( \frac{t_r}{W_h} \right)^2 \tag{2.6.1}$$

$$N = 16 \left( \frac{t_r}{W_b} \right)^2 \tag{2.6.1}$$

$$H = \frac{L}{N} \tag{2.6.2}$$

$$N_{eff} = 5.545 \left( \frac{t'_r}{W_h} \right)^2 \tag{2.6.3}$$

$$N_{eff} = 16 \left( \frac{t'_r}{W_b} \right)^2 \tag{2.6.3}$$

$$H_{eff} = \frac{L}{N_{eff}} \tag{2.6.3}$$

$$H_{min} \text{ (theoretical)} = r \sqrt{\frac{(11\,k^2 + 6\,k + 1)}{3\,(1+k)}} \tag{2.7}$$

$$\text{UTE\%} = \left[ \frac{H_{min} \text{ (theoretical)}}{H \text{ (actual)}} \right] \cdot 100 \tag{2.7}$$

## Separation

$$\alpha = \frac{k_2}{k_1} \tag{2.8}$$

## Resolution

$$R = 1.18 \left( \frac{t_{r2} - t_{r1}}{W_{h1} + W_{h2}} \right) \tag{2.9}$$

$$R = 2 \left( \frac{t_{r2} - t_{r1}}{W_{b1} + W_{b2}} \right) \tag{2.9}$$

$$\text{TZ} = \left( \frac{t_{r2} - t_{r1}}{W_{h1} + W_{h2}} \right) - 1 \tag{2.10}$$

$$\text{TZ} = \left( \frac{R}{1.177} \right) - 1 \tag{2.10}$$

**Carrier Gas**

$$H_{\mathrm{min}} = A + \frac{B}{\overline{u}} + C\,\overline{u} \qquad (6.4)$$

$$\overline{u} = \frac{L}{t_{\mathrm{m}}} \qquad (6.5.1)$$

$$t_{\mathrm{m}} = \frac{L}{60\,\overline{u}} \qquad (6.5.1)$$

$$\overline{F} = \frac{\pi\, r^2\, L}{t_{\mathrm{m}}} \qquad (6.5.2)$$

# Appendix C
# Mass, Volume and Length Unit Conversions

## Mass

| | Equals |
|---|---|
| 1 pg | 0.001 ng |
| 10 pg | 0.01 ng |
| 100 pg | 0.1 ng |
| 1000 pg | 1 ng |
| 1 ng | 0.001 µg |
| 10 ng | 0.01 µg |
| 100 ng | 0.1 µg |
| 1000 ng | 1 µg |
| 1 µg | 0.001 mg |
| 10 µg | 0.01 mg |
| 100 µg | 0.1 mg |
| 1000 µg | 1 mg |
| 1 mg | 0.001 gm |
| 10 mg | 0.01 gm |
| 100 mg | 0.1 gm |
| 1000 mg | 1 gm |
| 1 g | 0.001 kg |
| 10 g | 0.01 kg |
| 100 g | 0.1 kg |
| 1000 g | 1 kg |

## Volume

| | Equals |
|---|---|
| 1 µL | 0.001 mL |
| 10 µL | 0.01 mL |
| 100 µL | 0.1 mL |
| 1000 µL | 1 mL |
| 1 mL | 0.001 L |
| 10 mL | 0.01 L |
| 100 mL | 0.1 L |
| 1000 mL | 1 L |

## Length

| | Equals |
|---|---|
| 100 µm | 0.1 mm |
| 0.01 cm | 0.1 mm |
| 0.1 cm | 1 mm |
| 1 cm | 10 mm |
| 10 cm | 0.1 m |
| 100 cm | 1 m |
| 1000 cm | 10 m |
| 10,000 cm | 100 m |

# Appendix D
# Column Bleed Mass Spectra

These column bleed mass spectra are representative of ones typically obtained for each listed stationary phase. While the major ions will be the same for most GC/MS systems and similar columns, there will be differences in the identities and abundances of the minor ions. Column manufacturer, GC/MS parameters, column age and history, carrier gas purity, and system contaminants will all affect the minor ions.

The mass spectra were obtained using an Agilent 5973N Inert GC/MS system and at the isothermal temperature limit of each column. The helium carrier gas flow rate was 1.0 mL/min at the column exit.

### 100% Dimethylpolysiloxane

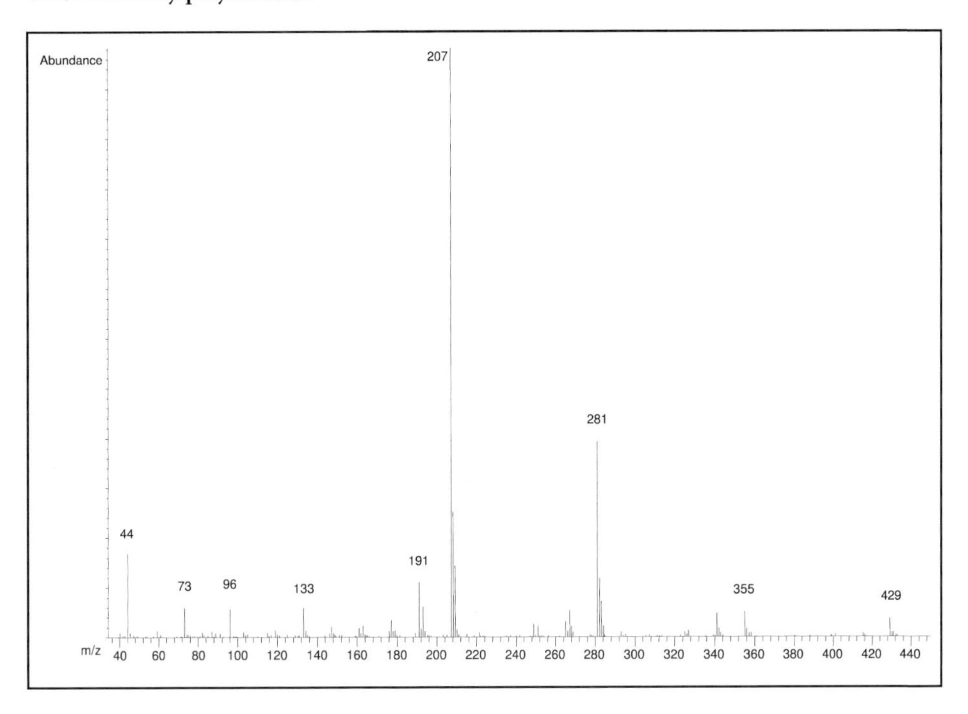

## 5% Diphenyl-95% dimethylpolysiloxane

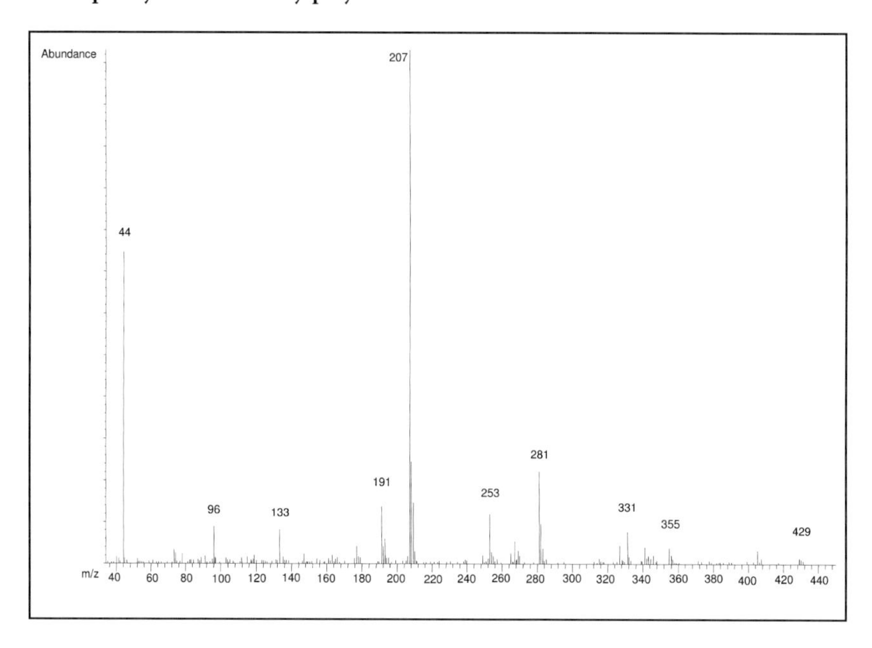


## 35% Diphenyl-65% dimethylpolysiloxane

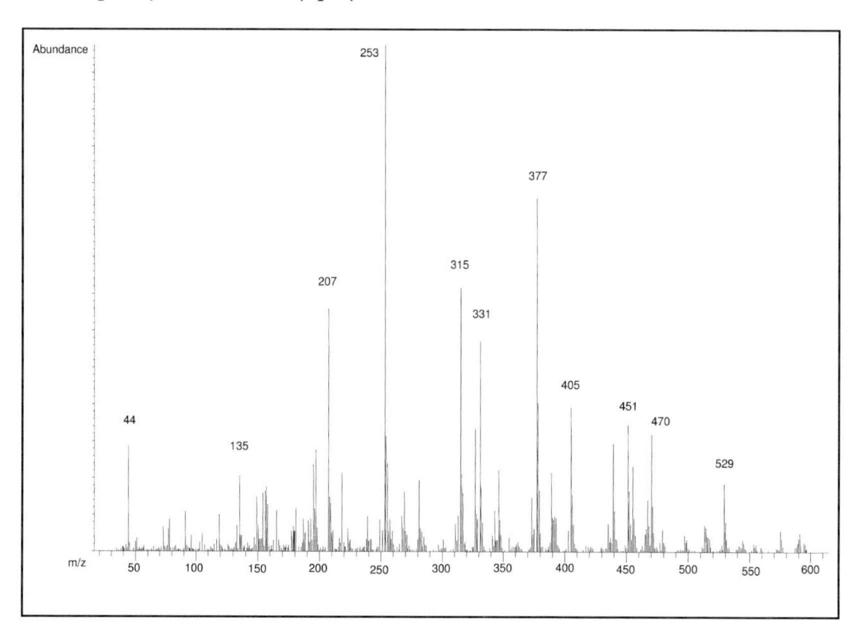

## 50% Diphenyl-50% dimethylpolysiloxane

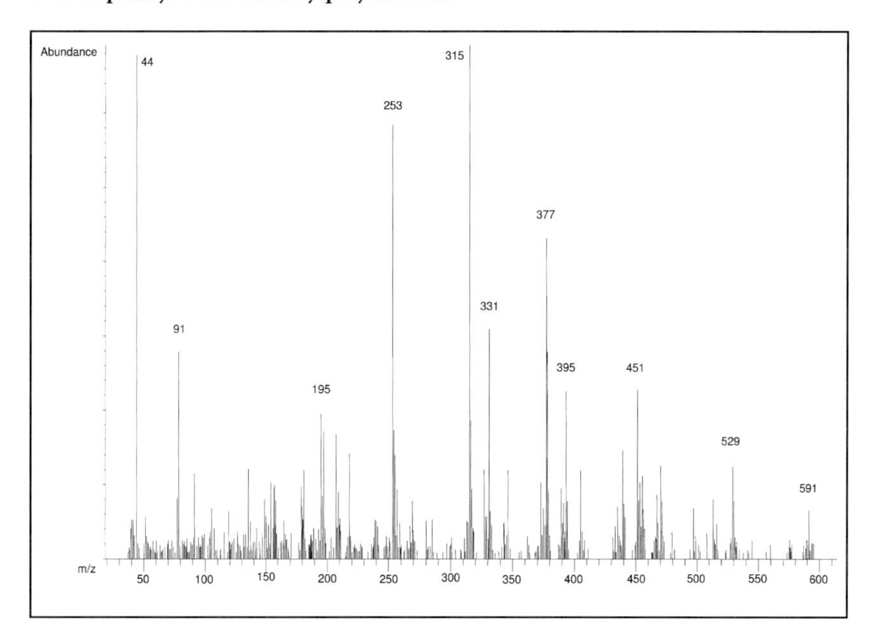


## 14% Cyanopropylphenyl-86% dimethylpolysiloxane*

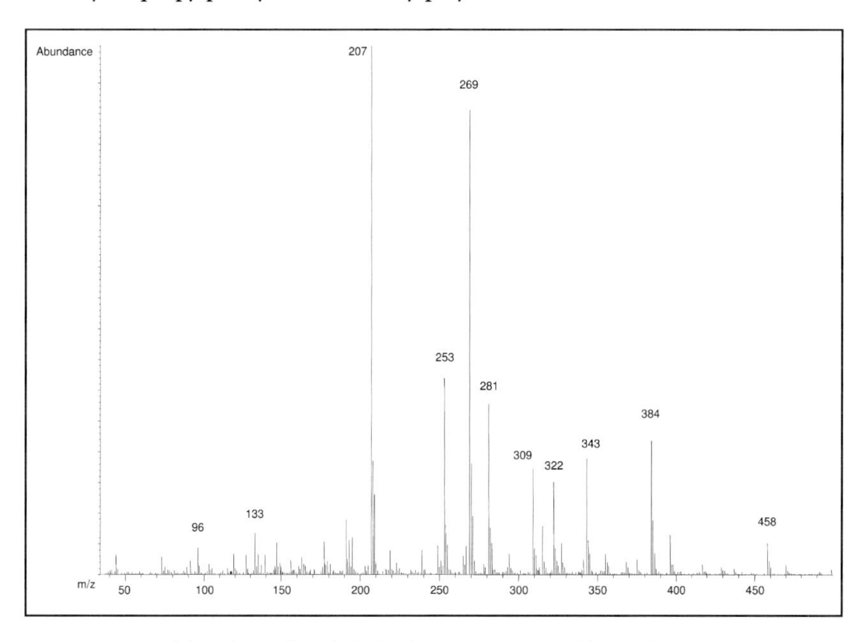


* 6% Cyanopropylphenyl-94% dimethylpolysiloxane spectrum will be similar

## 50% Cyanopropylphenyl-50% dimethylpolysiloxane

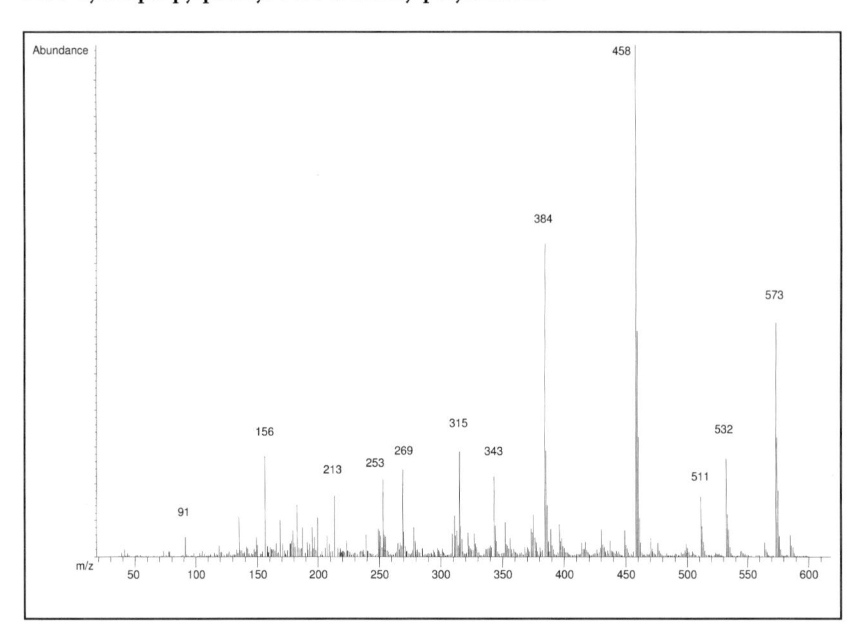


## 50% Cyanopropyl-50% dimethylpolysiloxane

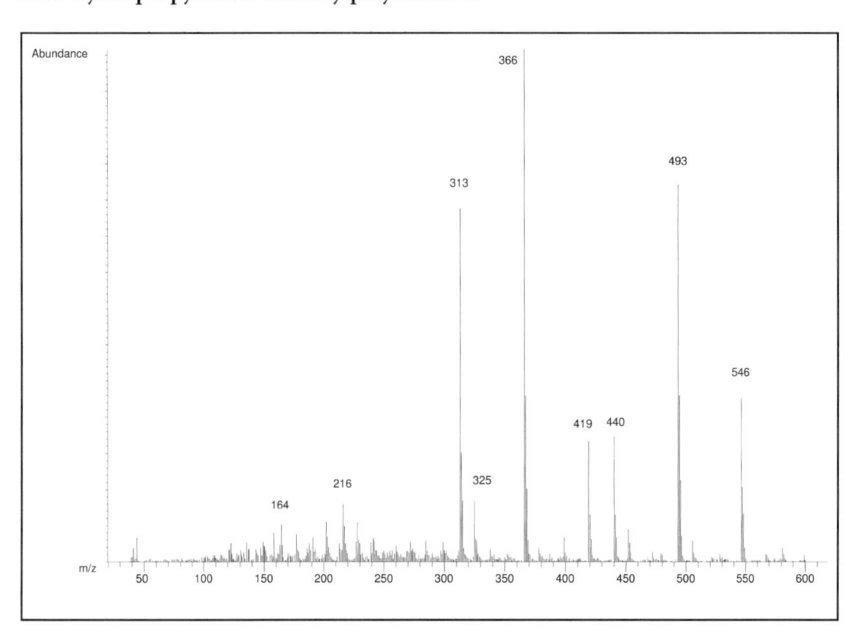

## ~95% Cyanopropyl-5% dimethylpolysiloxane

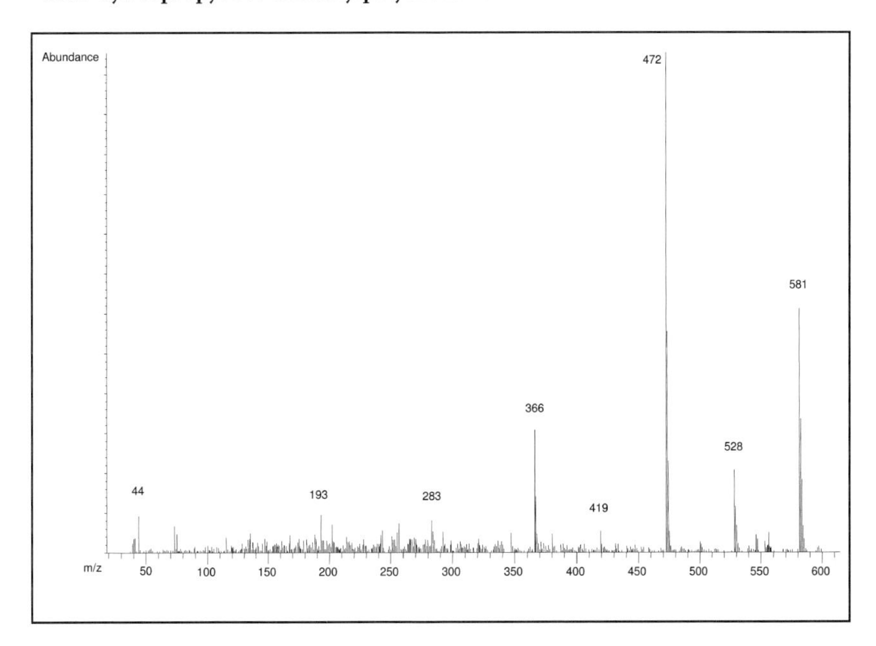


## 50% Trifluoropropyl-50% dimethylpolysiloxane

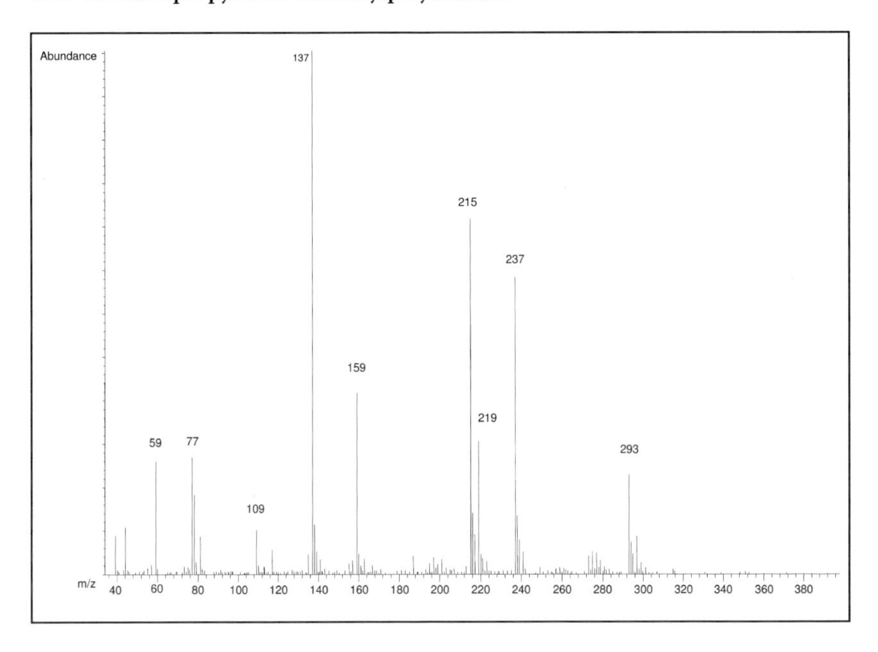

## 100% Polyethylene glycol (PEG)

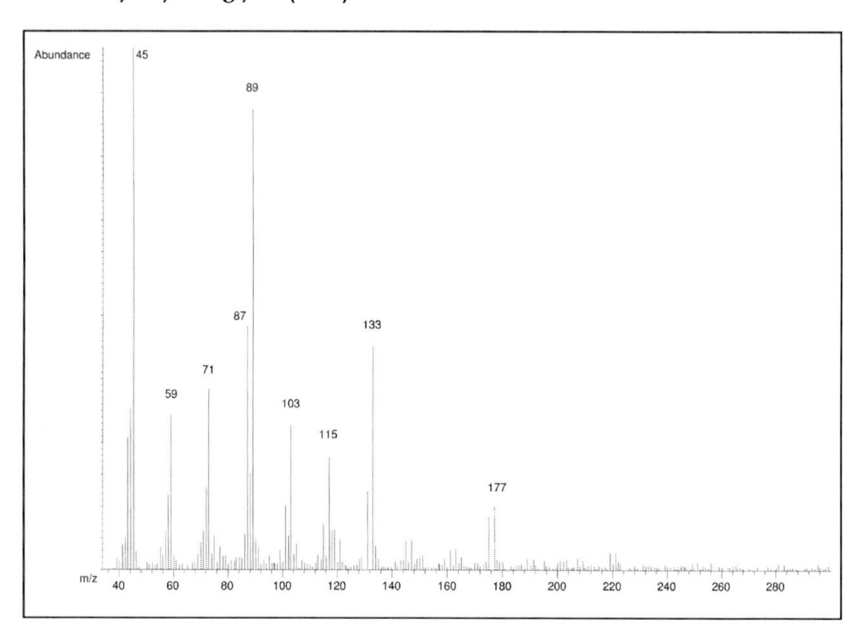


## FFAP

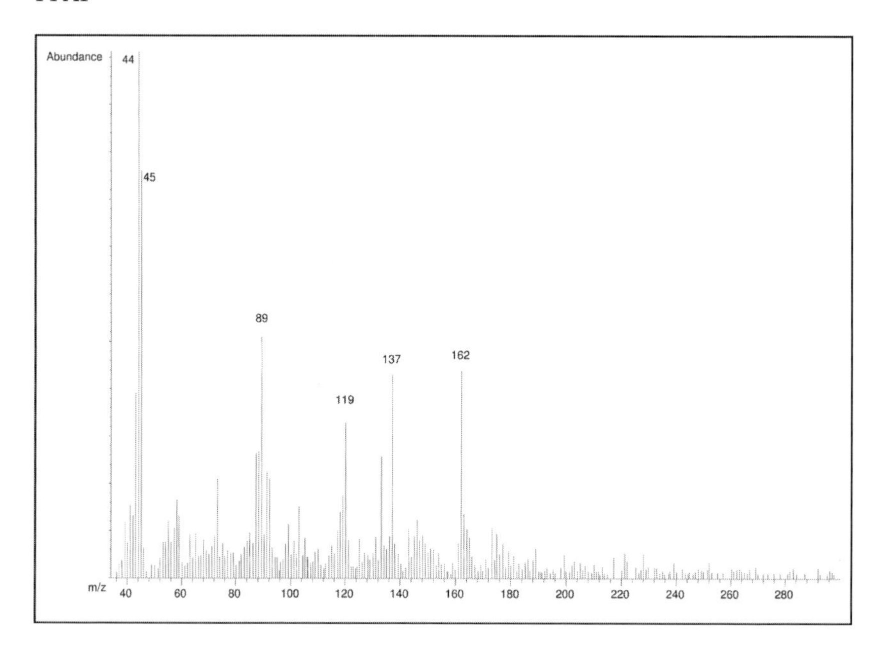

# Appendix E
# The Basics of High Speed GC Using Small Diameter Columns

## E.1
## Introduction

There are several approaches to reducing the run time of a GC analysis. Some require specialized equipment or significant changes to GC hardware and methods while others are less complex. This chapter focuses on the less complex approach of using small diameter columns to obtain reduced run times. Using small diameter columns is within the capabilities of many laboratories and usually does not require large investments in equipment. However, developing a high speed GC method or transferring to one from an existing conventional GC method may not be trivial, possible or easy to execute.

The intent of this chapter is provide a basic introduction to high speed GC using small diameter columns. It summarizes the requirements or considerations for running high speed GC using small diameter columns. All of the terms and concepts used in this chapter are previously explained in their corresponding chapters.

## E.2
## Column Considerations

The high efficiency of small diameter capillary GC columns largely enables the reduction in run times. While 0.15–18 mm i.d. columns may be represented or described as high speed columns, 0.10 mm i.d. columns provide the best balance between time reduction, ease of use and practicality. Very small diameter columns such as 0.05 mm i.d. are available, but they are impractical for most laboratories due to their difficult and demanding nature, exacting GC hardware requirements and other limitations.

The number of theoretical plates per meter (N/m) increases as capillary column diameter decreases, thus a shorter length of a smaller diameter column is needed to obtain the same total number of theoretical plates as compared to a larger diameter column. A 10 m × 0.10 mm i.d. column is a common replacement for the commonly used 30 m × 0.25 mm i.d. column. For example, total theoretical plates of ~125,000 compared to ~142,575, respectively, are obtained for a peak at $k = 5$. The shorter column length results in a decreased run time. As long as the phase ratio ($\beta$) of the smaller diameter column is the same or higher than the conventional diameter column, there will be no increase in retention ($k$) associated with the smaller column diameter. If the phase ratio is lower, any increases in retention can be counteracted by changes to the temperature program (Section E.6).

A decrease in column capacity occurs with a decrease in column diameter, thus overloaded peaks may occur for analytes at high concentrations. Since overloaded peaks are broad, the efficiency gains of a small diameter columns are negated. Increasing the column film thickness is an effective method to increase column capacity; however, increased retention also occurs which is counter to reducing run times. Some of the retention increase can be counteracted by changes to the temperature program (Section E.6). A film thickness of 0.1–0.2 μm is the most common for 0.10 mm i.d. columns when changing from an 0.25–0.32 mm i.d., 0.25 μm film thickness column. A film thickness of 0.3–0.4 μm is used when changing from an 0.25–0.32 mm i.d., 1.0 μm thickness column or when substantially higher column capacity is needed.

One assumption is the same or equivalent stationary phase will be used when changing to a smaller diameter column. Sometimes significant run time reductions are possible with only a change of the stationary phase. This option may be worth exploring if not previously considered. Using the same stationary phase may be required due to regulatory or practical factors. A change of stationary phase combined with a smaller diameter column can result in very large reductions in run time.

## E.3
## Carrier Gas Considerations

Relatively high carrier gas pressures are needed for small diameter columns. While shorter column length decreases the carrier gas pressure needed, the pressures are still substantially higher than for 0.25 mm i.d. and larger columns. Hydrogen is the preferred the carrier gas for high speed GC due to the lower pressures required to obtain the desired average linear velocities. The hydrogen van Deemter curves for 0.10 mm i.d. columns at high linear velocities are relatively flat and do not deviate far from the $\bar{u}_{opt}$ values except for analytes with high $k$ values. Operating at very high average linear velocities also helps to reduces analysis run times. While helium can be used as a carrier gas, analysis time reductions are not as large due to its lower $\bar{u}_{opt}$ values and higher carrier gas pressure are required which is usually undesirable.

For 0.10 mm i.d. columns, the optimal average linear velocity for hydrogen is 50–90 cm/sec with higher $k$ analytes favoring the lower end of the range. For a 10 m × 0.10 mm i.d. column at 100 °C, 34 psig is needed to obtain an average linear velocity of 60 cm/sec and 44 psig for 75 cm/sec. For a 20 m × 0.10 mm i.d. column at 100 °C, 73 and 93 psig is needed to obtain 60 and 75 cm/sec, respectively. Depending on the model of GC and its installed options, the higher carrier gas pressures may exceed the pressure limits of the GC. Many GCs have high pressure injectors or hardware retro-fits to enable very high carrier gas pressures. An increase in the delivery pressure of carrier gas to the GC may also be required.

## E.4
## Injector Considerations

Small diameter columns have specific injector requirements or require specialized injectors. The largest challenge is overcoming the low carrier gas flow rates used with small diameter columns. For a 10 m × 0.10 mm i.d. column, the typical hydrogen flow rate into the column is about 0.3 mL/min. Managing the higher carrier gas pressure can also be a challenge.

Conventional split injectors are most commonly used with 0.10 mm i.d. columns. A split ratio of 1 : 100 or higher is usually necessary to avoid severe loss of injector efficiency which would negate the high efficiency of small diameter columns. High split ratios are also necessary to avoid potential column overload with many samples. The use of smaller inner diameter injector liners (e.g., 2 mm) allows the use of lower split ratios (about 4 times lower compared to a 4 mm i.d. liner), but sample backflash in the injector can be a problem.

Conventional splitless injectors are rarely compatible with small diameter columns. The very slow transfer rate of the vaporized sample from the injector into the column requires very long purge activation times which results in extremely large solvent fronts and wide peaks. The very large solvent usually obscures the earlier eluting peaks. The large amount of sample introduced into the column with splitless injections usually severely overloads the column.

Specialized injectors suitable for use with small diameter columns are available. Most are some type of temperature programmable injector which do not have to be operated in the split mode. The most common utilize sample injection at a low injector temperature followed by an extremely rapid heating of the injector. Venting of the sample solvent may be involved to avoid swamping the head of the column with solvent. Some form of carrier gas flow programming during the sample introduction cycle may also be involved. Cryogenic interfaces are also available which cold trap the sample at the head of the column. All of these specialized injectors or interfaces are more complex and expensive than conventional split/splitless injectors.

Injectors capable of operating at higher pressures may be needed for small diameter columns. Carrier gas pressures of 100 psig or higher may be required especially for longer columns. Injector leaks are more common at higher pressures. More frequent septa changes are required since septa leaks occur for a lower number of syringe needle punctures when using high injector pressures. Better sealing septa may also be necessary to minimize the frequency of septa changes.

## E.5
## Detector and Data System Considerations

Peak widths of 0.5 sec and less are very common with small diameter columns. Detectors that immediately respond to analytes as they elute from the column (e.g., FID, NPD) are generally better suited for small diameter columns. Detectors with cells (e.g., ECD, TCD) or large volume detection zones may be less suited for small diameter columns. Rapid sweeping of the eluting compounds through the cell or detection zone is necessary to avoid peak broadening caused by long residence times in the detector. Higher auxiliary or make-up gas flow rates may be needed to rapidly sweep the detector volume; however, a loss of sensitivity may occur at the high flow rates. If available, micro cell design detectors are recommended for use with small diameter columns. For mass spectrometers (MS), the scan rates need to be high enough to obtain a minimum of 10 scans per peak with higher scans per peak being more accurate and desirable.

One of the parameters in data systems is sampling rate. Sampling rates of 50 Hz or higher are recommended for 0.10 mm i.d. columns to obtain an accurate retention time and area (or height) measurement of the peaks. Lower sampling rates often cause inaccurate or inconsistent integration results especially for very narrow or partially resolved peaks. High sampling rates generate large file sizes, thus sufficient hard drive capacity is needed to avoid data loss.

## E.6
## GC Oven Considerations

To take full advantage of the high speed capability of small diameter columns, a GC oven with a fast heating rate is needed. The maximum rate is dependent on the model of GC and the oven temperature. Maximum heating rates decreases as the oven temperature increases. Conventional GC uses heating or ramp rates of 5–25 °C/min with faster rates sometimes used during the initial portion of a temperature program. Newer models of GCs tend to have higher maximum heating rates than older models. Some GCs have a high power option allowing very fast heating rates. Aftermarket, add-on heating modules are also available to increase the heating rate of an existing GC.

Large increases in resolution when changing to the more efficient smaller diameter column are not obtained for samples with a large number of analytes with retention factors ($k$) above 7–10. The temperature program needs to be adjusted so the latest eluting analytes have $k$ values below 10. Very fast heating ovens are often necessary to get the $k$ values below 10. Heating rates of 40–50°/min are often needed for the fastest run times.

One other factor to consider is the cool down time of the GC oven. Depending on the initial and final program temperatures, it can take 3 to over 10 minutes for the GC to cool down which is significant compared to some high speed GC run times. If the high speed GC run is very fast, the cool down time can be close or greater than the run time. Newer model GCs tend to have faster cool down times. Add-on heating modules also tend to have faster cool down times.

## E.7
## Sample Considerations

High speed GC with small diameter columns is usually limited to more concentrated samples since split injectors at high split ratios are required; however, column overload can be a problem due to the low capacity of small diameter columns. A balance between sensitivity (detection limit) and column capacity often has to be carefully managed. Specialized or more complex injectors are often needed if low concentration samples are being analyzed and using a split injector does not provide the required method sensitivity.

Small diameter columns are more prone to contamination since they have a small inner surface area. A small amount of non-volatile residues at the front of a small diameter column can render it unusable. Compared to a conventional diameter column, additional or better sample cleanup may be necessary to prevent the rapid fouling of a smaller diameter column. To fully gain from high speed GC, rapid sample preparation is also needed. If the GC is idle while waiting for samples to be prepared, any analysis speed reductions are essentially wasted.

## E.8
## An Example of High Speed GC Using a Small Diameter Column

Figure E-1 shows a peppermint extract sample analyzed using a conventional and small diameter column. The run time for the high speed 10 m × 0.10 mm i.d. column is about half of the conventional 30 m × 0.25 mm i.d. column (7.5 vs. 15 minutes) while maintaining nearly equivalent peak resolution. The primary differences in the analysis conditions are the faster carrier gas average linear velocity (75 vs. 60 cm/sec) and faster ramp rate of the temperature program (20 vs. 7.5 °C/min) used for the 10 m × 0.10 mm i.d. column.

This example did not involve any GC hardware changes or modifications, thus this type of result is possible with a conventional GC. While a 50% reduction in the run time was obtained, further time reduction would have been possible with more changes to the GC and analysis conditions. A 0.2 µm film column was used to keep the column capacity close to the original 0.25 mm i.d. column. A 0.1 µm film column would have preserved the same phase ratio and further reduced the run time, but at the expense of column capacity. A 0.4 µm film column was tried, but the run time reduction was only 25% and determined to be too insignificant.

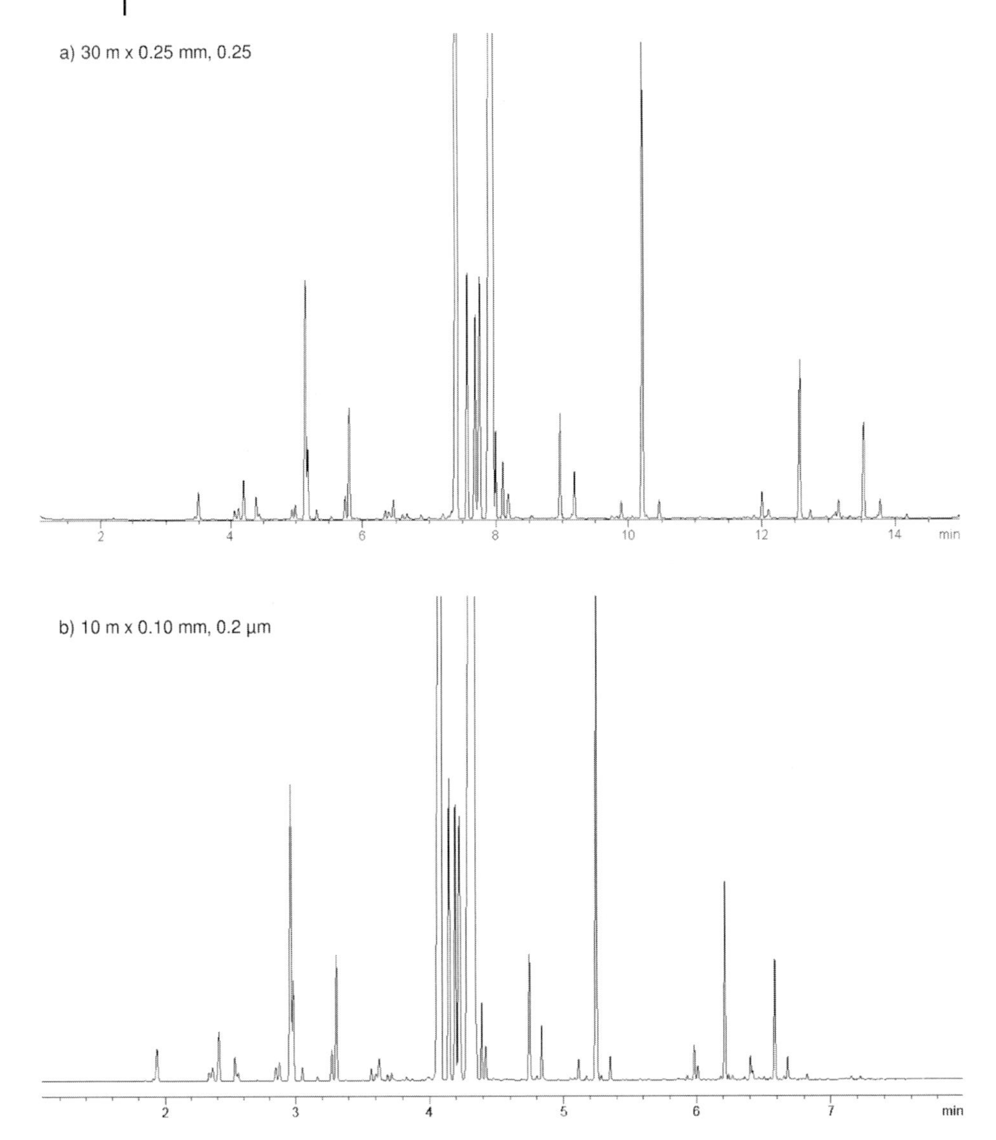


**Figure E-1** Comparison of conventional and high speed GC analyses.
Sample: 1% peppermint extract in ethanol

a) *Column:* DB-1, 30 m × 0.25 mm i.d., 0.25 µm
*Injector:* Split 1 : 100, 250 °C
*Detector:* FID, 300 °C
*Carrier gas:* Hydrogen at 60 cm/sec in the constant flow mode
*Oven:* 70 °C for 3 min, 70–160 °C at 7.5 °/min

b) *Column:* DB-1, 10 m × 0.10 mm i.d., 0.2 µm
*Injector:* Split 1 : 100, 250 °C
*Detector:* FID, 300 °C
*Carrier gas:* Hydrogen at 75 cm/sec in the constant flow mode
*Oven:* 70 °C for 2 min, 70–190 °C at 20°/min

## E.9
## High Speed GC Summary

High speed GC is not for the beginner or novice and requires more experienced operators with a good understanding of basic GC theory and hardware. Modifying a conventional GC method to a high speed method involves more than just installing a smaller diameter column. As previously discussed, there are numerous other factors to consider and adjustments to be made. Sometimes the adjustments or changes are straightforward such as for the example in Figure E-1. In other cases, significant effort is required to obtain the best results.

Table E-1 summarizes the major parameter differences between conventional and high speed GC. All of these factors need to be considered and weighed before proceeding to develop a high speed GC method using small diameter columns.

**Table E-1** Comparison of conventional and high speed GC parameters.

| Parameter | Conventional GC | High speed GC |
|---|---|---|
| **Column** | | |
| Diameter | 0.20–0.53 mm i.d. | 0.05–0.10 mm i.d. |
| Length | 12–100 m | 1–20 m |
| Film thickness | 0.1–5.0 µm | 0.05–0.40 µm |
| **Carrier gas** | | |
| Type | Helium or hydrogen | Hydrogen |
| Pressure | 2–60 psig | 30–150 psig |
| **Average linear velocity** | 20–60 cm/sec | 50–100 cm/sec |
| **Injector** | Conventional split, splitless, direct, on-column | Split or specialized (programmable; cryofocusing) |
| **Detector** | Conventional | Immediate detection or small volume cell; high scan rates (MS) |
| **Sampling rate** | < 50 Hz | ≥ 50 Hz |
| **Peak widths** | > 0.5 sec | ≤ 0.5 sec |
| **Oven heating rates** | 2–50 °C/min | 15–100 °C/min |

# Appendix F
# Basic Quantitative Capillary GC

## F.1
## Intentions

Quantitative results are affected by nearly every aspect of the GC and column. There are numerous variables that can adversely affect the accuracy and precision of quantitative results. The proper preparation and use of analytical standards is also required. Guidelines, requirements and recommendations concerning the GC and columns have been presented in the proceeding chapters. *The purpose and intent of this chapter is to provide some basic definitions, techniques and guidelines for the preparation and use of analytical standards along with an overview of a select number of basic calibration techniques. This section is not intended as an exhaustive treatment of quantitative GC techniques.* Instead of generic equations containing numerous variables, step-by-step instructions are provided along with illustrative examples where possible. The techniques presented within this chapter are not the only ones that can be used, but ones the author has found useful and reliable.

An assumption contained within this chapter is the reader is experienced and qualified in the proper use of measurement equipment such as balances and pipets along with the proper selection and use of suitable containers such as vials, bottles and caps. Following procedures and protocols for the safe handling, storage, labeling and disposal of hazardous compounds is critical and required before preparing or handling analytical standards or samples.

Gaseous samples are difficult to make and quantitatively measure since temperature, pressure and volume have to be measured and controlled. Quantitation and standard preparation techniques for gases are beyond the scope of this book and the following descriptions are limited to liquid and solid samples.

## F.2
## Definitions

Some of the terms used in discussing quantitative GC are confusing or have subtle differences. Also, some terms are commonly interchanged or used incorrectly.

For consistency and to avoid confusion, the following definitions are used in this chapter or provided as a reference.

1. *Concentration:* The amount (mass or volume) of a compound dissolved, mixed or suspended in a known amount (mass or volume) of another compound or substance.
2. *Analyte:* The compound of interest that is being measured. It is often called the target analyte. It may also be referred as a solute when used in the context of chromatographic processes. A sample may contain multiple analytes.
3. *Standard:* A solution containing one or more analytes at a known concentration. It is usually made up in a clean solvent, ideally the same as the sample solvent. Standards are either prepared in the laboratory or purchased from a commercial supplier. They are sometimes called analytical or working standards.
4. *Stock solution or standard:* A standard that is highly concentrated and used to prepared more dilute analytical standards.
5. *Sample:* Usually the substance or material being analyzed for the presence of analytes. It is often extracted or prepared for analysis. This term is often used to describe any solution being analyzed (e.g., blank or control sample).
6. *Matrix:* A more specific term describing any part of the sample extracted or derived from the original sample substance or material. Food products, soil, groundwater or biological tissues and fluids (e.g., blood, urine, plant) are examples of matrices. This term is more specific than sample and refers to any material present in the final sample after extraction, preparation or clean up. The analytes are not considered to be part of the matrix.
7. *Blank sample:* A sample composed of a clean, representative solvent that has been extracted using the same procedure as for actual samples. For example, saline (0.9% NaCl) may be used in place of urine or plasma to create a blank sample. A blank sample is used to determine if any peaks in the sample chromatogram are contributed by the materials (solvents, reagents, glassware, etc.) used in the sample extraction or preparation procedure.
8. *Control sample:* A sample of the same type as the sample of interest, but without any analytes. It is subjected to the same extraction or preparation procedure as used for actual samples. A control sample is used to determine whether the sample matrix chromatogram contains any peaks that may interfere or co-elute with the analyte peaks. Sample matrix materials may alter the effectiveness of the extraction or clean up procedure by their presence. Control samples are very useful when developing sample extraction or clean up methods. Control samples may be difficult to obtain or unavailable. A limited selection of control samples is available from government agencies (e.g., NIST) or commercial companies.
9. *Spike:* A control sample to which a known amount of analyte has been added. The spike sample is subjected to the same extraction or cleanup procedure as an actual sample then the amount of the analyte in the spiked sample is deter-

mined. The amount of analyte loss is used to correct the sample concentration, thus it compensates for any analyte loss during sample extraction or preparation. The amount in the spiked sample is considered "recovered" and the amount is often reported as the recovery percent (% recovered). Spiked samples are very useful when developing sample extraction or clean up methods. Spiked samples are also useful determining the affect of the matrix on GC behavior (e.g., response). If a suitable control sample is not available, adding analytes to a blank sample is an alternative, but a less desirable, option.

## F.3 Concentration

Concentration is used to denote the amount of a compound or analyte in a known amount of another substance, material or sample. Concentration is usually expressed with the numerator (i.e., matrix or sample portion) normalized to 1. For example, 125 mg of an analyte in 500 gm of a sample can be denoted as 125 mg/500 gm; however, 0.25 mg/gm or 250 gm/kg is a more conventional format. When possible, units are converted to avoid large numbers or ones with numerous decimal points. For example, 15,300 μg/kg would be written as 15.3 mg/kg. This is for ease of writing and reduction of errors due to miscounting of decimal points or zeros. The actual units selected often depend on sample sizes. For large samples (e.g., truck load), mg/kg may be a more relevant unit; ng/μL may be better when making μL sized GC injections.

Concentration is a relative measure and does not denote the absolute amount of the analyte. The concentration is constant regardless of the sample size. For example, an analyte concentration of 2.5 mg/mL is the same whether the total sample volume is 25 mL, 1 L or 40 L. The absolute or total amount of the analyte is determined by the total amount of the sample. For example, a sample with a concentration of 10 mg/L contains a total of 10 mg of the analyte in a 1 L volume where a 25 mL volume contains 0.25 mg (or 250 μg) of the analyte.

While volume is normally used to measure liquids, mass is also used especially when the liquid is the analyte. Conversion between mass and volume units may be necessary or desired for liquid analytes or matrices. Compound densities are needed and used to make the conversion (see Section F.4).

### F.3.1 Weight-to-Weight (w/w) and Weight-to-Volume (w/v)

Weight-to-weight (w/w) or weight-to-volume (w/v) are often used to denote concentration. An example of a weight-to-weight unit is 10 ng/mg; an example of a weight-to-volume measure is 5 mg/μL. Volume-to-volume (e.g., μL/L) or volume-to-weight (e.g., mL/kg) formats can also be used, but they are less common. The units usually make it obvious whether w/w or w/v is being used; however, it should be stated if there is any possibility of confusion or misinterpretation.

### F.3.2
### Parts per Million (ppm) and Parts per Billion (ppb)

Concentrations are sometimes expressed as a ratio of the analyte to the matrix. Part per million (ppm) is equal to 1 part of the analyte to 1 million parts of the matrix (i.e., 1 per $10^6$). Part per billion (ppb) is equal to 1 part of the analyte to 1 billion parts of the matrix (i.e., 1 per $10^9$). Part per trillion is denoted as ppt which should not be confused with parts per thousand. A part per thousand is usually better expressed as percentages (see Section F.3.3). Table F-1 lists some of the common units used when calculating ppm and ppb.

When solids dissolve in a liquid, there is usually a slight increase or decrease in the total volume of the final solution. The same effect is more pronounced when two liquids are mixed. When mixing two liquids, the final volume is not the sum of the two original volumes. For example, if water is added to 50 mL of ethanol (with careful mixing) until a final volume of 100 mL is obtained, 53.7 mL of water has to be added, thus the final mixture is not 50% ethanol. The consequence is making a 1 ppm solution by weight or volume rarely results in a solution that is exactly 1 ppm by volume. To be completely accurate, it is necessary to include w/w, w/v or v/v when using ppm, ppb or ppt as a concentration measurement; however, it is often omitted. Sometimes a "v" or "w" is used to denote whether volume or weight was used (e.g., ppmv).

There are several variants in the meaning of ppm, ppb and ppt, thus they can be ambiguous. The measures of ppm, ppb and ppt are not accepted by the National Institute of Standards and Technology (NIST) due to their country and language-dependent nature. Consult "The Guide for the Use of the International System of Units" (Document SP811) from NIST for more detailed information on this subject (www.physics.nist.gov). The use of weight-to-weight (w/w) or weight-to-volume (w/v) concentration measurements instead of ppm, ppb or ppt is recommended to avoid any confusion or ambiguity. The inaccuracy with ppm, ppb and ppt is often small, thus this recommendation is frequently ignored and these units are commonly seen and used.

**Table F-1** ppm and ppb conversions.

| w/w | | w/v | | v/v | |
|---|---|---|---|---|---|
| Equals | | Equals | | Equals | |
| ppm | mg/kg<br>µg/gm<br>ng/mg | ppm | mg/L<br>µg/mL<br>ng/µL | ppm | mL/kL<br>µL/L<br>nL/mL |
| ppb | µg/kg<br>ng/gm<br>pg/mg | ppb | µg/L<br>ng/mL<br>pg/µL | ppb | µL/kL<br>nL/L<br>pL/mL |

### F.3.3
### Percent (%)

Higher concentrations are often expressed as percents. One (1) percent is equivalent to 1 part per hundred. To prevent confusion, w/w or w/v should be included when using percentages as a concentration measure. Table F-2 lists some of the common units used when calculating percent concentrations.

**Table F-2** Percent conversions.

| w/v | | v/v | |
|---|---|---|---|
| | Equals | | Equals |
| % | gm/100 mL<br>mg/100 μL<br>10 gm/L<br>10 mg/mL | % | mL/100 mL<br>μL/100 μL<br>10 mL/L<br>10 μL/mL |

| w/w | | Misc. | |
|---|---|---|---|
| | Equals | | Equals |
| % | gm/100 gm<br>mg/100 mg<br>μg/100 μg<br>10 gm/kg<br>10 mg/gm<br>10 μg/mg | 1%<br>1%<br><br>0.1%<br>0.1% | 1 part per hundred<br>10,000 ppm<br><br>1 part per thousand<br>1000 ppm |

### F.3.4
### Molarity (M or mM)

Molarity is another concentration measure. It is rarely used to denote analyte concentration in GC samples. Molarity is defined as 1 mole of a compound (analyte) per liter of solution. One mole of a compound is equal to its formula weight in grams. For example, 1 mole of sodium chloride (NaCl) weighs 58.44 gm. A 1 molar (M) NaCl aqueous solution would be obtained by dissolving 58.44 gm of NaCl in 1 L of water; dissolving 5.844 gm in 1 L would result in an 0.1 M NaCl solution. Obviously, volumes other than 1 liter can be made by proportionally adjusting the amount of NaCl and water (e.g., 2.922 gm of NaCl in 50 mL makes a 1 M solution).

For more dilute solutions, millimolar (mM) is often used. It is 1000 times lower in concentration than molar (M), thus 1000 mM equals 1 M. For example, 58.44 mg of NaCl in 1 L of water makes a 1 mM solution. Millimolar is often used when working with a less than 1 M solution. For example, 50 mM could be used instead of 0.050 M.

## F.4
## Density (ρ)

Density (ρ; sometimes denoted as *d*) is the mass of a compound or substance per given unit volume (ρ = mass/volume). The standard unit is kg/m$^3$; however, gm/mL is an often used unit. Density is temperature dependent and 25 °C is commonly used as a reference or standard temperature. The values listed in Table F-3 are in gm/mL at 25 °C.

**Table F-3** Densities and boiling points of common solvents.

| Solvent | Density (gm/mL at 25 °C) | Boiling point (°C) |
|---|---|---|
| Acetone | 0.791 | 56 |
| Acetonitrile | 0.786 | 81–82 |
| Carbon disulfide | 1.266 | 46 |
| Chloroform | 1.492 | 60.5–61.5 |
| Cyclohexane | 0.779 | 80.7 |
| Dichloromethane (methylene chloride) | 1.325 | 40 |
| Diethyl ether | 0.706 | 34.6 |
| Dimethyl sulfoxide (DMSO) | 1.100 | 189 |
| *N,N*-Dimethylformamide (DMF) | 0.944 | 153 |
| Ethanol | 0.789 | 78 |
| Ethyl acetate | 0.902 | 76.5–77.5 |
| *n*-Hexane | 0.659 | 69 |
| Methanol | 0.791 | 65 |
| *i*-Octane (2,2,4-trimethylpentane) | 0.692 | 98–99 |
| *n*-Pentane | 0.626 | 35–36 |
| 1-Propanol | 0.804 | 97 |
| *i*-Propanol (2-propanol; IPA) | 0.785 | 82 |
| Tetrahydrofuran (THF) | 0.889 | 65–67 |
| Toluene | 0.865 | 110–111 |
| Water | 0.997 | 100 |

Knowing the densities of liquid compounds can be convenient when preparing analytical standards. If the volume and density of the liquid is known, the mass can be calculated. If the mass and density are known then the volume can be calculated. Instead of weighing a liquid, its carefully measured volume can be used to calculate the weight since mass = density × volume. For example, if 5 mL of acetone is accurately dispensed then its mass can be calculated as 3.955 gm (0.791 gm/mL × 5 mL = 3.955 gm). If the mass and density of a liquid is known, its volume can be calculated since volume = mass/density. For example, if 1.978 gm of acetone is weighed then its volume can be calculated as 2.5 mL (1.978 gm/0.791 gm/mL = 2.5 mL).

Densities are normally listed at 25 °C which is close to room temperature. Adjustments to compound densities due to a slight difference in room temperature are usually unnecessary since any density errors are insignificant when preparing smaller volumes of solutions. Using densities in the preparation of analytical standards is further described throughout Section F.7.

Specific gravity is the density of a substance relative to water. Be careful not to confuse density and specific gravity. The density of water is often rounded to 1.0 gm/mL even though the actual value at 25 °C is 0.997 gm/mL.

If the density of a liquid is unknown, it can be approximated by weighing a known volume. Using a tared vial with a cap, carefully dispense a measured volume of the liquid into the vial. Once the weight is obtained, the density can be calculated by dividing the measured weight by the volume. If necessary, convert the density unit into the desired values (e.g., mg/μL or gm/mL).

## F.5 Calibration for Quantitative Purposes

The basic premise of quantitative GC is fairly straightforward. A standard containing a known amount of the target analyte is injected in the GC. The sample is then injected using the *exact* same analysis conditions. If the sample contains a peak with the same retention time as the one in the standard, there is a high probability the sample contains the corresponding analyte. The relative areas or heights of the corresponding peaks are compared to determine the analyte amount or concentration in the injected sample. More details can be found in Section 1.1.4. Often there are additional calculations required to obtain the analyte amount in the original sample and several techniques are described in the following sections.

### F.5.1 Single and Multiple Point Calibration

Quantitative accuracy is often poor with only a single standard. In most cases, superior quantitative results are obtained using multiple standards at various concentrations. If extremely accurate quantitative is not needed, a single point calibration is usually sufficient.

A few examples where a single point calibration is often sufficient are:

1. An analysis where some quantitative error is acceptable or the error will have a relatively insignificant consequence.
2. A qualitative analysis where only a determination whether a compound is present.
3. An analysis to determine whether a compound is present above or below a threshold concentration, but knowing the actual amount is not critical or needed.
4. A quantitative analysis where an approximate compound amount is sufficient (e.g., low, moderate or high).

Multiple point calibrations require more effort and calculations. The data from the multiple standard injections are used to generate a calibration curve for quantitative purposes.

### F.5.2
### Calibration Curves

A calibration curve is generated by plotting the peak area or height of the analyte peak in the standard versus its corresponding amount. A calibration curve is generated for each of the target analytes. A representative calibration curve for a single compound is shown in Figure F-1.

After input of the chromatographic data from standard injections, most data systems automatically generate, store and use calibration curves. The exact

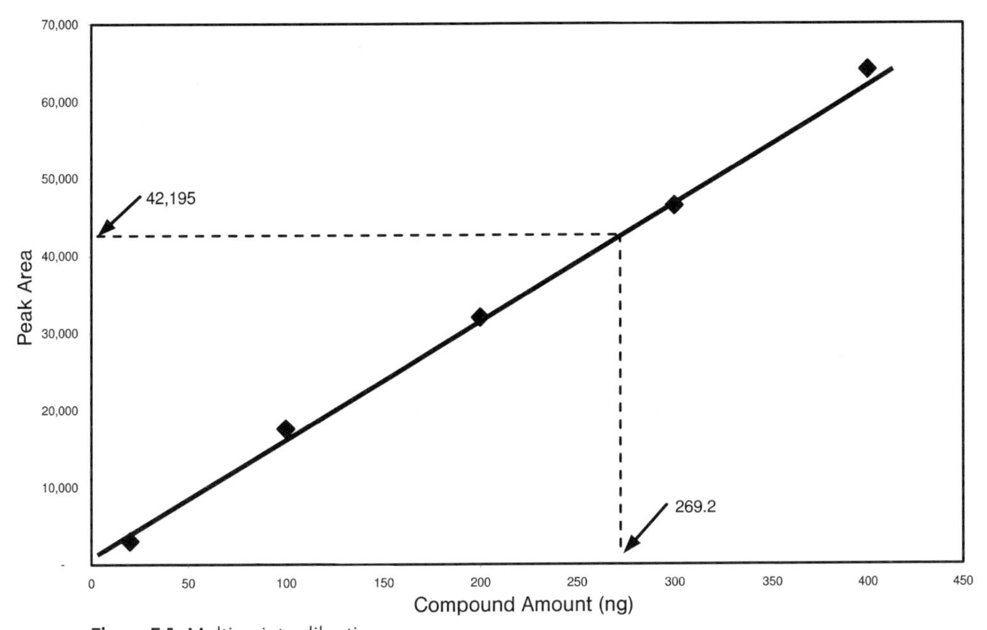


**Figure F-1** Multi-point calibration curve.

method is unique to each data system, thus detailed information can be found in the respective instruction manuals. Even though the data system constructs the calibration curve, it is recommended to understand the concept of calibration curve generation and use.

Once the calibration curve is generated and stored as part of the method, a peak in the sample positively identified by its retention time is compared to its respective calibration curve. The peak area or height is found on the peak size axis of the curve and the corresponding amount is located on the amount axis. Using the curve in Figure F-1 as an example, a sample peak with an area of 42,195 would correspond to 269.2 ng of the compound in the injected sample. This is done by finding 42,195 on the peak area axis ($y$-axis) and locating the corresponding point on the curve. The point on the compound amount axis ($x$-axis) corresponding to this point on the curve is located and used to determine the compound amount (269.2 ng).

It is generally accepted that a minimum of three points is needed to generate a good calibration curve. The lowest level standard should be at or slightly below the lowest amount expected in a sample. This amount may be at or slightly above the method detection limit when running trace level analysis. The highest level standard should be at or slightly above the highest amount expected in a sample. If a 3-point curve is being generated, the third standard should be about midway between the lowest and highest standards. If more than 3-point curve is being generated, the standards should be approximately evenly distributed between the lowest and highest level standards.

The number of points used in a calibration curve depends on several factors. One of the most important factors is the linear range of the detector (Section 8.2.6). If the detector is very linear over the range of the calibration curve, a fewer number of points can be used. If deviations from linear behavior are present, more points may be necessary. Some detectors such as an ECD have a relatively small linear range, thus more than one calibration curve may be necessary if the samples span a large concentration range. One curve is used for lower level samples and another is used for higher level samples. A 5-point calibration curve is used when more accurate or confident quantitative results are desired or when the sample concentrations span a large range.

A linear, best fit line forced through zero is probably the most common method used to generate calibration curves and usually provides the best results. The correlation coefficient is used to determine the quality of the linear fit. A coefficient of 0.99 or better is considered to be an excellent fit for quantitative purposes. If the correlation coefficient is too low, one or more of the data points may be erroneous, or the detector is not linear over the standard concentration range. A visual inspection usually exposes the problem data point; however, it can be difficult to find the actual problem point for three point calibration curves. Sometimes a more linear fit is obtained if the line is not forced through zero. Most data systems allow several choices of line fits, thus some trial and error may be necessary. For very large compound amount ranges and some detectors, a non-linear fit may be better.

Consistency in the generation of the calibration curves is very important. The exact same GC conditions must be used when generating the peak area or height data for each point in the curve. The only difference should be the concentration of the standards. Using different conditions such as injection volume or injector parameters (e.g., temperature, split ratio, purge times) will result in inaccurate calibration curves and quantitative errors. Samples also need to be analyzed using the exact same GC conditions as used with the standards or quantitative errors will occur.

More accurate calibration curves are obtained with multiple (repetitive) injections of each concentration standard. For example, multiple injections of the low, middle and high level standards are made for a three point calibration curve. The average of the peak areas or heights for each level standard is used to generate the calibration curves. Using the average peak areas or heights to generate the calibration curve helps to account for injection-to-injection variations. Ideally, the same number of injections at each level is used to obtain the averages.

New calibration curves should be generated at reasonable and logical intervals. A new set or group of samples or samples analyzed on different days are just some of the reasons for generating a new set of calibration curves. GC response changes over time due to a large number of reasons. Running frequent calibrations curves also aids in earlier detection or identification of some types of GC problems. Keeping a record of the calibration curves is useful in tracking the long-term performance of a GC. The calibration curve data is also useful in developing a preventative maintenance program for the GC.

## F.6
## Quantitation Calculations

Most samples require some type of dilution or extraction prior to GC analysis. If diluted or extracted, the sample injected into the GC does not reflect the actual analyte concentration in the original sample. Quantitation calculations have to account for every dilution or extraction step that alters the concentration, volume or weight of the original sample.

The use of external and internal standards is the most common GC quantitative methods. Many of the calculations are the same, but the internal standard method requires a few additional steps.

### F.6.1
### External Standard

The external standard technique is used when basic quantitation results are sufficient or when high accuracy is not needed. Unlike the internal standard technique, there are no corrections or adjustments made for sample losses, extraction inefficiencies, measurement inconsistencies or other sources of variation.

**Example**

1. 100 mL of solvent is added to 25 gm of a solid sample.
2. After thorough shaking followed by filtering, 50 mL of the extraction solvent is removed and concentrated to dryness.
3. The resulting dry residue is dissolved in 1 mL of solvent.
4. An injection of 2 μL of this solution is made.
5. Using a previously generated calibration curve, a target analyte amount of 192.2 ng is obtained.

*Step 1: Calculate the sample matrix concentration in the extraction solvent.*

Divide the sample weight by the volume of the extraction solvent.

$$\frac{25\ \text{gm}}{100\ \text{mL}} = 0.25\ \text{gm/mL}$$

*Step 2: Calculate the amount of the sample matrix contained in the solvent volume taken from the original extract.*

Multiply the sample matrix concentration and the volume removed.

$$0.25\ \text{gm/mL} \cdot 50\ \text{mL} = 12.5\ \text{gm}$$

Upon evaporating to dryness, the dry residue contains target analyte equivalent to the amount present in 12.5 gm of the original sample. The dry residue also contains any sample matrix components soluble in the extraction solvent corresponding to the amount present in 12.5 gm of the original (i.e., unextracted) sample. Sometimes the term "equivalents" is used to describe the amount of sample contained in an extracted or removed portion of a sample. In this example, the dry residue contains 12.5 gm equivalents and the extraction solvent contains 0.25 gm/mL sample equivalents.

*Step 3: Calculate the sample matrix concentration in the solvent used to dissolve the dried sample residue.*

Divide the dried sample residue amount by the solvent volume used to dissolve the dry sample residue.

$$\frac{12.5\ \text{gm}}{1\ \text{mL}} = 12.5\ \text{gm/mL} = 12.5\ \text{mg/μL}$$

*Step 4: Calculate the amount of sample matrix injected into the GC.*

Multiply the sample matrix concentration and the injection volume.

$$12.5\ \text{mg/μL} \cdot 2\ \text{μL} = 25.0\ \text{mg}$$

*Note:* This is not the amount of the target analyte.

*Step 5: Calculate the concentration of the target analyte in the injected sample.*

Divide the target analyte amount by the sample matrix amount injected into the GC.

$$\frac{192.2\ \text{ng}}{25.0\ \text{mg}} = 7.7\ \text{ng/mg}$$

For this example, the concentration of the target analyte is 7.7 ng/mg or 7.7 μg/gm. The original 25 gm of sample contained 192.5 μg of the target analyte (7.7 μg/gm · 25 gm).

F.6.2
### Internal Standard

Whenever the sample requires multiple extractions, splitting, drying and/or re-dissolution in solvents, using an internal standard is highly recommended. An internal standard is a compound added to the sample in a known amount at one point during the extraction or preparation process. It is usually added at the beginning before any processing steps occurs. The internal standard is used to measure or represent loss of target analytes during the sample extraction or preparation process. Failure to account for analyte loss will result in inaccurate and imprecise quantitative measurements. Variations in measurements such as pipeting and weighing, loss of sample and solvents due to evaporation or spillage, and analyte loss due to extraction technique limitations are some of the sources of quantitative inaccuracies. Using an internal standard aids in normalizing or correcting for these errors or variations. An internal standard is sometimes called a surrogate when it is added at the beginning of the sample preparation process.

The chemical and physical characteristics of the internal standard should be very similar to those of the analyte. The type and number of functional groups, molecular weight and volatilities of the analyte and internal standard should be as close as possible. In an ideal situation, each target analyte would have its own internal standard; however, this is rarely practical. Usually a single internal standard is used for multiple analytes of similar characteristics. If the characteristics of the target analytes differ by large amounts, multiple internal standards may be needed to account for the different analyte properties. Based on their properties, target analytes are associated with one of the multiple internal standards.

Selecting internal standards can sometimes be complicated or problematic. Some of the biggest challenges are compound availability and finding ones that do not co-elute with target analytes or sample matrix interferences. One strategy is to use internal standards that differ in one of the aliphatic components of their structures. For example, substituting an isopropyl for a propyl group, or adding a methyl group or double bond are often good routes to a satisfactory internal standard. The substitution or addition of a halogen is another approach. Substituting functional groups may not be a good approach since their chemical properties can be quite dissimilar. For example, using a ester as an internal standard for an alcohol may result in quantitative errors. For GC/MS analyses, a dueterated form of a target analyte is ideal if one is available.

Sometimes compromises are necessary especially for unusual or esoteric target analytes or for analysis with numerous target analytes of varying chemical structures and forms. Experimentation with several compounds may be required before an acceptable internal standard is found. This is done by adding known amounts of the target analytes and the various internal standards to blank or control samples followed by extraction and analysis. The recovery of the analytes and corresponding internal standards extracted from the control sample or blank should be similar. If the recovery amounts differ by more than 20%, the internal standard and analyte are not a good combination.

The calculations used for internal standards are more complicated than for external standards. There are a variety of quantitation calculation methods for using internal standards. Most data systems will perform the calculations if setup properly. The following example illustrates one approach. It is intended to show the concepts of using internal standards and not necessarily as the only or preferred method.

### Example

1. 100 mL of solvent is added to 25 gm of a solid sample.
2. 200 μL of a 600 ng/μL internal standard solution is added to the extraction solvent.
3. After thorough shaking followed by filtering, 50 mL of the extraction solvent is removed and concentrated to dryness.
4. The resulting dry residue is dissolved in 1 mL of solvent.
5. An injection of 2 μL of this solution is made.
6. Using a previously generated calibration curve, an analyte amount of 192.2 ng and an internal standard amount of 95.7 ng is obtained.

**Part A:** Determine the amount of sample matrix injected into the GC

*Step 1: Calculate the sample matrix concentration in the extraction solvent.*

Divide the sample weight by the volume of the extraction solvent.

$$\frac{25\ \text{gm}}{100\ \text{mL}} = 0.25\ \text{gm/mL}$$

*Step 2: Calculate the amount of the original sample matrix contained in the solvent volume taken from the original extract.*

Multiply the sample matrix concentration and the volume removed.

$$0.25\ \text{gm/mL} \cdot 50\ \text{mL} = 12.5\ \text{gm}$$

Upon evaporating to dryness, the dry residue contains target analyte equivalent to the amount present in 12.5 gm of the original sample. The dry residue also contains any sample matrix components soluble in the extraction solvent corresponding to the amount present in 12.5 gm of the original sample. Sometimes the term "equivalents" is used to describe the amount of sample contained in an extracted or removed portion of a sample. In this example, the dry residue contains 12.5 gm equivalents and the extraction solvent contains 0.25 gm/mL equivalents.

*Step 3: Calculate the sample matrix concentration in the solvent used to dissolve the dried sample residue.*

Divide the dried sample residue amount by the solvent volume used to dissolve the dry sample residue.

$$\frac{12.5\ \text{gm}}{1\ \text{mL}} = 12.5\ \text{gm/mL} = 12.5\ \text{mg/}\mu\text{L}$$

*Step 4: Calculate the amount of sample matrix injected into the GC.*

Multiply the sample equivalents concentration and injection volume.

$$12.5\ \text{mg/}\mu\text{L} \cdot 2\ \mu\text{L} = 25.0\ \text{mg}$$

*Note:* This is not the amount of the target analyte.

**Part B:** Determine the amount of the internal standard compound injected into the GC assuming there is no extraction loss (i.e., 100% recovery).

*Step 5: Calculate the amount of the internal standard compound added to the sample matrix.*

Multiple the concentration of the internal standard and the volume added.

$$600\ \text{ng/}\mu\text{L} \cdot 200\ \mu\text{L} = 120{,}000\ \text{ng} = 120\ \mu\text{g}$$

*Step 6: Calculate the internal standard concentration in the extraction solvent.*

Divide the internal standard amount by the volume of the extraction solvent.

$$\frac{120\ \mu\text{g}}{100\ \text{mL}} = 1.2\ \mu\text{g/mL}$$

*Step 7: Calculate the amount of the internal standard contained in the solvent volume taken from the original extract.*

Multiply the internal standard concentration and the volume removed.

$$1.2\ \mu\text{g/mL} \cdot 50\ \text{mL} = 60\ \mu\text{g}$$

*Step 8: Calculate the internal standard concentration in the solvent used to dissolve the dried sample residue.*

Divide the internal standard amount by the solvent volume used to dissolve the dry sample residue.

$$\frac{60\ \mu g}{1\ mL} = 60\ \mu g/mL = 60\ ng/\mu L$$

*Step 9: Calculate the amount of internal standard injected into the GC.*

Multiply the internal standard concentration and injection volume.

$$60\ ng/\mu L \cdot 2\ \mu L = 120\ ng$$

*Note:* This is the amount of internal standard compound assuming there is no loss during the extraction and preparation process.

**Part C:** Adjust the target analyte amount based on the recovery of the internal standard compound.

*Step 10: Calculate the amount of internal standard compound loss.*

Divide the amount of the internal standard compound detected by the amount if no loss occurred. This is the internal standard correction factor.

$$\frac{95.7\ ng}{120\ ng} = 0.798$$

*Note:* This result indicates that 20.2% of the external standard was lost during the extraction and preparation process ($1 - 0.798 = 0.202$). It is assumed the same amount of the target analyte was lost also.

*Step 11: Adjust the target analyte amount by the amount of the internal standard loss.*

Divide the target analyte amount by the internal standard correction factor.

$$\frac{192.2\ ng}{0.798} = 240.9\ ng$$

**Part D:** Determine the target analyte concentration in the original sample.

*Step 12: Calculate the concentration of the target analyte in the injected sample.*

Divide the adjusted target analyte amount by the sample matrix amount injected into the GC (from Step 4).

$$\frac{240.9\ \text{ng}}{25.0\ \text{mg}} = 9.6\ \text{ng/mg}$$

For this example, the concentration of the target analyte is 9.6 ng/mg or 9.6 μg/gm using an internal standard. Using the external standard method, a concentration of 7.7 ng/mg was obtained (from the example in Section F.6.1). The external standard method concentration is under reported by 1.9 ng/mg or 19.8%. Providing the internal standard compound was properly selected, the internal standard based result is more accurate and precise.

### F.6.3
### Modified Standard Addition

The use of quantitative standards assumes the analyte response is the same in the prepared sample (matrix) as in the standard. Sometimes matrix components may alter detector response or generates a peak that co-elutes with a standard peak, thus introducing quantitative inaccuracies. This is most common with complex samples where typical sample preparation techniques are insufficient or satisfactory sample cleanup is not feasible due to cost, time or accessibility. In these cases, using a form of the technique called standard addition may help.

The final sample is divided into a minimum of two portions prior to GC analysis. A known, but different, amounts of the target analyte is added to all but one portion. While one spiked sample is sufficient, two or more provides better accuracy. The same guidelines governing the concentration range of multiple standards (Section F.5.2) also applies to standard addition. The same volume of each respective spiking solution has to be added to each spiked sample. Clean solvent at the same volume is added to the unspiked sample. If the same volume can not be used for every sample, additional solvent can be added when necessary to make the total added volumes equal. It is critical that the same total volume is added to every sample (unless the sample volume is large) or concentration errors will occur. The additional added volume also has to be considered and adjusted for the quantitative calculations.

After the samples have been analyzed and data generated, a multi-point calibration curve is created (Section F.5.2). A difference with a regular calibration curve is the *x*-axis extends into the negative region. Unless the target analyte is not present in the sample, the *y*-intercept will be greater than zero and the *x*-intercept will be negative. The point where the curve intersects with the *x*-axis (i.e., $y = 0$) is the inverse of the target analyte amount in the sample.

An example three point standard addition curve is shown in Figure F-2. In this example, the target analyte amount in the injected sample is 15.5 ng. The same concentration calculations previously described in Sections F.6.1 and F.6.2 can be used to determine the analyte concentration in the original sample. Some data systems have a standard addition option as part of their calibration curve generation features.

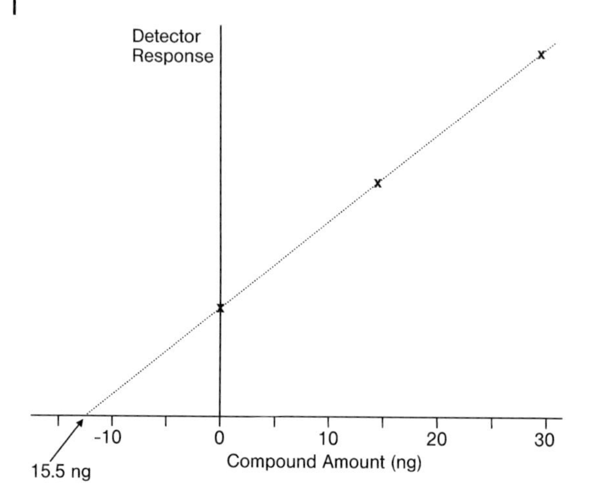


**Figure F-2** Modified standard addition curve.

Using standard addition has several drawbacks to consider. Standard addition requires a large amount of extracted sample, thus it is not well suited when very small or limited sample amounts are available. Additional sample handling is required since every sample requires accurate addition of the spiking standard or solvent. Since each sample is injected numerous times, a much larger number of GC injections per sample are required. The extra effort may be worthwhile in some cases since a modified standard addition technique provides the best quantitative results.

### F.6.4
### Relative Percent

Relative percent is often used for determining sample purities. A very small amount of the sample is analyzed and the areas of every peak is recorded. All of the peak areas are summed to determine the total peak area. The area of each individual peak is divided by the total peak area and converted into a percentage. The total should add up to 100%. Sometimes rounding results in a 0.1–0.2% error. Most data systems will generate a relative peak areas report.

Solid samples have to be dissolved in a solvent. The peak areas of the solvent peaks are not included in the calculations. Since the solvent is rarely 100% pure, a solvent blank is necessary to identify every peak corresponding to the solvent so their respective areas can be eliminated from the calculation. Using the proper integration settings or program will allow the data system to generate a relative peak area report for a solvent diluted sample (i.e., it disregards the solvent peaks).

Relative percent values may contain some errors due to a few assumptions. It is assumed that all sample compounds have the same amount of injector discrimination and detector response factors. Differences between the various compound functional groups results in response factor variations. For example, adding one or more oxygen, nitrogen and halogen groups to a hydrocarbon significantly changes its FID response especially for small molecular weight compounds. Selective detectors are usually not well suited for determining relative percent amounts due to their often extreme response differences between compounds. Another assumption is each compound amount is within the linear response range of the detector. If not, the peak areas for the highest and lowest amount compounds may be over or under reported. Compounds present in very high amounts may exceed the range of the detector. If any of the large peaks in the chromatogram have a flat or square top, the range has been exceeded and the corresponding peak area is too small. Increasing the detector range often comes at the expense of sensitivity, thus the ability to detect impurities at very low levels. The errors induced by accepting these assumptions can lead to errors of a 0.1% to over 10%. Finally, whenever determining purity levels by GC, it should be noted as part of the results. There may be impurities in the sample not detected by GC analysis, thus the reported purity level may not be absolutely correct without the qualifying statement.

## F.7
## Techniques for Preparation of Analytical Standards for GC

There are numerous methods to prepare standards for GC. They vary in accuracy, precision and complexity. The amount of accuracy and precision along with available glassware and measurement instruments may dictate which method is used. Working with toxic, highly volatile or unstable compounds may require special handling and safety precautions.

Pre-made standards are often available from commercial sources for regulated or standardized methods. Depending on the compounds and amounts used, it may be more cost effective to purchased pre-made standards. Specially prepared or custom standards are available, but the cost is usually high. In these cases, preparing the standards in the lab may be desired or necessary.

### F.7.1
### Standard Composition Considerations

A few decisions and simple calculations are necessary before proceeding with standard preparation or purchase. Many of these factors determine which preparation method, supplies and tools are required. Also, errors or wasted efforts may occur if some of these issues are ignored or forgotten.

1. *Compounds:* The physical and chemical properties and structures of each compound should be obtained and studied. Information such as physical form (e.g., viscous liquid, powder, crystals, etc.), solubility, reactivity, stability and toxicity are important qualities. This information may influence other decisions concerning the format and preparation of a standard.

2. *Number and combination of compounds:* Some compounds properties may require some to be in separate standards to avoid problems such as reactivity or insolubility. Some methods have numerous target analytes which can make standard preparation difficult or problematic if every one is put into a single standard. Preparing multiple standards each containing a subset of the compounds if often easier or more manageable.

3. *Concentration:* The concentration of each standard compound needs to be determined. Factors such as method detection limits, detector linear range, range of analytes in typical samples, injector and detector type, injection volume, detector sensitivity and selectivity, and compound solubilities have to be considered.

4. *Solvent:* All of the standard compounds need to be completely soluble in the solvent. If the standards will be stored in a refrigerator or freezer, the compounds also need to be soluble in the solvent at the lower temperatures. The solvent must not react with any of the compounds (e.g., acetone and primary amines). Extremely volatile solvents should be avoided due to their tendency to easily evaporate, thus increase the concentration of the standard over time. Some injection modes such as splitless and on-column require solvents with specific boiling point properties, thus instrument constraints have to be considered. Flammability and toxicity may be other solvent factors to take into account.

5. *Stability:* Some compounds degrade or react in solution with other compounds or the solvent. Some methods contain acidic and basic compounds which may react if present in the same solution, thus separate standards become necessary. Some standards may require special handling such as nitrogen purging or refrigeration.

6. *Volume:* The final volume of the standard may dictate the preparation method, supplies and tools. If stable over a long time and properly handled, preparing a large volume of a stock or analytical standard is usually better. This often reduces variation between standards unavoidably introduced with multiple preparations. Preparing smaller volumes of less stable standards may be required to ensure consistent concentrations and avoid unnecessary generation of chemical waste due to frequent disposal of degraded standards.

7. *Containers:* The volume of available bottles, vials or flasks may influence the volume of the standard. Reactive standard compounds may require the use of silylated glass vials or bottles. Teflon lined caps are usually needed to prevent extraction of impurities from the vial or bottle caps. Plastic bottle or vials may also leach materials into the standard.

8. *Equipment usage and capabilities:* Experience with accurate and precise operation of balances, pipets, micropipettors and related equipment is essential in obtaining the proper concentration standards. The small masses and volumes often encountered with standard preparation require capable balances and pipets along with good chemical handling techniques.

## F.7.2
## Preparing One Component Standards

### F.7.2.1 Using a Volumetric Flask

Using a volumetric flask is probably the most familiar method to prepare solutions and standards. The method is fairly simple and reliable providing proper weighing techniques are used.

#### Solid Analyte – Exact Weight Method

**Example**

1. 2.0 mg/mL standard concentration
2. 50 mL standard volume

*Step 1: Calculate the amount of the analyte compound to add to the flask.*

Multiply the standard concentration and the standard volume.

$$2.0\ \text{mg/mL} \cdot 50\ \text{mL} = 100\ \text{mg}$$

*Step 2: Add the analyte compound to the flask.*

Place the flask and stopper on the balance and tare (zero) the balance. Add 100 mg of the compound to the flask. Alternatively, the compound can be weighed out on weighing paper then transferred to a flask.

*Step 3: Add solvent to the flask.*

Fill the flask to the mark with the desired solvent, insert the flask stopper, thoroughly mix and label.

#### Solid Analyte – Approximate Weight Method

Weighing an exact amount into the volumetric flask or onto weighing paper can be difficult and trying. Also, accurately transferring all of the compound from weighing paper into a flask can also be problematic. Very small amounts of the compound usually have to be repeatedly added or removed from the flask or weighing paper to obtain the exact amount. An amount close to the target amount is used and the standard concentration is corrected accordingly.

**Example**

1. 2.0 mg/mL standard concentration
2. 50 mL standard volume

*Step 1: Calculate the amount of analyte compound to added to the flask.*

Multiply the standard concentration and the standard volume.

$$2.0\ \text{mg/mL} \cdot 50\ \text{mL} = 100\ \text{mg}$$

*Step 2: Add the analyte compound to the flask.*

Place the flask and stopper on the balance and tare (zero) the balance. Add approximately 100 mg of the compound to the flask. Alternatively, the compound can be weighed out on weighing paper then transferred to a flask.
For this example, 100.5 mg of the compound was added.

*Step 3: Add solvent to the flask.*

Fill the flask to the mark with the desired solvent, insert the flask stopper, thoroughly mix and label.

*Step 4: Calculate the concentration of the prepared standard solution.*

Divide the compound amount (from Step 2) by the volumetric flask volume.

$$\frac{100.5\ \text{mg}}{50\ \text{mL}} = 2.01\ \text{mg/mL}$$

#### Liquid Analyte – Exact Volume Method

Preparing a standard of a liquid compound requires an additional calculation. Adding and removing very small volumes of liquid compounds is difficult and they are prone to evaporation especially extremely volatile ones. This problem can be avoided by using a liquid compound's density. Once the desired weight of the liquid compound is known, the corresponding volume can be calculated using its density (Section F.4).

**Example**

1. 2.0 mg/mL standard concentration
2. 0.963 gm/mL compound density
3. 50 mL standard volume

*Step 1: Calculate the amount of analyte compound to add to the flask.*

Multiple the standard concentration and the standard volume.

$$2.5\ \text{mg/mL} \cdot 50\ \text{mL} = 125\ \text{mg}$$

*Step 2: Convert the density units to match the standard concentration and volume, if necessary.*

$$0.963\ \text{gm/mL} = 0.963\ \text{mg/µL}$$

*Step 3: Calculate the volume of the analyte compound equal to the mass determined in Step 1.*

Divide the compound amount by its density.

$$\frac{125\ \text{mg}}{0.963\ \text{mg/µL}} = 129.8\ \text{µL}$$

*Step 4: Add the compound to the flask.*

Pipet 129.8 µL of the analyte compound into the flask and immediately insert the flask stopper.

*Step 5: Add solvent to the flask.*

Fill the flask to the mark with the desired solvent, insert the flask stopper, thoroughly mix and label.

#### Liquid Analyte – Approximate Volume Method

It can be difficult to accurately pipet a small volume and obtain the weight calculated using the compound's density. More accurate results are obtained by weighing the liquid compound after it is added to the volumetric flask. The measured weight is then used to calculate the actual concentration of the standard.

**Example**

1. 2.0 mg/mL standard concentration
2. 0.963 gm/mL compound density
3. 50 mL standard volume

*Step 1: Calculate the amount of analyte compound to add to the flask.*

Multiple the standard concentration and the standard volume.

$$2.5 \text{ mg/mL} \cdot 50 \text{ mL} = 125 \text{ mg}$$

*Step 2: Convert the density units to match the standard concentration and volume, if necessary.*

$$0.963 \text{ gm/mL} = 0.963 \text{ mg/}\mu\text{L}$$

*Step 3: Calculate the volume of the analyte compound equal to the mass determined in Step 1.*

Divide the compound amount by its density.

$$\frac{125 \text{ mg}}{0.963 \text{ mg/}\mu\text{L}} = 129.8 \text{ }\mu\text{L}$$

*Step 4: Add the compound to the flask.*

Place the flask and stopper on the balance and tare (zero) the balance. Remove the flask from the balance. Pipet 129.8 µL of the compound into the flask and immediately re-insert the flask stopper.

*Step 5: Determine the weight of the analyte compound.*

Placed the flask with the inserted stopper on the still tared balance. Record the weight of the analyte compound.
For this example, 127.5 mg of the compound was added.

*Step 6: Calculate the concentration of the standard.*

Dividing the analyte compound amount by the flask volume.

$$\frac{127.5\ \text{mg}}{50\ \text{mL}} = 2.55\ \text{mg/mL}$$

### F.7.2.2 Using Vials and an Exact Measurement Technique

In general, working with whole numbers or numbers easy to mathematically manipulate is desired. While concentrations such as 2.01 or 2.55 mg/mL are useable, they are somewhat unwieldy numbers to use in calculations or for dilutions. There is a different standard preparation method that gives the concentration as originally intended.

This method uses a vial or bottle instead of a volumetric flask. A volumetric flask can be used, but the fill line or level is ignored. Before proceeding, the final volume of the standard needs to be known.

#### Solid Analyte

**Example**

1. 1.5 mg/mL standard concentration
2. 10 mL standard volume

*Step 1: Calculate the amount of analyte compound to add to the vial.*

Multiply the standard concentration and the standard volume.

$$1.5\ \text{mg/mL} \cdot 10\ \text{mL} = 15\ \text{mg}$$

*Step 2: Add the analyte compound to the vial.*

Place a vial along with its cap on the balance and tare (zero) the balance. Add approximately 15 mg of the analyte compound to the vial.
For this example, 16.2 mg was being added.

*Step 3: Calculate the volume of solvent to add to the vial.*

Divide the analyte compound amount by the standard concentration.

$$\frac{16.2\ \text{mg}}{1.5\ \text{mg/mL}} = 10.8\ \text{mL}$$

*Step 4: Add solvent to the vial.*

Add the solvent volume calculated in Step 3 to the vial, cap, thoroughly mix and label.
For this example, the desired concentration of 1.5 mg/mL is obtained when 10.8 mL of solvent is added to the vial containing 16.2 mg of the analyte compound.

### Liquid Analyte

#### Example

1. 5.0 mg/mL standard concentration
2. 1.05 gm/mL compound density
3. 100 mL standard volume

*Step 1: Calculate the amount of analyte compound to add to the vial.*

Multiply the standard concentration and the standard volume.

$$5.0 \text{ mg/mL} \cdot 100 \text{ mL} = 500 \text{ mg}$$

*Step 2: Convert the density units to match the standard concentration and volume, if necessary.*

$$1.05 \text{ gm/mL} = 1.05 \text{ mg/}\mu\text{L}$$

*Step 3: Calculate the volume of the liquid compound equal to the mass determined in Step 1.*

Divide the analyte compound amount by its density.

$$\frac{500 \text{ mg}}{1.05 \text{ mg/}\mu\text{L}} = 476.2 \text{ }\mu\text{L}$$

*Step 4: Add the compound to the flask.*

Place the vial and its cap on the balance and tare (zero) the balance. Remove the vial from the balance. Pipet 476.2 µL of the analyte compound into the vial and immediately re-cap.

*Step 5: Determine the weight of the analyte compound.*

Placed the vial on the still tared balance. Record the weight of the analyte compound.
For this example, 498.0 mg of the compound was added.

*Step 6: Calculate the volume of solvent to add to the vial.*

Divide the analyte compound amount by the standard concentration.

$$\frac{498.0 \text{ mg}}{5.0 \text{ mg/mL}} = 99.6 \text{ mL}$$

*Step 7: Add solvent to the vial.*

Add 99.6 mL of solvent to the vial, cap, thoroughly mix and label.
For this example, the desired concentration of 5.0 mg/mL is obtained when 99.6 mL of solvent is added to the vial containing 498.0 mg of the compound.

## F.7.3
## Preparing Multi-Component Standards

Nearly every GC analysis is intended for multiple analytes, thus analytical standards containing multiple compounds are needed. Usually a concentrated stock solution for each compound is prepared. Stock solutions can be prepared using any of the methods described in Section F.7.2. Stock solutions are mixed in the appropriate amounts to obtain an analytical standard containing the analyte compounds at their desired concentrations. A stock solution can contain more than one compound provided it can be easily and accurately prepared.

There are numerous methods to calculate and prepare multi-component standards. The two methods described below are designed to keep the math simple by using numbers that are easy to manipulate and error check.

### F.7.3.1 Equal Volume Method

This method is designed so the same volume of each stock solution is used to prepare the analytical standard. Several decisions or assumptions have to be made before or during the preparation process.

#### Using a Volumetric Flask

**Example**

1. Standard containing 4 compounds
2. 20 mL standard volume
3. Analyte compound concentrations
   a) Compound 1 at 100 µg/mL
   b) Compound 2 at 200 µg/mL
   c) Compound 3 at 50 µg/mL
   d) Compound 4 at 50 µg/mL

*Step 1: Calculate the amount of each analyte compound in the final standard.*

Multiply each compound concentration and the standard volume.

Compound 1: $100\ \mu g/mL \cdot 20\ mL = 2000\ \mu g$
Compound 2: $200\ \mu g/mL \cdot 20\ mL = 4000\ \mu g$
Compound 3: $50\ \mu g/mL \cdot 20\ mL = 1000\ \mu g$
Compound 4: $50\ \mu g/mL \cdot 20\ mL = 1000\ \mu g$

*Step 2: Select the stock solution volume used to make the analytical standard.*

A volume of 100 µL was selected in this example due to the ease of calculations and the availability of accurate measurement tools (e.g., digital pipettor). As long as they are equal, volumes other than 100 µL can be used if desired; however, the sum of the stock solution volumes can not exceed the desired volume of the analytical standard.

*Step 3: Calculate the concentration of each stock solution.*

Divide the compound amount in the standard (from Step 1) by the selected volume (from Step 2).

$$\text{Compound 1:} \quad \frac{2000\ \mu g}{100\ \mu L} = 20\ \mu g/\mu L = 20\ mg/mL$$

$$\text{Compound 2:} \quad \frac{4000\ \mu g}{100\ \mu L} = 40\ \mu g/\mu L = 40\ mg/mL$$

$$\text{Compound 3:} \quad \frac{1000\ \mu g}{100\ \mu L} = 10\ \mu g/\mu L = 10\ mg/mL$$

$$\text{Compound 4:} \quad \frac{1000\ \mu g}{100\ \mu L} = 10\ \mu g/\mu L = 10\ mg/mL$$

*Step 4: Select the volume of each stock solution to prepare.*

For this example, a stock solution volume of 5 mL was selected.

*Step 5: Calculate the amount of each analyte compound to weigh or dispense.*

Multiply the stock solution concentration (from Step 3) and volume (from Step 4).

Compound 1: $20\ mg/mL \cdot 5\ mL = 100\ mg$
Compound 2: $40\ mg/mL \cdot 5\ mL = 200\ mg$
Compound 3: $10\ mg/mL \cdot 5\ mL = 50\ mg$
Compound 4: $10\ mg/mL \cdot 5\ mL = 50\ mg$

*Step 6: Prepare each stock solution.*

Use any of the techniques described in Section F.7.2 to prepare 5 mL of each stock solution at their respective concentrations.

*Step 7: Add each stock solution to the volumetric flask.*

Accurately pipet 100 µL of each stock solution into a 20 mL volumetric flask.

*Step 8: Add solvent to the volumetric flask.*

Fill the flask to the mark with the desired solvent, insert the flask stopper, thoroughly mix and label.

**Using a Vial or Bottle**

If a suitable volumetric flask is not available, an alternative method using a vial or bottle can be employed.

**Example**

1. Standard containing 4 compounds
2. 20 mL standard volume
3. Analyte compound concentrations
   a) Compound 1 at 100 μg/mL
   b) Compound 2 at 200 μg/mL
   c) Compound 3 at  50 μg/mL
   d) Compound 4 at  50 μg/mL

*Steps 1–6 are the same as for the volumetric flask method.*

*Step 7: Add each stock solution to the vial or bottle.*

Accurately pipet 100 μL of each stock solution into a the vial or bottle.

*Step 8: Calculate the total volume of stock solution added to the vial or bottle.*

Sum the volumes of the stock solutions added to the vial or bottle.

$$100 + 100 + 100 + 100 = 400\ \mu L = 0.40\ mL$$

*Step 9: Calculate the volume of solvent to add to the vial or bottle.*

Subtract the total stock solution volume (Step 8) from the final standard volume.

$$20\ mL - 0.40\ mL = 19.6\ mL$$

*Step 10: Add solvent to the vial or bottle.*

Add the calculated volume of solvent (Step 9) to the vial or flask, cap, thoroughly mix and label.

#### F.7.3.2 **Equal Concentration Method**

This method is designed so the same concentration of each stock solution is used to prepare the analytical standard. Several decisions or assumptions need to be made before or during the process.

**Using a Volumetric Flask**

**Example**

1. Standard containing 4 compounds
2. 20 mL standard volume
3. Analyte compound concentrations
   a) Compound 1 at 100 μg/mL
   b) Compound 2 at 200 μg/mL
   c) Compound 3 at  50 μg/mL
   d) Compound 4 at  50 μg/mL

*Step 1: Calculate the amount of each analyte compound in the final standard.*

Multiply each compound concentration and the standard volume.

Compound 1: 100 µg/mL · 20 mL = 2000 µg
Compound 2: 200 µg/mL · 20 mL = 4000 µg
Compound 3: 50 µg/mL · 20 mL = 1000 µg
Compound 4: 50 µg/mL · 20 mL = 1000 µg

*Step 2: Calculate the concentration of the stock solutions.*

Find the compound with the lowest concentration in the analytical standard.

Multiply its concentration by a value between 100 and 500. Usually a value evenly divisible by 100 works the best (e.g., 100, 200, etc.). Some trial and error may be necessary to obtain a suitable concentration that works in Step 4. This value represents the dilution factor used to make the analytical standard from the lowest concentration stock solution.

For this example, compounds 3 or 4 are the lowest concentration standards. A multiplier of 400 was selected.

$$50\ \mu g/mL \cdot 400 = 20{,}000\ \mu g/mL$$
$$20{,}000\ \mu g/mL = 20\ \mu g/\mu L = 20\ mg/mL$$

*Step 3: Select the volume of each stock solution to prepare.*

Any reasonable volume can be used. For this example, a stock solution volume of 5 mL is being used.

*Step 4: Calculate the amount of each analyte compound to weight or dispense.*

Multiply the stock solution concentration (Step 2) and volume (Step 3).

$$20\ mg/mL \cdot 5\ mL = 100\ mg$$

*Step 5: Prepare each stock solution.*

Use any of the techniques described in Section F.7.2 to prepare 5 mL of each stock solution.

*Step 6: Calculate the volume of each stock solution needed to prepare the analytical standard.*

Divide the amount of each analyte compound in the analytical standard (Step 1) by the stock solution concentration (Step 2).

Compound 1: $$\frac{2000\ \mu g}{20\ \mu g/\mu L} = 100\ \mu L$$

Compound 2: $$\frac{4000\ \mu g}{20\ \mu g/\mu L} = 200\ \mu L$$

Compound 3: $\dfrac{1000\ \mu g}{20\ \mu g/\mu L} = 50\ \mu L$

Compound 4: $\dfrac{1000\ \mu g}{20\ \mu g/\mu L} = 50\ \mu L$

If any of the volumes obtained in this step are too small or large, adjust the selected concentration of the stock solution obtained in Step 2. Increase the concentration if the volume is too large and decrease the concentration if the volume is too small. If an adjustment is not possible, the equal concentration method can not be used. The unequal volume and concentration method will have to be used (Section F.7.3.3).

*Step 7: Add each stock solution to the volumetric flask.*

Accurately pipet the corresponding volume of each stock solution (Step 6) into a 20 mL volumetric flask.

*Step 8: Add solvent to the volumetric flask.*

Fill the flask to the mark with the desired solvent, insert the flask stopper, thoroughly mix and label.

#### Using a Vial or Bottle

If a suitable volumetric flask is not available, an alternative method using a vial or bottle can be employed.

**Example**

1. Standard containing 4 compounds
2. 20 mL standard volume
3. Analyte compound concentrations
   a) Compound 1 at 100 μg/mL
   b) Compound 2 at 200 μg/mL
   c) Compound 3 at 50 μg/mL
   d) Compound 4 at 50 μg/mL

*Steps 1–6 are the same as for the volumetric flask method.*

*Step 7: Add each stock solution to the vial or bottle.*

Accurately pipet the corresponding volume of each stock solution (Step 6) into a vial or bottle (100, 200, 50 and 50 μL).

*Step 8: Calculate the total volume of stock solution added to the vial or bottle.*

Sum the volumes of the stock solutions added to the vial or bottle.

$$100 + 200 + 50 + 50 = 400\ \mu L = 0.40\ mL$$

*Step 9: Calculate the volume of solvent to add to the vial or bottle.*

Subtract the total stock solution volume (Step 8) from the final standard volume.

20 mL – 0.40 mL = 19.6 mL

*Step 10: Add solvent to the vial or bottle.*

Add the calculated volume of solvent (Step 9) to the vial or flask, cap, thoroughly mix and label.

#### F.7.3.3 **Unequal Volume and Unequal Concentration Method**

Sometimes using equal stock solution volumes or concentrations is not possible or desirable. A desired stock solution concentration may not be obtained to due limited solubility of a compound in the selected solvent or some stock solution volumes are too small or large to handle. Perhaps the desired concentrations may not be available from the supplier or vendor when using prepared stock solutions.

This method is used when the stock solution concentrations are not equal and the stock solution volumes have to unequal, thus allowing the use of standard concentrations that may not ideal.

**Using a Volumetric Flask**

**Example**

1. Standard containing 4 compounds
2. 20 mL standard volume
3. Analyte compound concentrations
   a) Compound 1 at 100 µg/mL
   b) Compound 2 at 200 µg/mL
   c) Compound 3 at 50 µg/mL
   d) Compound 4 at 50 µg/mL

*Step 1: Calculate the amount of each analyte compound in the final standard.*

Multiply each compound concentration and the standard volume.

Compound 1: 100 µg/mL · 20 mL = 2000 µg
Compound 2: 200 µg/mL · 20 mL = 4000 µg
Compound 3: 50 µg/mL · 20 mL = 1000 µg
Compound 4: 50 µg/mL · 20 mL = 1000 µg

*Step 2: Calculate the concentration of each stock solution.*

Multiply the concentration of each analyte compound in the desired analytical standard by a value between 100 and 500. The exact number is not critical; however,

a value evenly divisible by 100 usually works the best (e.g., 100, 200, etc.). The goal is to end up with a stock solution concentration that results in a volume that is easy to accurately measure in Step 5. Some trial and error may be necessary to find a suitable or optimal concentration, thus this is an iterative process.
For this example, the following stock solution concentrations are used.

Stock solution 1: 4 μg/μL
Stock solution 2: 10 μg/μL
Stock solution 3: 5 μg/μL
Stock solution 4: 2 μg/μL

*Step 3: Prepare or obtain the stock solutions.*

Any of the methods in Section F.7.2 can be used to make the stock solutions. If purchased stock solutions are being used, the concentration will be supplied by the vendor. Dilution of the vendor supplied solution may be necessary to avoid volumes too low to accurately dispense.

*Step 4: Calculate the volume of each stock solution to add to the analytical standard.*

Divide the amount of each analyte compound in the analytical standard (Step 1) by each respective stock solution concentration (Step 2).

Compound 1: $$\frac{2000\ \mu g}{4\ \mu g/\mu L} = 500\ \mu L$$

Compound 2: $$\frac{4000\ \mu g}{10\ \mu g/\mu L} = 400\ \mu L$$

Compound 3: $$\frac{1000\ \mu g}{5\ \mu g/\mu L} = 200\ \mu L$$

Compound 4: $$\frac{1000\ \mu g}{2\ \mu g/\mu L} = 500\ \mu L$$

*Step 5: Add each stock solution to the volumetric flask.*

Accurately pipet the corresponding volume of each stock solution (Step 4) into a 20 mL volumetric flask.

*Step 6: Add solvent to the volumetric flask.*

Fill the flask to the mark with the desired solvent, insert the flask stopper, thoroughly mix and label.

#### Using a Vial or Bottle

If a suitable volumetric flask is not available, an alternative method using a vial or bottle can be employed.

**Example**

1. Standard containing 4 compounds
2. 20 mL standard volume
3. Analyte compound concentrations
   a) Compound 1 at 100 μg/mL
   b) Compound 2 at 200 μg/mL
   c) Compound 3 at 50 μg/mL
   d) Compound 4 at 50 μg/mL

*Steps 1–4 are the same as for the volumetric flask method.*

*Step 5: Add each stock solution to the vial or bottle.*

Accurately pipet the corresponding volume of each stock solution (Step 4) into a vial or bottle (500, 400, 200 and 500 μL).

*Step 6: Calculate the total volume of stock solution added to the vial or bottle.*

Sum the volumes of the stock solutions added to the vial or bottle.

$$500 + 400 + 200 + 500 = 1600\ \mu L = 1.6\ mL$$

*Step 7: Calculate the volume of solvent to add to the vial or bottle.*

Subtract the total stock solution volume (Step 6) from the final standard volume.

$$20\ mL - 1.6\ mL = 18.4\ mL$$

*Step 8: Add solvent to the vial or bottle.*

Add the calculated volume of solvent (Step 7) to the vial or flask, cap, thoroughly mix and label.

### F.7.4
### Serial Dilution

To generate multi-point calibration curves, multiple standards at different concentrations are needed. Instead of making completely separate standards, a technique called serial dilution can be used. Serial dilution results in better accuracy and precision especially when three or more standards are being used.

Serial dilution involves starting with a high concentration standard and a portion is diluted to make a lower concentration standard. The resulting lower concentration standard is used to make an even lower concentration standard. This process is repeated until the desired number of analytical standards is obtained.

The most concentrated solution can be used as the first (i.e., highest concentration) analytical standard or used as a stock solution to prepare all of the analytical standards. It is recommended to prepare a separate stock solution for further dilution since this helps to maintain the integrity of the highest concentration standard. Also, additional analytical standards can be easily and repeatedly made from the already prepared stock solution. Sometimes using a stock solution may not be possible due to situations such as solubility limits or limited concentration choices of commercially supplied standards.

A number of calculations are required to determine the volume of each analytical standard to prepare. The calculations start with the lowest concentration analytical standard and progresses in order towards the stock solution. These calculations need to be completed before preparing any of the analytical standards or stock solution.

**Example**

1. Four analytical standards and one stock solution
2. Stock solution: 1000 µg/mL
3. Standard 1: 20 mL at 200 µg/mL
4. Standard 2: 10 mL at 100 µg/mL
5. Standard 3: 10 mL at 50 µg/mL
6. Standard 4: 20 mL at 20 µg/mL

**Part A:** Determining the Volume of Standard 4

The volume of Standard 4 is already known since it is stated in the original information (20 mL).

**Part B:** Calculating the Total Volume of Standard 3 to Prepare

*Step 1: Calculate the amount of the analyte compound in Standard 4.*

Multiply the concentration of Standard 4 and the volume of Standard 4.

$$20\ \mu g/mL \cdot 20\ mL = 400\ \mu g$$

*Step 2: Calculate the volume of Standard 3 containing the analyte compound amount in the prepared volume of Standard 4.*

Divide the analyte compound amount in Standard 4 (Step 1) by the concentration of Standard 3.

$$\frac{400\ \mu g}{50\ ng/\mu L} = \frac{400\ \mu g}{50\ \mu g/mL} = 8\ mL$$

*Note:* This is only the volume of Standard 3 needed to prepare Standard 4.

*Step 3: Calculate the minimum volume of Standard 3 to prepare.*

Add the volume of Standard 3 needed to make Standard 4 (Step 2) to the final volume of Standard 3.

$$8\ \text{mL} + 10\ \text{mL} = 18\ \text{mL}$$

*Note:* A larger volume can be made for ease of preparation or calculation.
For this example, 20 mL of Standard 3 was prepared and used in the following calculations.
After 8 mL is removed to make Standard 4, the remaining 12 mL is greater than the desired or minimum final volume of Standard 3 (10 mL).

**Part C:** Calculating the Total Volume of Standard 2 to Prepare

*Step 4: Calculate the amount of the analyte compound in the prepared volume of Standard 3.*

Multiply the concentration of Standard 3 and the total volume of Standard 3 (Step 3).

$$50\ \mu\text{g/mL} \cdot 20\ \text{mL} = 1000\ \mu\text{g}$$

*Step 5: Calculate the volume of Standard 2 containing the analyte compound amount in the prepared volume of Standard 3.*

Divide the analyte compound amount in Standard 3 (Step 4) by the concentration of Standard 2.

$$\frac{1000\ \mu\text{g}}{100\ \text{ng}/\mu\text{L}} = \frac{1000\ \mu\text{g}}{100\ \mu\text{g/mL}} = 10\ \text{mL}$$

*Note:* This is only the volume of Standard 2 needed to prepare Standard 3.

*Step 6: Calculate the minimum volume of Standard 2 to prepare.*

Add the volume of Standard 2 needed to make Standard 3 (Step 5) to the final volume of Standard 2.

$$10\ \text{mL} + 10\ \text{mL} = 20\ \text{mL}$$

After 10 mL is removed to make Standard 3, the remaining 10 mL is equal to the desired final volume of Standard 2.

**Part D:** Calculating the Total Volume of Standard 1 to Prepare

*Step 7: Calculate the amount of the analyte compound in the prepared volume of Standard 2.*

Multiply the concentration of Standard 2 and the total volume of Standard 2 (Step 6).

$$100\ \mu\text{g/mL} \cdot 20\ \text{mL} = 2000\ \mu\text{g}$$

*Step 8: Calculate the volume of Standard 1 containing the analyte compound amount in the prepared volume of Standard 2.*

Divide the analyte compound amount in Standard 2 (Step 7) by the concentration of Standard 1.

$$\frac{2000\ \mu g}{200\ ng/\mu L} = \frac{2000\ \mu g}{200\ \mu g/mL} = 10\ mL$$

*Note:* This is only the volume of Standard 1 needed to prepare Standard 2.

*Step 9: Calculate the minimum volume of Standard 1 to prepare.*

Add the volume of Standard 1 needed to make Standard 2 (Step 8) to the final volume of Standard 1.

$$10\ mL + 20\ mL = 30\ mL$$

After 10 mL is removed to make Standard 2, the remaining 20 mL is equal to the desired final volume of Standard 1.

Standard 1 can be used as the stock solution. It is recommended to prepare a more concentrated stock solution instead of using the first standard (i.e., most concentrated analytical standard) as the stock solution.

**Part D:** Calculating the Volume of the Stock Solution to Prepare

The volume of the stock solution is more of choice or decision; however, there is minimum volume needed to prepare the highest concentration analytical standard. The following calculations provide that value.

There is a large amount of leeway in the choice of the stock solution concentration. It has to be more concentrated than the highest concentration analytical standard (Standard 1 in this example). Ideally, the concentration allows for easy preparation and use. A concentration 5–10 time higher than the highest concentration analytical standard is convenient. Higher stock solution concentrations can be used. A stock solution concentration two times higher than the highest concentration analytical standard should be a minimum limit.

*Step 10: Calculate the amount of analyte compound in the prepared volume of Standard 1.*

Multiply the concentration of Standard 1 and the prepared volume of Standard 1 (Step 9).

$$200\ \mu g/mL \cdot 30\ mL = 6000\ \mu g$$

*Step 11: Calculate the volume of the stock solution containing the analyte compound amount in the prepared volume of Standard 1.*

Divide the analyte compound amount in Standard 1 (Step 10) by the concentration of the Stock Solution.

$$\frac{6000\ \mu g}{1000\ \mu g/mL} = 6\ mL$$

*Note:* This is only the volume of the Stock Solution needed to prepare Standard 1.

*Step 12: Determine the volume of stock solution to prepare.*

To facilitate the preparation of multiple sets of analytical standards, making a volume greater than 6 mL is recommended. For more stable stock solutions or when larger volumes are quickly consumed, a larger volume can be made. For less stable stock solutions or when small volumes are slowly consumed, a smaller volume should be made.

For this example, 30 mL is selected so 4–5 sets of analytical standards can be made from the same stock solution.

#### Summary of the Volume Calculations

The previous calculations provide the volume of each standard to prepare and the amount to use for each dilution. For this example, the total volume of each analytical standard and stock solution to prepare is:

Stock solution: 30 mL
Standard 1: 30 mL
Standard 2: 20 mL
Standard 3: 20 mL
Standard 4: 20 mL

The volume of each analytical standard and the stock solution needed to prepare the next lower concentration analytical standard is:

Stock solution: 6 mL
Standard 1: 10 mL
Standard 2: 10 mL
Standard 3: 8 mL
Standard 4: 0 mL

The final volume of each analytical standard and the stock solution after serial dilution is:

Stock solution: 30 mL – 6 mL = 14 mL
Standard 1: 30 mL – 10 mL = 20 mL
Standard 2: 20 mL – 10 mL = 10 mL
Standard 3: 20 mL – 8 mL = 12 mL
Standard 4: 20 mL – 0 mL = 20 mL

These volumes are the same or slightly greater than the desired final volumes (as stated in the original example information).

### Preparation of the Stock Solution and Analytical Standards

The stock solution and analytical standards can be prepared using any of the methods described in Sections F.7.2 for a single compound standard or Section F.7.3 for multiple compound standards.

# References

These references are recommended for more detailed explanations and discussions of gas chromatographic theory and practice.

W. Jennings, E. Mittlefehldt, P. Stremple: Analytical Gas Chromatography, 2nd edition. Academic Press (1997).

L. S. Ettre, J. V. Hinshaw: Basic Relationships of Gas Chromatography. Advanstar (1993).

J. V. Hinshaw, L. Ettre: Introduction to Open-Tubular Column Gas Chromatography. Advanstar (1994).

J. V. Hinshaw: Getting The Best Results From Your Gas Chromatograph. John Wiley & Sons, Inc. (2005).

R. L. Grob, E. F. Barry (Eds.): Modern Practice of Gas Chromatography, 4th edition. John Wiley & Sons, Inc. (2004).

M. Klee: GC Inlets – An Introduction. Hewlett Packard Co. (1994).

K. Grob: Split and Splitless Injection for Quantitative Gas Chromatography, 4th edition. Wiley-VCH (2001).

K. Grob: On-Column Injection in Capillary Gas Chromatography. Hüthig Buch Verlag (1991).

S. A. Stafford (Ed.): Electronic Pressure Control in Gas Chromatography. Hewlett Packard Co. (1993).

H. H. Hill, D. G. McMinn: Detectors for Capillary Chromatography. John Wiley & Sons, Inc. (1992).

This paper is the more in-depth source of the information used in the simplified explanation of stationary phase interactions in Section 4.4.

W. Burns, S. J. Hawkes: "Choice of Stationary Phase in Gas Chromatography". Journal of Chromatographic Science 15, 185 (1977).

*The Troubleshooting and Maintenance Guide for Gas Chromatographers, Fourth Edition.* Dean Rood

ISBN: 978-3-527-31373-0

# Subject Index



*The Troubleshooting and Maintenance Guide for Gas Chromatographers, Fourth Edition.* Dean Rood

ISBN: 978-3-527-31373-0

## *Related Titles*

R. L. Grob, E. F. Barry (Eds.)

### Modern Practice of Gas Chromatography

2004

ISBN 978-0-471-22983-4

K. Grob

### Split and Splitless Injection for Quantitative Gas Chromatography

**Concepts, Processes, Practical Guidelines, Sources of Error**

2001

ISBN 978-3-527-29879-2

H. M. McNair, J. M. Miller

### Basic Gas Chromatography

1998

ISBN 978-0-471-17261-1